T0136787

FLORA ZAMBESIACA

Flora terrarum Zambesii aquis conjunctarum

VOLUME NINE: PART ONE

FLORA ZAMBESIACA

MOZAMBIQUE

MALAWI, ZAMBIA, ZIMBABWE

BOTSWANA

VOLUME NINE: PART ONE

Edited by
E. LAUNERT
on behalf of the Editorial Board:
G. Ll. LUCAS
Royal Botanic Gardens, Kew
E. LAUNERT
British Museum (Natural History)
M. L. GONÇALVES
Centro de Botânica, Instituto de Investigação Cientifíca Tropical, Lisboa

Published by the Managing Committee on behalf of the contributors to Flora Zambesiaca
1988

©*Flora Zambesiaca Managing Committee, 1988*

Typeset by TND Serif Ltd Hadleigh, Ipswich, Suffolk
Printed in Great Britain by Halesworth Press Ltd

ISBN 0 9507682 3 5

CONTENTS

LIST OF FAMILIES INCLUDED IN VOLUME IX, PART 1

130. Plantaginaceae
131. Nyctaginaceae
132. Amaranthaceae
133. Chenopodiaceae
134. Basellaceae
135. Phytolaccaceae

LIST OF NEW TAXA PUBLISHED IN THIS WORK

Sarcocornia mossambicensis Brenan, Spec. nov.

130. PLANTAGINACEAE

By G. Lehmann

Annual or perennial, terrestrial or aquatic (not in Flora Zambesiaca area) herbs. Leaves mostly spirally arranged and all radical, less often cauline and spiral, alternate or opposite, simple, often sheathing at the base, sometimes reduced, exstipulate. Inflorescence usually a spike, rarely capitate racemes, terminating a scape. Each flower subtended by a bract. Flowers small, hermaphrodite or unisexual (not in Flora Zambesiaca area), usually 4-merous, actinomorphic or rarely so, sessile, bracteate. Calyx herbaceous, persistent, 4-lobed, the lobes more or less free or sometimes the lower pair more or less united. Corolla gamopetalous, scarious, 3–4 lobed; tube ampulliform or cylindric, lobes imbricate, equal, spreading. Stamens (1–2) 4, inserted in the corolla tube and alternating with the leaves, rarely hypogynous, usually exserted; filaments usually long, filiform; anthers versatile, 2-thecous, the thecae dehiscing longitudinally. Ovary superior, 1–4 -locular, with 1-many ovules in each locule, axile or basal; style single, long, usually filiform, simple, often exserted. Fruit a capsule, membranous, circumscissile or 1-seeded, hard, indehiscent. Seeds peltately attached, often mucilaginous when wet, 1-many; embryo straight, rarely curved, enclosed by the fleshy endosperm.

A large family of 3 genera and comprising between 250 and 300 species widely distributed throughout both hemispheres. Mostly in temperate and subtemperate regions but extending into the tropics.

PLANTAGO L.

Plantago L., Sp. Pl.: 112 (1753); in Gen. Pl., ed. 5: 52 (1754).

Annual or perennial, terrestrial, caulescent or mostly acaulescent herbs. Leaves usually in radical rosettes or opposite or alternate on branched stems, various in shape, usually entire. Inflorescence a pedunculate, cylindrical spike or head, with inconspicuous flowers, each subtended by a bract. Flowers almost always hermaphrodite, 4-merous. Calyx about as long as the corolla tube. Corolla 4-lobed, somewhat scarious; tube campanulate-tubular or urceolate; lobes lanceolate or ovate, shorter than the tube, usually with membranous margins or deflexed. Stamens 4, inserted in the corolla tube, alternating with the corolla lobes, exserted; filaments slender. Disk rare. Ovary globose or nearly so, bilocular, ovules 1-many in each locule; style usually much exserted and villous, more rarely included; stigma simple. Fruit a membranous capsule, circumscissile. Seeds 1-many in each locule, usually more or less boat-shaped with ventral hilum, mucilaginous when wet; albumen fleshy; embryo straight or curved; radicle inferior.

A genus of world-wide distribution comprising over 250 species, several of them occurring as weeds; three occur in the Flora Zambesiaca area.

1. Leaf lamina palmatilobed; bracts keeled - - - - - - - - 1. *palmata*
– Leaf lamina entire; bracts dorsally rounded or flat - - - - - - - - 2
2. Leaf lamina linear-lanceolate to lanceolate or spathulate, tapering into a petiole or sometimes sessile; inferior sepals connate for more than half their length; seeds 2 - - - - - - - - - - - - - 2. *lanceolata*
– Leaf lamina ovate, broadly elliptic or elliptic, always abruptly petiolate; inferior sepals free for more than half their length; seeds 4 - - - - - - - 3. *major*

1. **Plantago palmata** Hook. f. in Journn. Linn. Soc. Bot. **6**: 19 (1861); in op. cit. **7**: 213 (1864).—Engl. Pl. Ost-Afr. **C**: 374 (1895).—Baker in F.T.A. **5**: 504 (1900).—Thonner, Blütenpfl. Afr., t. 142 (1908).—Pilg. in Engl. Pflanzenr. IV **269**: 77 (1937).—Hepper in F.W.T.A., ed. 2, 2: 306, fig. 271 (1963).—Verdc. in Journ. E. Afr. Nat. Hist. Soc. **24** (108): 60 (1964). TAB **1** fig. C. Type from Bioko.

Perennial, pilose or glabrescent herb with short, straight rhizome and numerous fleshy roots. Leaves in a basal rosette, spirally arranged; lamina 6–13

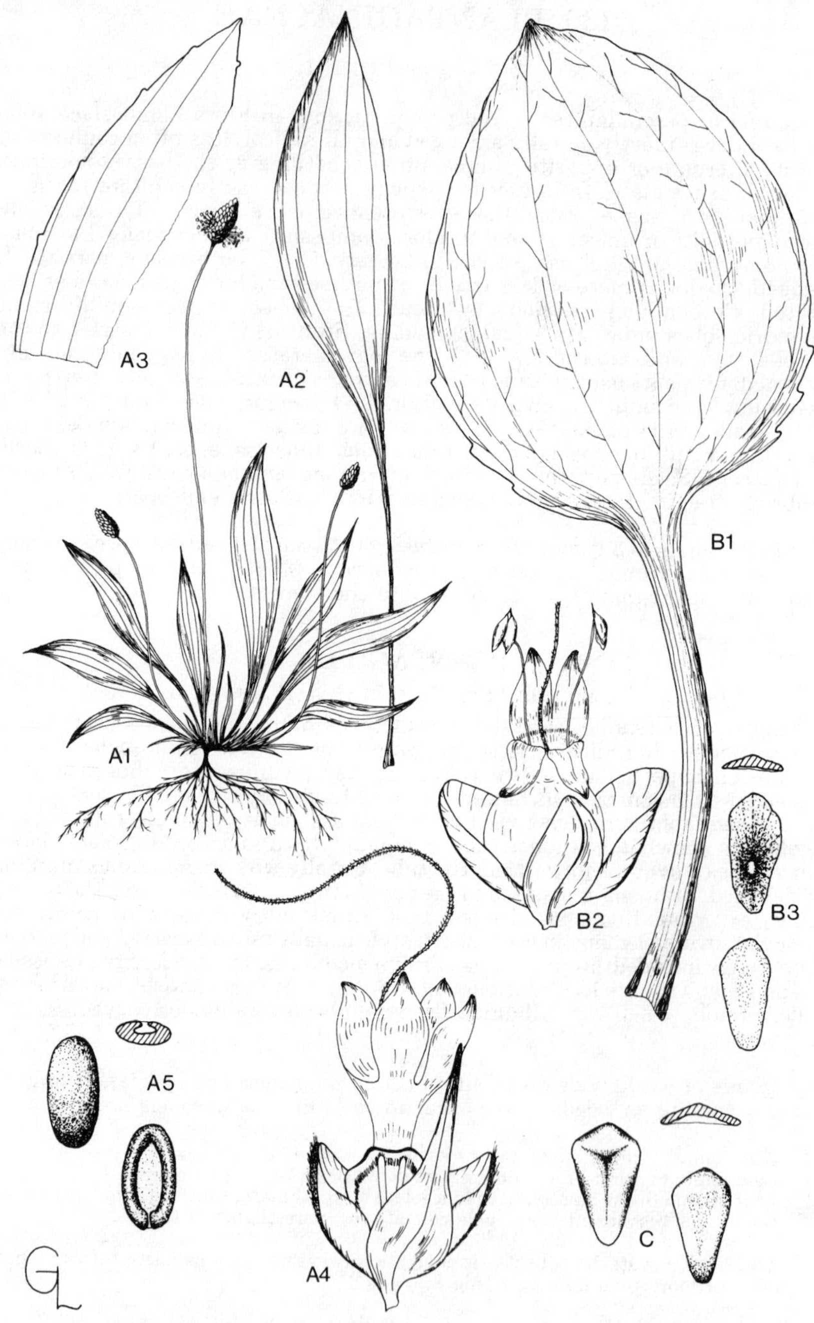

Tab. 1. A.—PLANTAGO LANCEOLATA. A1, habit (×$\frac{1}{4}$); A2, leaf (×$\frac{1}{2}$); A3, part of leaf showing margin (×1); A4, flower with bract (×10); A5, seed in cross section, dorsal and ventral view (×7), A1—5 from *Wild* 2498. B.—PLANTAGO MAJOR. B1, leaf (×$\frac{1}{2}$); B2, flower (×10); B3, seed in cross section, dorsal and ventral view (×7), B1—3 from *Garside* 1340. C.—PLANTAGO PALMATA. seed in cross section, dorsal and ventral view (×8) *Wild* 2512.

× 5–11 cm., distinctly palmatilobed, cordate-ovate to subcircular in outline, (3) 5 (7) veined at the base, main nerves fairly distinct with secondary nerves, usually glabrous sometimes slightly pilose; lobes 2.3–6 × 1.6–6, 1.4–3 × 1–2 cm. narrowly to broadly ovate with the apex subobtuse to rounded. Petiole 5–31 cm. long, 2–4 mm. in diam., glabrous, sometimes somewhat pubescent, dilated at the base; slightly cauliculate. Scape (5) 6.5–27 cm. long, 0.5–1.0 mm. in diam., erect or ascending, usually 5-sulcate, rarely pubescent but usually densely pilose beneath the spike. Inflorescence spicate; spikes cylindric to globose, 1–11 (17) cm. long, 3–5 mm. in diam., usually rather dense (at first much longer and less dense in fruit). Bracts 2–2.5 (3.5) mm. long, slightly boat-shaped, ovate, obtuse, glabrous, midrib slightly keeled. Sepals c. 3 mm. long, almost equal, slightly keeled midrib, obtuse, glabrous. Corolla tube c. 25 mm. long, glabrous, tube somewhat constricted at the mouth; lobes c. 15 mm. long, lanceolate to ovate, acute to subobtuse, usually spreading or reclined. Stamens exserted; anthers yellowish or cream coloured. Ovary ellipsoid to globose; pistil scarcely exceeding the lobes. Capsule globose or nearly so, 2–4 mm. long, usually 2-seeded. Seeds c. 2 mm. long, ellpisoid or ellipsoid-trigonous.

Zimbabwe. E: Inyanga, eastern slopes of Inyangani Mt., c. 1676 m., fl. 9.vii.1961, *Wild* 5512 (K; SRGH).

Also in Kenya and Uganda. Along streams in submontane forest.

2. **Plantago lanceolata** L., Sp. Pl.: 113 (1753).—Baker in F.T.A. 5: 503 (1900).—Pilg. in Engl. Planzenr. IV **269**: 313 (1937).—Levys in Adamson & Salter, Fl. Cape Penins.: 730 (1950).—Verdc. in Journ. E. Afr. Nat. Hist. Soc. **24** (108): 60 (1964).—Sagar & Harper in Journ. Ecol. **52**: 211–218 (1964).—Chater in Fl. Europaea **4**: 42 (1976). TAB. **1** fig. A. Type specimen of *P. angustifolia* Major in Hort. Cliff. p. 36, No. 3 (BM, lectotype).

Perennial herb, extremely variable, glabrous, pubescent or more rarely densely pilose, growing from a more or less erect stout short rhizome; roots terete, 0.1–0.75 mm. in diam.; stem sericeous. Leaves in a basal rosette, spirally arranged, petiolate, rarely sessile. Lamina (2) 8–25 (45) × 0.5–3.5 (8) cm., linear-lanceolate, ovate-lanceolate or spathulate, gradually narrowed into the petiole, entire or revolutely and shallowly dentate (3) 5 (7) veined, glabrous, appressed-pubescent or villous, apex acute or acuminate; petiole 2–21 cm. or longer, caulicate, flexible, glabrous or rarely pubescent. Scape erect or ascending (4) 10–80 (120) cm. long, 1–3 mm. in diam. 5-sulcate, more or less densely appressed pilose. Inflorescence spicate; spikes cylindric to globose, (0.3) 0.5–5 (10) cm. long, 0.5–1.0 mm. in diam., very dense. Bracts 2.5–3.5 mm. long, ovate-acuminate, glabrous or shortly hairy; midrib distinct, brownish. Sepals 2.5–3.5 mm. long, the anteriors adnante for most of their length but their midrib separate, ovate, apex slightly retuse, lateral sepals free, ovate, keeled often shortly hairy, usually ciliate along the keel above. Stamens exserted, anthers white with a yellow tinge. Ovary ellipsoid to globose; pistil about twice as long as the flower, hairy. usually 2-seeded. Capsule ovoid, 3–4 mm. long, seeds 2 (2.5) mm. long, oblong in outline, blackish dorsally convex, ventrally concave.

Botswana. SE: Mahalapye Experimental Station, on cultivated imported grass plots, c. 975 m., fr. v.1959, *De Beer* 893 (K; LISC; SRGH). **Zimbabwe**. C: Harare, Spicer's Farm, c. 5 km. above Makabusi-Hunyani confluence on Mukabusi, c. 1401 m., fl. 1.ix.1959, *Phipps* 2183 (BM; K; SRGH). E: Nyanyadzi, Sabi R., fl. 3.ii.1948, *Wild* 2498 (K; SRGH). **Malawi**. S: Thuchila, near Mulanje, fl. 9.ix.1952, *Jackson* 975 (K).

Also from Tanzania and Cape Prov. In open woodland and grassland, in irrigated land and long roads.

3. **Plantago major** L., Sp. Pl.: 112 (1753).—Baker in F.T.A. 5: 503 (1900).—Pilg. in Engl. Pflanzenr. IV **269**: 41 (1937).—Verdc. in Journ. E. Afr. Nat. Hist. Soc. **24** (108): 60 (1964).—Sagar & Harper in Journ. Ecol. **52**: 189-205 (1964). TAB. **1** fig. B. Type from Sweden.

Perennial herb, very variable, usually robust, glabrous or pubescent, growing with a short straight erect stem, with numerous adventitious roots. Leaves in a basal rosette spirally arranged, petiolate; lamina (1.5) 5–30 (40) × (0.5) 3–10 (17) cm., ovate to elliptic or rarely subcircular, entire sinuate or irregularly dentate, 3–9-veined at the base, usually glabrous or rarely pubescent, with the apex rounded to subobtuse, abruptly narrowed into the petiole; petiole 3–38 cm. long, usually as long as the lamina or shorter, cauliculate, flexible, glabrous or rarely pubescent. Scape (1) 7–15 (47) cm. long, 1–4 mm. in diam., erect or ascending, striate, glabrous or sometimes with short, appressed or ascending hairs. Inflorescence spicate, the spikes nearly always cylindrical, almost invariably simple, (0.5) 10–15 (70) cm. long, 3–8 mm. in diam.; flowers densely to loosely arranged. Bracts 1–2 (2.5) mm. long, ovate, glabrous, midrib distinct. Sepals 1.5–2 (2.5) mm. long, almost boat-shaped, usually equal, glabrous, green with narrow scarious margins, midrib prominent. Corolla tube c. 2 mm. long, glabrous; lobes 1–1.25 mm. long, lanceolate to ovate or narrowly triangular, acute to subacute, glabrous. Stamens 2–3 mm. long, exserted; anthers at first lilac, later whitish or yellowish. Ovary globose; pistil longer than the flower, hairy. Capsule 2–3 (5) mm. long, ovoid to globose or subconical, (3) 8–16 (34) seeded. Seeds 1–1.7 mm. long, ellipsoid or ellipsoid-trigonous, with the ventral side more convex than the dorsal.

Zambia. W: Ndola, fr. 21.viii.1973, *Fanshawe* 16 (K; NDO). **Zimbabwe**. W: Bulawayo, Municipal Park, c. 1372 m., 30.iv1958, *Drummond* 5519 (K; LISC; SRGH). C: Harare, Waterfalls, bank of Makabusi R., fr. 9.ix.1971, *Biegel & Pope* 3602 (K; LISC; SRGH). E: Vumba, fl. & fr. iii.1942, *Farrar* 4761 (SRGH).

Also all over tropical Africa and in S. Africa. Ranging naturally throughout Europe and northern and central Asia, but now naturalised throughout most of the world. Along streams and ditches, usually in moist sandy soil.

131. NYCTAGINACEAE

by B.L. Stannard

Trees, shrubs, herbs or climbers, sometimes spiny. Leaves opposite, alternate or in fascicles, simple, exstipulate, usually petiolate. Inflorescences axillary or terminal, cymose, umbellate, glomerulate, verticillate or thyrsoid. Flowers hermaphrodite or unisexual, actinomorphic, bracteate; bracts free or connate into an involucre, sometimes brightly coloured. Perianth gamophyllous, lower part persistent, usually green, upper part often petaloid and coloured, sometimes caducous after anthesis. Stamens 1-many, hypogynous; filaments free or connate at base, subequal to unequal; anthers dithecous, dehiscing longitudinally. Ovary superior, unilocular, sessile or stipitate; ovule 1, basal, erect; style 1, slender, more or less equal to or longer than stamens; stigma linear, capitate, peltate or penicillate. Fruit indehiscent, usually an anthocarp formed by the persistent accrescent basal part of the perianth enclosing an achene, sometimes winged or variously ribbed, sometimes glandular. Seed 1; endosperm present or absent; embryo straight or curved.

A family of about 30 genera and approximately 300 species, mainly in the tropics, mainly America.

1. Unarmed herbs - - - - - - - - - - - - - - - - 2
— Armed shrubs or spiny, woody climbers - - - - - - - - - - 4
2. Flowers more than 3 cm. long, conspicuous; bract connate forming, calyx-like involucre - **1. Mirabilis**
— Flowers less than 2 cm. long, relatively inconspicuous; bracts small, free - 3
3. Upper portion of perianth infundibuliform; stamens and styles exserted. Leaves in (sub) equal pairs. Anthocarps obscurely more or less 10-ribbed, with conspicuous, viscid, sometimes stipitate glands; plants often woody towards base and with scrambling or climbing stems - - - - - - - - **2. Commicarpus**
— Upper portion of perianth campanulate; stamens and styles exserted or included. Leaves often in unequal pairs. Anthocarps deeply 3–5 (6)-ribbed, without glands (although sometimes glandular pubescent); plant erect or diffuse - **3. Boerhavia**
4. Woody climber with recurved axillary spines; leaves alternate to sub-opposite, broad, petiolate; anthocarps 4–5-angled with longitudinal rows of viscid, stipitate glands on each angle - - - - - - - - - - - - - - - **4. Pisonia**

– Shrubs with spinescent branchlets; leaves usually in fascicles, linear, sessile or subsessile, anthocarps with 4 parchment-like wings - - - - 5. **Phaeoptilum**

In addition to the above, various species and forms of the South American shrub genus *Bougainvillea* Comm. ex Juss. are cultivated in the Flora Zambesiaca area. They can be recognised by their usually scrambling or scandent habit and the three large, brightly coloured bracts adnate to the pedicels of the triads of flowers.

1. MIRABILIS L.

Mirabilis L., Sp. Pl. **1**: 177 (1753); in Gen. Pl. ed. **5**: 82 (1754).

Erect perennial herbs, tuberous-rooted, much-branched. Leaves opposite. Inflorescences cymose, 1-many flowered. Involucre of connate bracts calyx-like, 5-lobed. Flowers conspicuous, hermaphrodite, ephemeral. Perianth variable in form but often tubular with expanded 5-lobed limb, lower portion constricted above ovary, persistent in fruit, upper portion petaloid, coloured, caducous after anthesis. Stamens 3–6, unequal, usually exserted. Ovary sessile or subsessile, 1-ovulate; style long exserted; stigma globose, usually with densely packed papillae or fimbriate. Fruit enclosed in hardened perianth base (anthocarp). Anthocarp ellipsoid, globose or claviform, often angled or ribbed, smooth or tuberculate.

A genus of about 60 species, mostly from the warmer regions of America, introduced and cultivated elsewhere.

Mirabilis jalapa L., Sp. Pl. **1**: 177 (1753).—Choisy in DC., Prodr. **13**, 2: 427 (1849).—Heimerl in Engl. & Prantl, Pflanzenfam. 3, **1B**: 24, fig. 8 E–H (1889); ed. 2, **16 C**: 108 (1934).—Baker & C.H. Wright in F.T.A. **6**, 1: 2 (1909).—Cooke in Fl. Cap. **5**, 1: 393 (1910).—F.W.T.A. ed. 1, **1**: 153 (1927); ed. 2, **1**: 178 (1954).—F.W. Andr., Fl. Pl. Anglo-Egypt. Sudan **1**: 152 (1950).—Balle in F.C.B. **2**: 90 (1951).—Cavaco, Fl. Madag., Nyctaginaceae: 4 (1954). TAB. 2. Type from India.

Herb up to 1.5 m. tall. Stems glabrescent or pubescent, pubescence often concentrated in longitudinal strips down stem. Leaves ovate, 3.5–13 × 2–8 cm., entire, acuminate, base rounded, truncate or cordate, glabrous to puberulent (particularly along veins above), margins often ciliate, rhaphides usually conspicuous beneath, superior leaves frequently subsessile, inferior leaves petiolate; petioles up to 4.5 cm. long, pubescent along superior surface. Inflorescences terminal. Pedicels pubescent. Involucre 5–13 mm. long, divided more or less halfway into 5 triangular to ovate, acute to acuminate lobes, glabrescent to pubescent or pilose particularly along margins. Perianth tubular 4–5.5 cm. long, lower portion green, upper portion spreading 2–3 cm. diam., purple, red, yellow, white or variegated. Stamens 5–6, exserted, 2.75–6 cm. long; filaments filiform, united into fleshy cup at base; anthers 1.2–2 mm. long, oblong-ellipsoid. Ovary ellipsoid or ovoid, 1–1.5 mm. long; style 4–6.5 cm. long, filiform; stigma with stalked papillae. Anthocarp ellipsoid to subglobose, 7–9 mm. long, ribbed or angled, tuberculate between the ribs (in dried state), hard, black.

Zambia. W: Alongside railway line, S. of Kitwe Station, fl. 12.ii.1961, *Linley* 70 (SRGH). S: Magoye Agric. Experiment Station S. of Mazabuka, fl. 7.v.1961, *Angus* 2865 (K; LISC; SRGH). **Zimbabwe**. N: Lomagundi Distr., Chinhoyi Dichwe Forest, fl. 16.ii.1965, *West* 6351 (SRGH). W: Bulalima Mangwe, fl. 28.xi.1954, *Meara* 93 (SRGH). C: Kwekwe, 1250 m., fl. & fr. 26.ii.1966, *Biegel* 859 (K; LISC; SRGH). E: Inyanga Distr., Pienaar's Farm, Juliasdale, fl. & fr. 12.ii.1961, *Rutherford-Smith* 512 (K; LISC; PRE; SRGH). **Malawi**. N: Nyika Plateau, fl. & fr. ii-iii.1903, *McClounie* 123 (K). C: Ngara Hill, Lilongwe, 1250 m., fl. & fr. 7.ii.1959, *Robson* 1506 (BM; K; LISC; PRE; SRGH). S: Zomba, banks of Mlunguzi R. between Govt. Hostel and Old Naisi Rd., 990 m., fl. 29.iii.1977, *Brummitt* 15016 (K). **Mozambique**. Z: Chamo, mouth of R. Shire, fl. v.1862, *Kirk* s.n. (K). T: *Angonia* Distr., outskirts of Ulongue, fl. & fr. 2.iii.1980, *Macuàcua & Mateus* LM 1153 (LMA). M: Maputo near the Laboratorios dos Estudos Gerais Universitarios in the Avenida de Mozambique, fl. 25.ix 1965, *Marques* 652 (LMU).

Introduced from tropical America, now widely cultivated and naturalised in all tropical regions.

Tab. 2. MIRABILIS JALAPA. 1, habit (×⅓), from cultivated specimen; 2, longitudinal section of flower (×1), from cultivated specimen; 3, fruiting involucre (×2) 4, fruit (×2), 3–4 from *Robson* 1506.

2. COMMICARPUS Standley

Commicarpus Standley in Contr. U.S. Nat. Herb. **12**: 373 (1909).

Perennial herbs or subshrubs. Stems long, slender, erect, reclining or climbing, often much branched. Leaves opposite in (sub) equal pairs, petiolate, often fleshy, usually entire or repand, rhaphides often visible on inferior surface in dried state. Inflorescences umbellate, verticillate, irregularly branched or mixed, pedunculate. Flowers small, fugacious, hermaphrodite, usually pedicellate; each pedicel bracteate. Bracts caducous. Perianth often with rhaphides clearly visible over surface, lower portion constricted above ovary with glands over surface often particularly around apex, greenish, persistent, upper portion petaloid, coloured, infundibuliform, shallowly lobed, sometimes with distinct basal tube, caducous after anthesis. Stamens 2–5 (6), exserted; filaments filiform, often unequal, connate into short sheath at the base; anthers dithecous. Ovary usually ellipsoid, often stipitate, 1-ovulate; style filiform, exserted; stigma capitate. Anthocarps cylindrical, fusiform, clavate or turbinate, more or less 10-sulcate, with variously arranged, large, viscid, sessile or stipitate glands.

A genus of about 20 species, mainly from Africa but also in southern Spain, Burma, Malaysia, southern China, Australia and warmer parts of the New World. It has a preference for arid habitats.

This account largely follows Meikle's treatment of the genus in his "Key to Commicarpus" in Notes Roy. Bot. Gard. Edin. **36**, 2: 235–249 (1978).

Although often confused with *Boerhavia*, in which it was once included, *Commicarpus* can be clearly distinguished from that genus by its often more suffruticose or scrambling/climbing habit, the shape of the perianth, and the large viscid or mucilaginous glands on the anthocarps. A poor specimen from Botswana (*Jobson* 5 [SRGH]) with "chewed" leaves and a few very young flowers on an undeveloped inflorescence resembles material of *Commicarpus fallacissimus* (Heimerl) Heimerl from Namibia - the stems have the same brownish grey to silvery white colour, the leaves seem to have a similar coarse, leathery texture, the indumentum matches and the flowers have just two stamens. More and better material is required, however, to establish with certainty whether this is indeed a representative of that species in the Flora Zambesiaca area.

1. Perianth widely infundibuliform, with a very short inconspicuous basal tube; flower (including anthocarp) less than 7 mm. long - - - - - - - - - - - 2
– Perianth narrowly infundifuliform with a well-developed basal tube; flowers (including anthocarp) more than 7 mm. long - - - - - - - - - - - - - 3
2. Mature fruits turbinate, tapering markedly from apex to base; anthocarps with sessile viscid glands scattered over surface and 5 stipitate viscid glands spreading stellately around apex; stems glabrous, scabridulous or pubescent - - - 1. *helenae*
– Mature fruits clavate, tapering at both ends; anthocarps with prominent but sessile glands scattered over surface; stems minutely but distinctly glandular pubescent - - - - - - - - - - - - - - - - - - - 2. *pilosus*
3. Whole plant conspicuously glandular pilose - - - - - 6. *grandiflorus*
– Plant glabrous to pubescent but not glandular pilose - - - - - - - 4
4. Flowers white (very rarely possibly tinged yellowish, pinkish or mauvish); anthocarps with more or less slender stalked viscid glands concentrated mainly around apex - - - - - - - - - - - - - - - - 3. *plumbagineus*
– Flowers pink, reddish, mauve or purple; anthocarps with sessile glands or with shortly stalked glands but these not concentrated around apex - - - - - 5
5. Inflorescences umbellate (rarely of mixed umbels and verticels). Anthocarps covered with prominent, sometimes shortly stalked viscid glands - - - - - - - - - - - - - 4. *chinensis* subsp. *natalensis*
– Inflorescences verticillate or irregularly branched. Anthocarps with 5 larger prominent or more or less thick stalked viscid glands around apex and smaller less conspicuous glands over rest of surface - - - - - - - - - - 5. *pentandrus*

1. **Commicarpus helenae** (J.A. Schultes) Meikle in Hook., Ic. Pl. **37**: t. 3694, p. 1 (1971).—Meikle in Notes Roy. Bot. Gard. Edin. **36**, 2: 246, fig. 3N (1978). TAB. 3 fig. C. Type from St. Helena.

Boerhavia helenae J.A. Schultes in Roem & Schult., Syst. Veg. 1, Mantissa **1**: 73 (1822) (as *Boerhaavia*). Type as above.

Boerhavia stellata Wight, Ic. Pl. Ind. Or. **3**, 2: 6, t. 875 (1843) (as *Boerhaavia*).—Choisy in DC., Prodr. **13**, 2: 454 (1949).—Berhaut in Bull. Soc. Bot. Fr. **100**: 51 (1953). Type from India.

Commicarpus verticillatus sensu Heimerl in Engl. & Prantl, Pflanzenfam. ed. 2, **16** C: 117 (1934) et auctt. mult non *C. verticillatus* (Poir.) Standl. in Contr. U.S. Nat. Herb. **18**: 101 (1916).—sensu Baker & C.H. Wright in F.T.A. **6**, 1: 6 (1909).—sensu F.W.T.A. ed. 1, **1**: 153 (1927); ed. 2, **1**: 177 (1954).—sensu Balle in F.C.B. **2**: 86, t. 7 (1951).

Commicarpus stellatus (Wight) Berhaut in loc. cit. Type as for *B. stellata*.

Herbs from a woody rootstock. Stems up to 1—1.5 m. long, slender, erect, decumbent or scrambling, branching, pubescent, scabridulous or glabrous. Leaves 1.5—5 × 0.7—3 cm., ovate, more or less fleshy, pubescent to glabrous, base cordate, rounded or more or less truncate, slightly attenuate along petiole, apex rounded to acute, apiculate, margins entire or sometimes repand; petioles 5—15 mm. long, sparsely pubescent. Inflorescence narrow, verticillate, long pedunculate, peduncles sparsely pubescent. Bracts linear-lanceolate, 1—2 mm. long, pubescent. Flowers sessile to very shortly pedicellate. Perianth 4—5 (6.5) mm. long, lower portion sulcate with 5 prominent viscid glands around apex, upper portion 2—3 (4.5) mm. long, widely infundibuliform with very short inconspicuous tube, purple, mauve, magenta, pink, white or yellow, pubescent. Stamens 2—3, 4—5 (6) mm. long; anthers 0.3—0.5 mm. long, transverse elliptic to rounded. Ovary 0.5—0.6 mm. long, ellipsoid, shortly stipitate, glabrous; style 3—3.5 (5.5) mm. long. Anthocarps turbinate, tapering markedly from apex to base, 3.5—7 × 1—2.5 mm., sessile viscid glands scattered over surface, 5 stipitate viscid glands spreading stellately around apex.

Botswana. N: Tsigara Pan, 48 km. W. of mouth of Nata R., 893 m., fl. & fr. 26.iv.1957, *Drummond & Seagrief* 5247 (K; SRGH). SE: Orapa, fl. & fr. 16.iii.1975, *Kerfoot* 7748 (PRE).

Also in Iran, S. Arabia, S. Palestine, Egypt, Canary Is., India, St. Helena and widespread in tropical Africa. Mostly in dry sandy or rocky areas but also on calcareous, black alluvial and volcanic soils, in grassland, beside rivers and in coastal regions; 0—1400 m.

2. **Commicarpus pilosus** (Heimerl) Meikle in Notes Roy. Bot. Gard. Edin. **36**, 2: 245, 249, fig. 3K (1978). TAB. **3** fig. D. Type from S. Africa.

Commicarpus fallacissimus f. *pilosus* Heimerl in Bothalia **3**: 233 (1937). Type as above.

Herb to 70 cm., woody towards base. Stems procumbent, branching, shortly but distinctly more or less densely glandular pubescent. Leaves 0.7—5 × 0.7—5 cm., ovate to more or less subcircular, more or less fleshy; very shortly and evenly glandular pubescent on both surfaces, base rounded, more or less truncate or broadly cuneate, apex rounded to acute, apiculate, margins entire or often repand; petioles 5—25 mm. long, glandular pubescent. Inflorescences verticillate, long pedunculate, peduncles glandular-pubescent. Bracts linear-lanceolate, 2.5—4 mm. long, glandular pubescent particularly along margins. Pedicels 0.5—5 mm. long elongating up to 11 mm. in fruit, glandular pubescent. Perianth 4—6 mm. long, spreading up to 7 mm., lower portion sulcate, with prominent globose viscid glands around apex, glabrous, upper portion 2.5—4 m. long, widely infundibuliform with very short inconspicuous tube, purple, mauve or pink, glabrescent to glandular-pubescent. Stamens 2—3, 6—10 mm. long; anthers 0.5—0.8 mm. long, transverse elliptic to subcircular, Ovary 0.65—0.75 mm. long, ellipsoid, stipitate, glabrous; style 5—7 mm. long. Anthocarps 6—7.5 × 1.5—2 mm., clavate, tapering at both ends, with sessile viscid glands scattered over surface, 5 (6) gland-warts remaining distinctly prominent around apex in dried state.

Botswana. SE: Ilatamabele-Mosu area, near Soa Pan, fl. & fr. 9.i.1974, *Ngoni* 279 (K; MO; SRGH). **Zimbabwe.** S: Beitbridge, fl. & fr. 10.i.1961, *Leach* 10664 (K; MO; PRE; SRGH).

Also in S. Africa. In savanna, in dry areas and on stony ground; 450—900 m.

3. **Commicarpus plumbagineus** (Cav.) Standley in Contr. U.S. Nat. Herb. **18**: 101 (1916).—Meikle in F.W.T.A. ed. 2, **1**: 177 (1954).—Meikle in Notes Roy. Bot. Gard. Edin. **36**, 2: 244, fig. 2J (1978). TAB. **3** fig. A. Type from Spain.

Boerhavia plumbaginea Cav., Ic. Pl. 2: 7, t. 112 (1793) (as *Boerhaavia*).—Baker & C.H. Wright in F.T.A. **6**, 1: 6 (1909).—Burtt Davy, Fl. Pl. Ferns Transv. **I**: 109 (1926).—F.W.T.A. ed. 1, **1**: 152, 153 (1927). Type as above.

Boerhavia verticillata Poiret in Lam., Encycl. Meth. Bot. **5**: 56 (1804). (as *Boerhaavia*). Type from Senegal.

Boerhavia commersonii Baillon in Bull. Soc. Linn. Paris **1**: 484 (1885). (as *Boerhaavia*). Type from Madagascar.

Commicarpus verticillatus (Poiret) Standley in Contr. U.S. Nat. Herb. **18**: 101 (1916). Type as for *B. verticillata*.

Commicarpus commersonii (Baillon) Cavaco in Bull. Soc. Bot. Fr. **100**: 297 (1953). Type as for *B. commersonii*.

Commicarpus africanus auct. non *C. africanus* (Lour.) Dandy in F.W. Andr., Fl. Pl. Anglo-Egypt. Sudan **1**: 152 (1950). The latter is based on *Boerhaavia africana* Lour., Fl. Cochinch. **1**: 16 (1790) from Mozambique and according to Meikle (Notes Roy. Bot. Gard. Edin. **36**, 2: 246 (1978)), is possibly a *Boerhavia* but not a *Commicarpus*.

Herbs from woody rootstock. Stems procumbent or scandent to 4 (10) m, woody towards base, much branched, glabrous, puberulous or pubescent. Leaves 1.5–12 × 0.7–8 cm., ovate, slightly fleshy, pubescent to glabrescent on both surfaces, base cordate, more or less truncate or rounded, apex acute, apiculate, margins entire or repand; petioles 1.5–40 mm. long, pubescent to glabrescent. Inflorescences irregular or of mixed umbels and verticels; peduncles pubescent to glabrescent. Bracts 1.5–4 mm. long, linear-lanceolate, pubescent. Pedicels 1–5 mm. long, pubescent. Perianth 8–15 mm. long, lower portion with viscid glands over the surface (particularly around the apex), glabrous to puberulous, upper portion 7–12 mm. long, spreading 5–9 mm. wide, narrowly infundibuliform with distinct basal tube, white (possibly very rarely with yellowish, pinkish tinges), often densely pubescent to glabrescent. Stamens 3–5 long exserted, 12–18.5 mm. long; anthers 0.6–0.8 × 0.8–1 mm., thecae reniform. Ovary 0.75–1 mm. long, ellipsoid, stipitate, glabrous, style 15–18 mm. long. Anthocarps 7–11 × 1–2 mm., cylindrical, fusiform to claviform, glabrous to pubescent with numerous usually more or less slender-stalked viscid glands concentrated towards apex but sometimes scattered sparsely among sessile viscid glands over rest of surface.

Botswana. N: Gwetshaa Isl., fl. & fr. 3.v.1973, *Smith* 569 (K; LISC; LMU, SRGH). SE: 72 km. N.E. of Zanzibar, fl. 15.x.1977, *Woollard* 297 (SRGH). **Zambia.** B: c. 13 km. N. of Nangweshi, 1036 m., fl. & fr. 23.vii.52, *Codd* 1771 (BM; K; PRE; SRGH). N: Mbala Distr., Cassava Sands, Lake Tanganyika, 780 m., fl. & fr. 17.ii.1959, *Richards* 10939(b) (K). C: Chilanga, Makulu stream near Mt. Makulu Res. St., fl. & fr. 12.iv.1958, *Angus* 1893 (K; PRE; SRGH). ?E: Luangwa R., fl. & fr. 30.v.1958, *Fanshawe* 4473 (K). S: Zambesi River, Katambora, fl. & fr. 25.xi.1949, *Wild* 3211 (K; SRGH). **Zimbabwe.** N: Mazoe, Iron Mask Hills, 1555 m., fl. & fr. iv.1906, *Eyles* 333 (BM; MO; SRGH). W: Hwange Distr., Victoria Falls, near Ministry of Water Development Pump-house, 8880 m., fl. & fr. 8.iii.1978, *Mshasha* 42 (K; MO). E: Mutare Distr., N. slope of Victory Hill, Commonage, 1158 m., fl. & fr. 1.xi.52, *Chase* 4694 (BM; MO; SRGH). S: Mwenezi Distr., Malangwe River, SW. Mateke Hills, 625 m., fl. & fr. 6.v.1958, *Drummond* 5650 (K; LISC; SRGH). **Malawi.** N: 32 km. S. of Karonga, 480 m., fl. & fr. 23.iv.75, *Pawek* 9534 (K; MO; PRE; SRGH). C: Dedza Distr., Nkata-taka, Chipoka Rd., fl. 10.iv.1969, *Salubeni* 1311 (SRGH). S: Sharpevale area, 500 m., fl. & fr. 1.ii.1959, *Robson* 1413 (BM; K; PRE; LISC; SRGH). **Mozambique.** Z: Morrumbala, between M'Gaza and Muriri, fl. 21.v.1943, *Torre* 5348 (LISC). T: Zumbo, Melause Rd., c. 3 km. from Zumbo, fl. & fr. 19.iv.1972, *de Aguiar Macedo* 5221 (LISC; LMA; LMU; SRGH). MS: Báruè, 40 km. from Changara towards Vila Gouveia 2 km. along path towards Catunguinenes kingdom (lower Luenha), c. 400 m., fl. & fr. 28.v.1971, *Torre & Correia* 18696 (LISC; LMU). GI: Gaza, Canicado, 12 km. from Lagoa Nova towards Chimai, fl. & fr. 23.vii.1969, *Correia & Marques* 1023 (BM; COI; LISC; LMU; PRE; SRGH). M: Maputo, Rd. to Porto Henrique, fl. & fr. 18.ix.1955, *Lemos* 108 (LISC; LMA).

Widespread in tropical Africa and also in S. Spain, Palestine, S. Arabia, Namibia and S. Africa. In forest and grassland, often beside rivers and water courses on a variety of soils; up to 1800 m.

Heimerl described a var. *trichocarpa* (Bull. Herb. Boiss. 5, App. 3: 68 (1897)), distinguishing it by the sparse covering of very short eglandular trichomes on the perianth, ovary and anthocarps. Meikle in his "Key to Commicarpus" (Notes Roy. Bot. Gard. Edin. 36 2: (1978)) maintains Heimerl's variety for some material from Botswana and Namibia, separating it on the basis of its stems being "closely crisped pubescent or scabridulous" as opposed to the "glabrous or at most very sparsely scabridulous or puberulous" stems of the type variety. Schreiber, on the other hand, does not accept the distinction (Merxm., Prodr. Fl. SW. Afr. 25: 5 (1969), under *C. africanus* (Lour.) Dandy). As the differences do not appear at all clear-cut, it would seem more practical to follow the latter author's view.

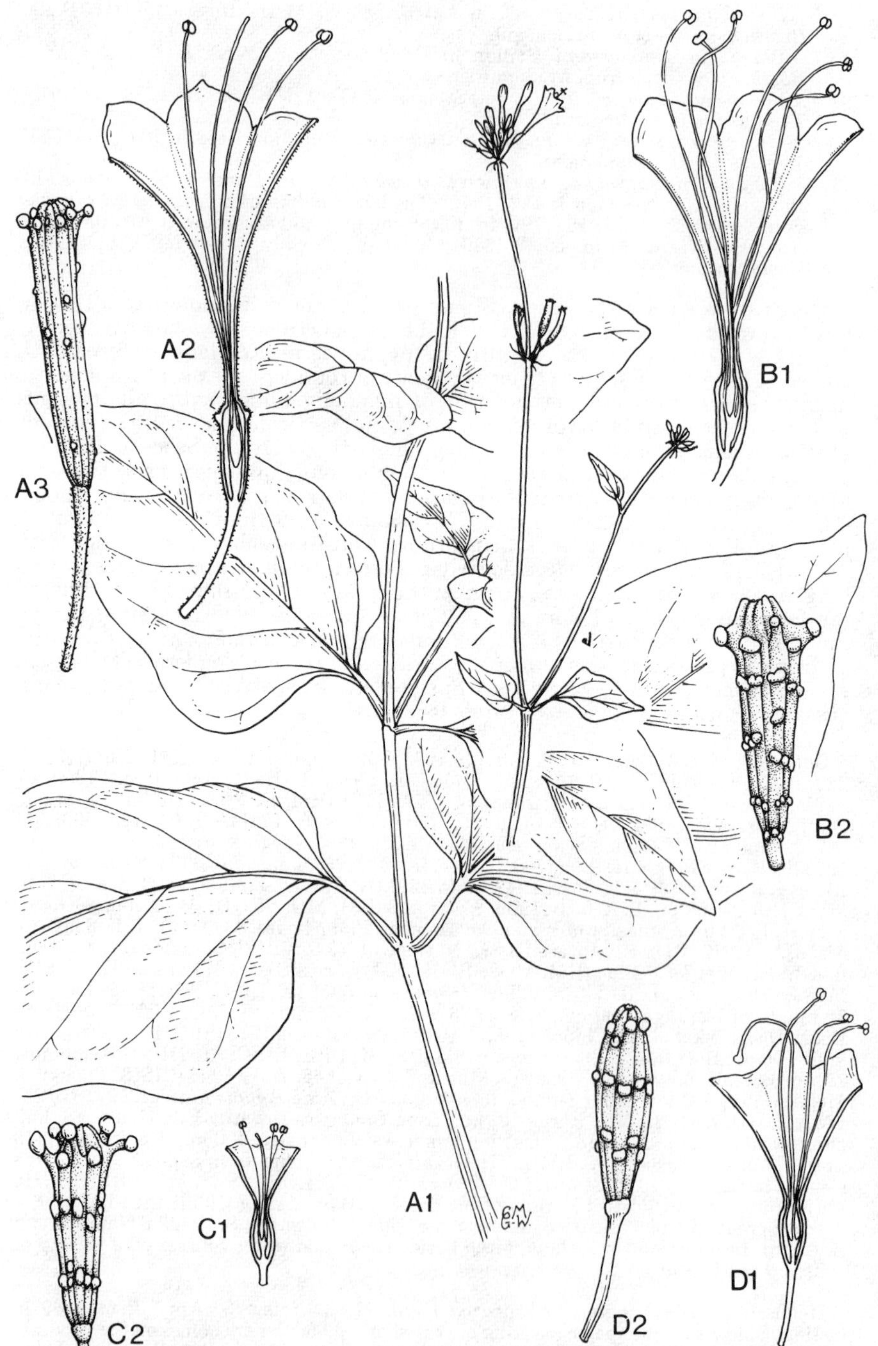

Tab. 3. A.—COMMICARPUS PLUMBAGINEUS. A1, habit (×⅔); A2, longitudinal section of flower (×4); A3, fruit (×4), A1–3 from *Torre & Correia* 17943. B.—COMMICARPUS PENTANDRUS. B1, longitudinal section of flower (×4); B2, fruit (×4) *Torre* 7415. C.—COMMICARPUS HELENAE. C1, longitudinal section of flower (×4) *Kerfoot* 7748; C2, fruit (×4) *Pearson* 2191. D.—COMMICARPUS PILOSUS. D1, longitudinal section of flower (×4); D2, fruit (×4) D1–2 *Smuts & Gillett* 4058.

4. **Commicarpus chinensis** (L.) Heimerl in Engl. & Prantl, Pflanzenfam. ed. 2, **16** C: 117 (1934). Type "in China. Osbeck" (LINN).
Valeriana chinensis L., Sp. Pl. **1**: 33 (1753). Type as above.
Boerhavia chinensis (L.) Aschers. & Schweinf., Beitr. Fl. Aethiop. **1**: 167 (1867). Type as above.

Subsp. **natalensis** Meikle in Notes Roy. Bot. Gard. Edinb. **36**, 2: 243, 247 (1978). Type from S. Africa.

Herbs. Stems scandent or prostrate, sparsely puberulous to glabrous. Leaves 1.6–7.0 × 1.4–6 cm., ovate, base shallowly cordate, apex acute, margins entire, sometimes slightly repand, sparsely pubescent (especially along veins on both surfaces and margins) to glabrescent; petioles 5–35 mm. long, pubescent to glabrescent. Inflorescences umbellate, very rarely of umbels and verticels; peduncles sparsely pubescent to glabrous. Bracts 2–5 mm. long, linear-lanceolate, pubescent. Pedicels 3–5 mm. long extending up to 15 mm. in fruit, sparsely pubescent to glabrous. Perianth 8–18 mm. long, lower portion striate, covered with viscid gland-warts, glabrous, upper portion 3.5–10 mm. long, narrowly infundibuliform with well-developed basal tube, opening out up to 10 mm. wide at the mouth, sparsely puberulous to glabrous, pink, pinkish red or purple. Stamens 2–3, long exserted, 14–24 mm. long; anthers 0.5–0.6 × 0.8–0.9 mm., transversely elliptic. Ovary 0.8–1.2 mm. tall, ellipsoid, stipitate, glabrous; style 20–28 mm. long, long exserted. Anthocarps 9–11 × 2–3 mm., fusiform, truncate at apex, covered with prominent, sometimes shortly stalked viscid glands.

Mozambique. M: Maputo, Bela Vista, Ponta de Ouro, fl. & fr. 8.iv.1968, *Balsinhas* 1200 (COI; LMA).
Also in S. Africa (Natal). In coastal dune forest and thickets; c. sea-level.

C. chinensis subsp. *chinensis* is found in India, Pakistan, S. China, Malay Peninsula and Islands, Thailand and Vietnam.

5. **Commicarpus pentandrus** (Burch.) Heimerl in Engl. & Prantl, Pflanzefam., ed. 2, **16** C: 117 (1934).—Pohnert & Schreiber in Merxm., Prodr. Fl. SW. Afr. **25**: 4 & 6 (1969).—Meikle in Notes Roy. Bot. Gard. Edin. **36**, 2: 241, fig. 2B (1978). TAB. **3** fig. B. Type from S. Africa.
Boerhavia pentandra Burch., Trav. Int. S. Afr. **1**: 432 (1822) (as *Boerhaavia*).—Cooke in Fl. Cap. 5, 1: 396 (1910) in part.—Marloth Fl. S. Afr. **1**: 192, fig. 85, 86, plate 47D (1913).—Burtt Davy, Fl. Pl. Ferns Transv. **1**: 209 (1926). Type as above.
Boerhavia burchellii Choisy in DC., Prodr. **13**, 2: 455 (1849). (as *Boerhaavia*). Types from S. Africa.
Boerhavia transvaalensis Gandoger in Bull. Soc. Bot. Fr. **66**: 221 (1919) (as *Boerhaavia*). Type from S. Africa.

Herbs forming mats up to 2 m. across from woody rootstock. Stems procumbent, branching, crisped pubescent to glabrescent. Leaves 1–4 (6) × 0.7–3.5 (6) cm., ovate, broadly elliptic or subcircular, fleshy, crisped pubescent on both surfaces (often particularly along veins) to more or less glabrous, base cordate to rounded, apex acute to rounded, often apiculate, margins irregularly serrulate; petioles 1–12(15) mm. long, crisped pubescent. Inflorescences verticillate or irregularly branched, long pedunculate, peduncles crisped pubescent to glabrescent, stout. Bracts 3–9 mm., linear-lanceolate, pubescent (often particularly along margins). Pedicels 1–10 mm. (terminal flowers sometimes subsessile), crisped pubescent to glabrescent. Perianth 9–15 mm. long, lower portion sulcate, with usually 5 prominent viscid glands around apex and smaller less conspicuous glands over rest of surface, glabrous, upper portion 8–12 mm. long, spreading 8–12 mm. wide, narrowly infundibuliform with distinct basal tube, purple, magenta or pinkish red, pubescent to glabrescent. Stamens (4) 5 (6), 11–17 mm. long, anthers 0.7–0.8 x 1.2–1.5 mm., transverse elliptic. Ovary 0.75–1 mm. long, ellipsoid, stipitate, glabrous; style 14–20 mm. long. Anthocarps 7–10 × 1.25–2.5 mm., clavate, glabrous to puberulous, with 5 larger prominent or more or less thick stalked viscid glands around apex, smaller less prominent viscid glands alternating with these at apex and over rest of surface.

Botswana. SW: Kang Pan, 300 km. W. of Gaborone, 1065 m., fl. & fr. 4.ii.77, *Mott* 1127 (K; MO; SRGH). SE: 60 km. NW. of Serowe, fl. & fr. 25.iii.1965, *Wild & Drummond* 7298 (K; LISC; SRGH). **Zimbabwe**. W: 25 km. S. of Bulawayo, fl. & fr. 18.xii.1971, *Plowes* 3487 (LISC). C: Gweru, fl. i.1905, *Gardner* 20 (K; SRGH). **Mozambique**. M: S. of Save at the fork of the Rds. to Namaacha and Moamba, c. 12.5 km. from Moamba, fl. & fr. 6.v.1952, *Myre & de Carvalho* 1199 (K; LISC; LMA; SRGH).

Also in Namibia and S. Africa. In grassland, forest clearings and on flood plains on a variety of soils; up to c. 1400 m.

Reported as being good food for livestock (Burtt Davy, Fl. Pl. Ferns Transv. 1: 209 (1926); Marloth, Fl. S. Afr. 1: 192 (1913)).

6. **Commicarpus grandiflorus** (A. Rich.) Standl. in Contrib. U.S. Nat. Herb. **18**: 101 (1916). Types from Ethiopia.

Boerhavia grandiflora A. Rich., Tent. Fl. Abyss. 2: 209 (1850). Type as above.

Herbs, whole plant sticky. Stems scandent or prostrate to 2 m., more or less densely glandular-pilose. Leaves 1.5–10 × 1–7 cm., ovate, rarely more or less subcircular base truncate to rounded, apex acute to acuminate, rarely rounded, margin entire, sometimes slightly repand, glandular-pilose. Inflorescences mostly umbellate but sometimes a few of mixed umbels and verticels on the same plant; peduncles densely glandular-pilose. Bracts 2–3 mm. long, linear-lanceolate, glandular pilose. Pedicels 2–10 mm. long, extending up to 15 mm. in fruit, densely glandular-pilose. Perianth 8–13 mm. long, lower portion covered with viscid gland warts, glandular-pilose to glabrescent, upper portion 6–10 mm. long, narrowly infundibuliform with well developed basal tube, opening out up to 10 mm. wide at the mouth, glandular-pilose, pink, mauve, magenta or purple. Stamens (?) 2, 3–4 long exserted, 12–17 mm. long; anthers 0.3 × 0.75 mm., transversely elliptic to oblong. Ovary 0.5–0.75 mm. tall, ellipsoid, stipitate, glabrous; style 12–20 mm. long, long exserted. Anthocarps 5–6.5 × 1–2.5 mm., clavate, covered with prominent, sessile to very shortly stalked viscid glands, glandular-pilose.

Malawi. N: Rumphi Distr., St. Patrick's Mission, 1.6 km. up Chelinda R., 1067 m., fl. & fr. 14.viii.1977, *Pawek* 12878 (K; PRE; SRGH).

Also in S. Sahara, N.E. and E. tropical Africa, S. Arabia, W. India. Rocky slopes and grasslands, light woodland and thickets, sometimes near water-courses; 850–2400 m.

3. BOERHAVIA L.

Boerhavia L., Sp. Pl.: 3 (1753); in Gen. Pl., ed. 5: 9 (1754).

Annual or perennial herbs, often from thick woody or fleshy rootstock, usually many-stemmed, sometimes viscid, raphides often present in all parts. Stems slender, erect, decumbent or ascending, sometimes woody towards base, terete, often much branched. Leaves opposite, commonly in pairs of unequal size at each node, sometimes fleshy, entire, sometimes repand, petiolate. Inflorescences terminal and axillary, paniculate, umbellate or cymose, sometimes diffuse, pedunculate. Bracteoles minute, caducous. Flowers small, solitary or glomerate, fugacious, hermaphrodite, subsessile to pedicellate. Perianth: lower portion constricted above ovary, cylindrical to obconical, upper portion petaloid, coloured, campanulate, slightly 5-lobed, caducous after anthesis. Stamens 1–4(6), exerted or included; filaments filiform, shortly united a base; anthers dithecous, usually subglobular. Ovary ellipsoid, shortly stipitate, 1-ovulate; style filiform, exserted or included; stigma capitate. Anthocarps fusiform, clavate or turbinate, conspicuously 3–5 sulcate, glabrous or glandular pubescent, often viscid; ribs rounded, acute or winged, usually mucous when wet. Fruit usually more or less filling anthocarp cavity.

A genus of about 20 species widely distributed in the tropics and subtropics, several being cosmopolitan weeds. Often found in waste places. The boundaries between three of the species treated as separate here (*B. diffusa* L., *B. coccinea* Mill., *B. repens* L.) are not at all clear. The whole genus is in much need of thorough revision. The name *Boerhavia elegans* Choisy has been misapplied to some material, from the Flora Zambesiaca area. This species occurs in N.E. tropical Africa, through Arabia and Iran and into Pakistan.

1. Anthocarps eglandular, glabrous, turbinate, apex truncate or broadly obtuse. Leaves often distinctly reddish brown punctate - - - - - - - - 1. *erecta*
– Anthocarps glandular pubescent, fusiform, apex tapering or rounded - - 2
2. Inflorescences terminal, leafless, ultimately of large, diffuse, lax, cymose panicles with slender, glabrous branches - - - - - - - - - - 2. *diffusa*
– Inflorescences terminal and axillary, usually leafy, of simple or aggregated cymes or pseudo-umbels - - - - - - - - - - - - - 3
3. Plants sprawling or ascending, glandular pubescent, less commonly glabrescent, often with long septate hairs, often viscid. Leaves mostly more than 2.5 cm. long. Inflorescences terminal and axillary, often aggregated, usually branching with two or more cymes of 3-many flowers, well exceeding leaves (repeated branching of stems and reduction of superior leaves sometimes gives a paniculate appearance) - - - - - - - - - - - - - - - 3. *coccinea*
– Plants prostrate, often mat-forming, pubescent, puberulous or glabrescent, not viscid. Leaves rarely more than 2.5 cm. long, often much less. Inflorescences axillary, simple, unbranched with one cyme of 2–4 flowers (7 distally), shorter than or only slightly exceeding leaves - - - - - - - - - - - - 4. *repens*

1. **Boerhavia erecta** L., Sp. Pl. **1**: 3 (1753).—Choisy in DC., Prodr. **13**, 2: 450 (1849).—Heimerl in Pflanzenfam. ed. 2, **16C**: 118 (1934).—Meikle in F.W.T.A. ed. 2, **1**: 178 (1954).—Codd in Bothalia **9**, 1: 115 (1966).—Agnew, Upland Kenya Wild Fl.: 162 (1974). TAB. **4** fig. B. Type from Mexico (Vera Cruz).

Annual or perennial herbs, branching freely from base, spreading. Stems prostrate, decumbent or ascending, glabrescent to pubescent, often with scattered longer septate hairs particularly around nodes, sometimes viscid. Leaves 1.3–6.5 × 0.4–6 cm., narrowly to broadly ovate, base subcordate, rounded, more or less truncate or broadly cuneate, apex acute to obtuse, sometimes apiculate, glabrescent to sparsely pubescent, discolourous, usually distinctly reddish-brown glandular punctate on both surfaces; petioles up to 2 cm. long, pubescent or sometimes pilose. Inflorescences terminal, cymose, ultimately diffusely paniculate; peduncles glabrous to puberulous, sometimes glandular. Bracteoles 1–1.5 mm. long, lanceolate, ciliate. Flowers usually 2–5 (10) per cluster, sessile or with pedicels up to 2 mm. long. Perianth 1.5–3 mm. long, glabrous, lower portion, 5-ribbed, upper portion 0.8–1.5 mm. long, campanulate, magenta, pink, mauve or white. Stamens 2–3, 2–2.5 mm. long, subequal, slightly exerted. Ovary 0.4–0.5 mm. long, glabrous; style 1.25–2 mm. long, slightly exerted. Anthocarps 3–4 × 1–1.2 mm., turbinate, apex broadly obtuse or truncate, eglandular, glabrous; ribs 5, acute.

Botswana. SE: Gaberone Village, 975 m., fl. & fr. 29.i.75, *Mott* 595 (K; SRGH). **Zambia**. C: railway embankment at the Kafue rail bridge, fl. & fr. 16.iii.63, *van Rensburg* 1680 (K; SRGH). E: Luangwa Valley, Kakumbi, 610 m., fl. & fr. 4.iii.69, *Astle* 5561 (K; SRGH). S: Monze, fl. & fr. 7.vi.62, *Fanshawe* 6861 (K; MO; SRGH). **Zimbabwe**. W: Hwange Colliery, fr., 19.x.1977, *Hill* in SRGH 260 244 (SRGH). **Malawi**. N: Karonga Distr., St. Anne's, 3.2 km. N. of Chilumba, 580 m., fr. 23.iv.69, *Pawek* 2295 (K). S: Liwonde National Park, near Park Office, fl. & fr. 17.iv.80, *Blackmore, Brummitt & Banda* 1245 (BM). **Mozambique**. N: Nampula Prov., Malema, Mutuali, C.I.C.A. Experimental St., fl. & fr. 25.iv.1961, *Balsinhas & Marrime* 449 (LISC; LMA; PRE). Z: Mocuba, fl. & fr. 13.iii.1943, *Torre* 4927 (LISC). T: Cahobra Bassa, left bank of Zambezi R., 230–330 m., fl. & fr. 13.iv.1972, *Pereira & Correia* 2010 (BM; COI; LISC; LMU; MO; PRE; SRGH). MS: Ancueza, Chiou, C.I.C.A. Experimental St., fl. & fr. 12.iv.1960, *Lemos & Macuacua* 65 (BM; COI; K; LISC; LMA; PRE; SRGH). M: Maputo, near the Laboratorios dos Estudos Gerais Universitarios in the Avenida de Moçambique, fl. & fr. 20.iii.1965, *Marques* 371 (BM; COI; K; LISC; LMU; MO; PRE; SRGH).

A widespread weed in Old and New World tropics. Commonly on disturbed and waste-ground; 110–1100 m.

Two specimens from Botswana (*Mitchison* 34 and 48) have fruits approaching those of *B. pterocarpa* S. Wats., a New World species found also in S. Africa. Reported to be good forage for animals.

2. **Boerhavia diffusa** L., Sp. Pl. **1**: 3 (1753).—F.W.T.A. ed. 1, **1**: 153 (1927) pro parte.—Meikle in F.W.T.A. ed. 2, **1**: 178 (1954).—Agnew, Upland Kenya Wild Flowers: 162 (1974). TAB. **4** fig. A. Type from India.

Boerhavia paniculata Rich., Actes Soc. Hist. Nat. Paris **1**: 105 (1792). Type from Cayenne.

Boerhavia adscendens Willd., Sp. Pl. **1**: 19 (1797).—Baker & C.H. Wright in F.T.A. **6**, 1: 4 (1909).—Cooke in Fl. Cap. **5**, 1: 395 (1910). Type from Guinea.

Boerhavia repens var. *diffusa* (L) Hook. f. in Fl. Brit. India **4**, 2: 709 (1885).—Baker & C. H. Wright in F.T.A. **6**, 1: 5 (1909).—Cooke in Fl. Cap. **5**, 1: 394 (1910). Type from S. Africa.

Perennial herbs, usually much branched from base. Stems spreading, decumbent with ascending flowering stems to 1 m., diffusely branched, glabrous to puberulous, often with long septate hairs particularly around nodes. Leaves 1.5−6 × 0.8−5 cm., ovate, elliptic or subcircular, base subcordate, rounded, more or less truncate or broadly cuneate, apex acute to rounded, sometimes apiculate, usually more or less glabrous with long septate hairs mainly along veins below and margins; petioles to 4 cm., glabrous to puberulous, often with long septate hairs. Inflorescences elongating greatly after start of flowering, ultimately lax, terminal, leafless, cymose panicles, occasionally a few scattered simple axillary cymes, branches glabrous. Bracteoles more or less 1 mm. long, ovate to lanceolate, acuminate, margins fimbriate. Flowers usually 2−7(11) per cluster, sessile or with pedicels to 1 mm. long. Perianth 1.6−2.5 mm. long, lower portion 5-ribbed, glandular-pubescent, upper portion 0.5−1.2 mm. long, campanulate, whitish, pink, carmine or purple. Stamens 1−3, subequal, slightly exserted 1.5−2.5 mm. long,. Ovary more or less 0.5 mm. long, glabrous; style 1−1.5 mm. long, slightly exserted. Anthocarps 3−3.5 × 1−1.75 mm., clavate, apex rounded, glandular pubescent.

Botswana. SE: Gaberone, Content Farm, fl. & fr. 28.xi.1972, *Kelaole* A73 (SRGH). **Zambia**. B: Mankoya, near resthouse Kaoma fl. & fr. 20.xi.1959, *Drummond & Cookson* 6654 (K; SRGH). N: Mbala Distr., Lake Tanganyika, Mpulungu, 880 m., fl. & fr. 20.xii.1951, *Richards* 110 (K). W: Kitwe, fl. & fr. 16.xi.67, *Mutimushi* 2365 (K; SRGH). C: Lusaka, by fence of railway reserve, 1220 m., fl. & fr. (no date given), *Best* 364 (MO; SRGH). E: Luangwa Bridge, Gt. East Rd., c. 300 m., fl. & fr. 7.x.1958, *Robson & Angus* 5 (BM; K; LISC; PRE; SRGH). S: Livingstone, fl. & fr. vi.1937, *Obermeyer* 36522 (PRE). **Zimbabwe**. N: Gokwe, fl. & fr. ii.1964, *Bingham* 1071 (LMU; SRGH). W: Hwange Nat. Park, near Kennedy Siding, c. 32 km. S.E. of Main Camp, fl. & fr. 12.xii.68, *Rushworth* 1358 (PRE; SRGH). E: Sabi Valley, Nyanyadzi, 550 m., fl. & fr. 3.ii.1948, *Wild* 2493 (K; SRGH). S: Chiredzi, Nyajena T.T.L., Mabagwashe, fl. & fr. 18.v.1971, *Taylor* 181 (K). **Malawi**. N: Nkhata Bay, White Fathers' Beach, 488 m., fl. & fr. 6.vi.1976, *Pawek* 11358 (K; PRE; SRGH). C: Salima Distr., Senga Bay, Grand Beach Hotel, fl. & fr. 18.iii.1977, *Grosvenor & Renz* 1296 (K; SRGH). S: Nsanje Distr., S. of Lilanje R., c. 90 m., fl. & fr. 25.iii.1960, *Phipps* 2709 (K). **Mozambique**. N: Nampula, fl. & fr. 11.viii.1936, *Torre* 633 (COI; LISC). Z: Muemba, fl. & fr. 12.iii.1943, *Torre* 4927 (LISC). T: Luenya R., 8 km. from Zimbabwe border, 488 m., fl. & fr. 27.ix.1948, *Wild* 2644 (K; SRGH). MS: Maringua, Sabi River, 183 m., fl. & fr. 24.vi.1950, *Chase* 2592 (K; BM; SRGH). GI: Baixo Limpopo, Inhamissa, Hidraulica Agricola camp, fl. & fr. 16.xii.1957, *Aguiar Macedo* 26 (K; LISC; LMA; SRGH). M: old Santaca to Catuane Rd., England Village, on bank of R. Maputo, fl. & fr. 2.ix.1948, *Gomes e Sousa* 3817 (COI; K; LMA; PRE; SRGH).

A pantropical weed. Commonly in waste or disturbed ground, often on sandy soil; 0−1220 m.

3. **Boerhavia coccinea** Miller, Gard. Dict. ed. 8, **4**: (1768).—Agnew, Upland Kenya Wild Flowers: 162 (1974). TAB. **4** fig. C. Type from Jamaica.

Boerhavia viscosa Lag. & Rodr. in Anal. Cienc. Nat. **4**: 256 (1801). Type from South America.

Boerhavia diffusa var. *hirsuta* Heimerl in Bot. Jahrb. **10**: 9 (1888).—Codd in Bothalia **9**, 1: 117 (1966). Type from S. Africa.

Boerhavia diffusa var. *viscosa* (Lag. & Rodr.) Heimerl. in Beitr. Syst. Nyct.: 27 (1897).—Meikle in F.W.T.A. ed. 2, **1**: 177 (1954).—Codd in Bothalia 9, **1**: 118 (1966). Type as above.

Boerhavia bracteata Cooke in Kew Bull. **1909**: 421 (1909).—Cooke in Fl. Cap. 5, 1: 394 (1910).—Burtt Davy, F.P.F.T. **1**: 209 (1926). Types from S. Africa.

Annual or perennial herbs, sometimes viscid. Stems sprawling or ascending to 2−3 (4) m. with erect flowering stems, diffusely branched, sparsely to densely glandular-pubescent, often with interspersed long, septate hairs. Leaves 0.7−5.5 × 0.4−5 cm., ovate, elliptic or subcircular, base subcordate to rounded, apex rounded to acute, often apiculate or mucronate, glandular-pubescent mainly along the veins or more commonly more or less glabrous except for long, septate hairs particularly along veins and margins; petioles up to 4 cm. long, glandular-

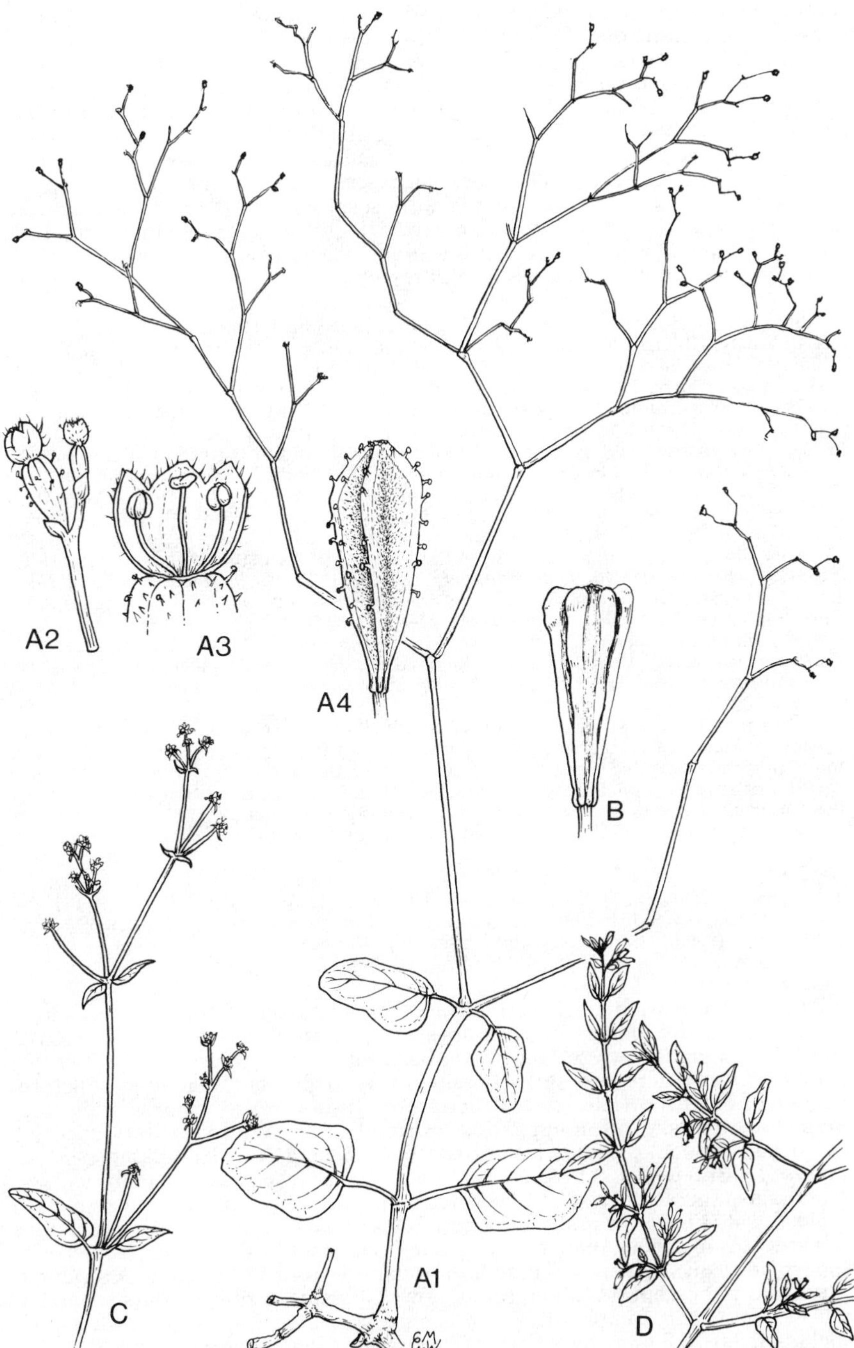

Tab. 4. A.—BOERHAVIA DIFFUSA. A1, habit ($\times\frac{2}{3}$), A2, inflorescence ($\times$8); A3, longitudinal section of apical part of perianth ($\times$18); A4, fruit ($\times$8), all from *Robson & Angus* 5. B.—BOERHAVIA ERECTA. B, fruit ($\times$8) *Balsinhas & Marrime* 449. C.—BOERHAVIA COCCINEA. C, habit ($\times\frac{2}{3}$) *Leach & Brunton* 9993. D.—BOERHAVIA REPENS. D, habit ($\times\frac{2}{3}$) *Gossweiler* 215 B.

pubescent. Inflorescences terminal and axillary, cymose, often aggregated, repeated branching distally and reduction of superior leaves sometimes giving a paniculate appearance, glandular-pubescent, more rarely glabrescent, often viscid; peduncles often stout. Bracteoles 0.6−2 mm. long, ovate, acuminate, margin fimbriate. Flowers 3-many per cluster, subsessile or with pedicels up to 1 mm. long. Perianth 1.75−3.5 mm. long, glandular-pubescent, lower portion 5-ribbed, upper portion 0.5−2.5 mm. long, campanulate, white, pink, magenta or mauve. Stamens (1) 2−3 (4), subequal, exserted 1.5−4.5 mm. long, subequal, exserted. Ovary 0.4−0.6 mm. long, glabrous; style 1.3−4.0 mm. long, exserted. Anthocarps 3−4 × 0.8−1 mm., narrowly fusiform to clavate-fusiform, apex obtuse, glandular- pubescent, rarely glabrescent, ribs 5, smooth, rounded, glandular-pubescent, more rarely glabrescent.

Botswana. N: Boteti floodplain, fl. & fr. 23.ii.1980, *Smith* 3142 (K; MO; SRGH). SW: Okwa Valley, Ghanzi-Lobatsi Rd., fl. & fr. 11.v.69, *Brown* 6073 (K; SRGH). SE: Orapa, Managers house, fl. & fr. 28.i.75, *Allen* 250 BCD (K; PRE). **Zambia**. N: Mbala, Mpulungu, Lake Tanganyika, c. 792 m., fl. & fr. 22.x.1967, *Simon & Williamson* 1152 (LISC; PRE; SRGH). C: near Mumbwa, fl. & fr. 1911, *Macaulay* 357 (K). E: Gt. East Rd., between Hofmeyr turn-off and Kachalolo, c. 650 m., fl. & fr. 12.xii.1958, *Robson* 908 (BM; K; LISC). S: Zeze, Sinazongwe, 600 m., fl. & fr. 28.xii.1958, *Robson* 987 (K; LISC). **Zimbabwe**. N: Urungwe Distr., 35 km. S. of Chirundu, fl. & fr. 11.vi.1960, *Leach & Brunton* 9993 (K; LISC; PRE). W: Matobo Distr., Maleme Dam, c. 1160 m., fl. & fr. i.1960, *Miller* 7076 (MO; SRGH). E: Chipinge Distr., E. Sabi, Mutungawuna, c. 427 m., fl. & fr. 24.i.1957, *Phipps* 154 (COI; LISC; SRGH). S: Beitbridge Distr., fl. & fr. 26.vii.1977, *Phillips* 2637 (K; MO; SRGH). C: Salima Boma, near the Tavern, fl. 9.vii.1980, *Salubeni* 2790 (MO; SRGH). S: Farringdon Rd., near Mangochi, fl. 14.iii.1955, *Exell, Mendonça & Wild* 866 (BM; LISC; SRGH). **Mozambique**. N: between Memba and Nacala, fl. & fr. 17.v.1937, *Torre* 1419 (COI; LISC). T: Cahobra Bassa, Mecangadzi R., left bank, 330 m, fr. 18.x.1973, *Correia, Marques & Pereira* 3515 (LMU). MS: Gorongosa, Chitengo, fl. & fr. 22.x.1965, *Balsinhas* 983 (COI). M: right bank of Incomati R., fl. & fr. iv.1893, *Quintas* 142 (COI).

A pantropical weed. On a variety of soils; up to 800 m.

A very polymorphic species which may well include more than one taxon. Sometimes the inflorescences resemble those of *B. repens* but they have longer peduncles and usually more flowers. Repeated branching distally and reduction of upper leaves can also sometimes give the appearance of one large terminal, paniculate inflorescence, causing possible confusion with *B. diffusa*. Some authors (Codd in Bothalia **9**, 1: 115 (1966); Porcher in Castanea **43**, 3: 172 (1978)) have considered this species and *B. diffusa* to be conspecific.

4. **Boerhavia repens** L., Sp. Pl. **1**: 3 (1753).—Baker & C.H. Wright in F.T.A. **6**, 1: 4 (1909) excl. vars.—Meikle in F.W.T.A. ed. 2, **1**: 177, 178 (1954).—Codd in Bothalia **9**, 1: 121, fig. 2, 3 (1966).—Agnew, Upland Kenya Wild Fl.: 162 (1974).—Fosberg in Smithson. Contrib., Bot. **39**: 8 (1978). TAB. **4** fig. D. Type from Egypt.

Annual herbs with thickened roots. Stems prostrate, sparingly to much branched, slender, puberulous or pubescent to glabrescent. Leaves 0.5−2.5(4) × 0.15−1.2 cm., ovate to lanceolate, base rounded to truncate, apex obtuse to acute, often mucronate, sparsely puberulous to glabrous, glandular punctate, subsessile or with petioles up to 0.7 cm. long. Inflorescences cymose or pseudo-umbellate, axillary, often appearing crowded out of axil by axillary shoot, 1 cyme per peduncle; peduncles up to 10 mm. long, slender, glabrescent to glabrous. Bracteoles 0.5−1.2 mm. long, linear-lanceolate, usually ciliate, glandular-punctate, crisped-puberulous (particularly around margins) to glabrescent. Flowers usually 2−4 per cluster sometimes up to 6−7 in cymes towards tips of shoots, more rarely solitary; pedicels to 1.5 mm. long (sometimes longer in fruit), glabrescent to glabrous. Perianth 1.5−2.5(3) mm. long, glandular-pubescent to glabrescent, lower portion 5-ribbed, upper portion 0.5−1.0 mm. long, campanulate, white, pink or mauve. Stamens 1−2, subequal, included, 0.6−0.8 mm. long. Ovary 0.3−0.5 mm. long, glabrous; style 0.6−1.2 mm. long. Anthocarps 3−3.5 × 1.5−2 mm., clavate to fusiform, apex rounded, glandular pubescent; ribs 5, smooth, more or less glabrous.

Botswana. SW: Bokspits, edge of Molopo valley, fl. & fr. 11.iii.1980, *Timberlake* 2197 (SRGH). **Zimbabwe**. S: Beitbridge, fl. & fr. 16.ii.1955, *Exell, Mendonca & Wild* 449 (BM; LISC; SRGH). Locally common in tropical Asia and Africa on dry soils.

One specimen from Zimbabwe (*Taylor* 29) has larger leaves than average (up to 4 × 2.5 cm.), probably the result of better growing conditions. The distinction between this species and *B. coccinea* is not clear and requires further investigation. C. Jarvis (BM(NH)) has pointed out to me that the specimen LINN 9.8, which *L.E. Codd* (in Bothalia **9**, 1: 121 (1966)) treats as the type of *B. repens*, is annotated "*? repens*". The question mark would possibly indicate that Linnaeus was trying to determine this specimen working from an earlier source and that he was not sure of the correct identity. In view of the fact that the geographical information cited in Sp. Pl. **1**: 3 (1753) is identical with that given in Vaillant's "Structure des Fleurs" (1718) (Nubia, entre Mocho & Tangos), it could be that the latter was this earlier source and that it was on Vaillant's description that Linnaeus based his own description in Sp. Pl. rather than on any specimen. This being the case (and no other authentic material having survived), it would be necessary to neotypify the species using LINN 9.8.

4. PISONIA L.

Pisonia L., Sp. Pl. **1**: 1026 (1753); in Gen. Pl. ed. **5**: 451 (1754).

Scandent or erect shrubs or trees, unarmed or with axillary spines. Leaves opposite, alternate or verticillate, petiolate or sessile, simple, entire. Inflorescences paniculate cymes, axillary or terminal, dense or lax, subsessile or pedunculate. Flowers small, dioecious, rarely monoecious or hermaphrodite, male and female flowers sometimes of different shape; bracts 2–4, not involucrate. Perianth infundibuliform, campanulate, urceolate or tubular, 5–10 lobed or toothed; lobes erect, spreading or reflexed. Stamens variable in number usually 5–10 mostly exserted, rudimentary in female flowers; filaments connate below. Ovary sessile or stipitate, elongate, rudimentary in male flowers; style filiform, included or exserted; stigma lobed-capitate, multifid. Fruit enclosed in hardened, elongated, oblong or clavate, smooth or 4–5-angled perianth base (anthocarp); angles sometimes furnished with longitudinal rows of viscid stipitate glands.

A genus of about 50 species, cosmopolitan in the tropics, mainly America.

Pisonia aculeata L., Sp. Pl. **2**: 1026 (1753).—Choisy in DC., Prodr. **13**, 2: 440 (1849).—Baker, Fl. Maurit. & Seychelles: 263 (1877).—Heimerl. in Engl. & Prantl, Pflanzenfam. **3**, 16: 29 (1889); ed. 2, **16 C**: 127 (1934).—Baker & C.H. Wright in F.T.A. **6**, 1: 8 (1909).—Cooke in Fl. Cap. **5**, 1: 397 (1910).—F.W.T.A. ed. 1, **1**: 152 (1927); ed. 2, **1**: 177 (1954).—Cavaco in Fl. Madag., Nyctaginaceae: 10, fig. 2 (1954). TAB. 5. Type from South America.

Large scandent shrubs, armed with recurved axillary spines (abortive shoots). Trunk attaining 150 cm. diam. Stems slender, terete, spreading; younger shoots ferruginous pubescent to tomentose. Leaves alternate to subopposite, 1.5–5 × 1–3 cm., elliptic to more or less circular, obtuse-acuminate to rounded at apex, broadly attenuate to more or less rounded at base, glabrous to pubescent (sometimes concentrated along main veins), subcoriaceous; petioles 0.3–2 cm long, pubescent. Inflorescences axillary, female laxer than male, branches densely pubescent to tomentose; pedicels much elongated in fruit; bracts 2–3(4), attached near base of perianth, 0.3–0.8 × 0.3–0.4 mm, ovate-oblong, pubescent. Flowers pale yellowish or greenish white, scented. Male flowers: perianth 2.5–3 mm. long, infundibuliform, pubescent, 5 larger lobes alternating with 5 smaller teeth, lobes triangular, spreading or recurved; stamens 6–8, exserted, 3.5–6 mm. long; filaments, 3–5.5 mm. long, filiform; anthers 0.5–0.6 mm. diam., orbicular. Female flowers: 2–2.5 × 1.3–1.5 mm., perianth tubular to urceolate, 5 larger lobes alternating with 5 smaller ones, pubescent, 4–5 longitudinal rows of black glands developing as perianth matures; ovary 1.5–1.7 × 0.5–0.75 mm., elliptic-ovate, glabrous, 1-ovulate; style 0.5–0.8 mm. long, exserted; stigma multifid. Anthocarps 10–25 × 2–4 mm., 4–5-angled, 2–3 longitudinal rows of viscid stipitate glands along each angle, tomentellous to tomentose on flat surfaces between glands.

Mozambique. M: Maputo, fl. 28.iii.1947, *Hornby* 2615 (LMA; PRE; SRGH).

Possibly introduced from tropical America; widespread in Old and New World tropics. In a wide range of habitats; sea level- 1400 m.

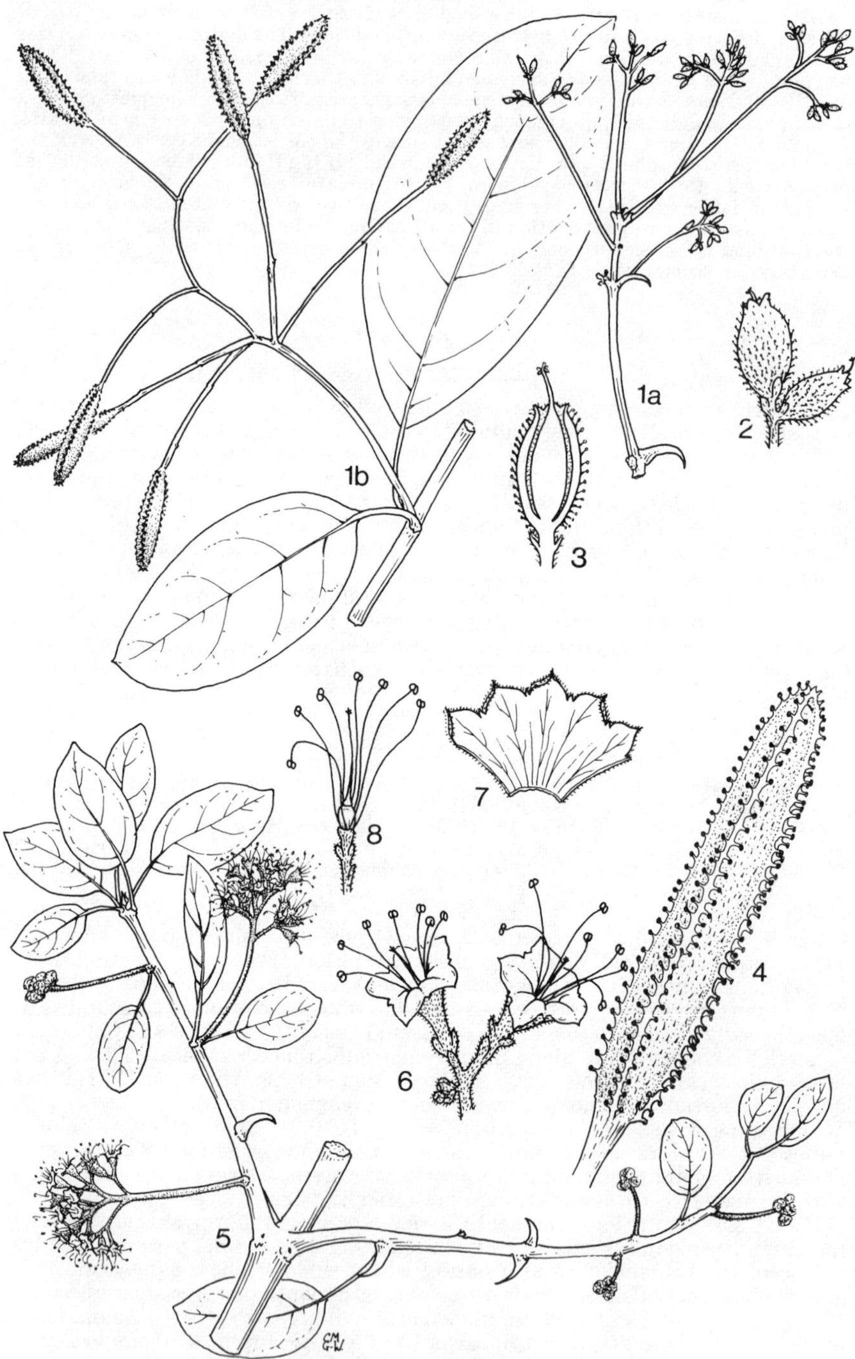

Tab. 5. PISONIA ACULEATA. 1a, habit, female flowers (×$\frac{2}{3}$) *Eggeling* 2821; 1b, fruiting habit (×$\frac{2}{3}$) *Leeuwenberg* 7329; 2, female flowers (×4); 3, longitudinal section of female flower (×4), 2–3 from *Eggeling* 2821; 4, fruit (×3) *Leeuwenberg* 7329; 5, habit, male flowers (×$\frac{2}{3}$); 6, male flowers (×4), 7, perianth of male flower (×4); 8, male flower with perianth removed (×4), 5–8 from *Hornby* 2615;

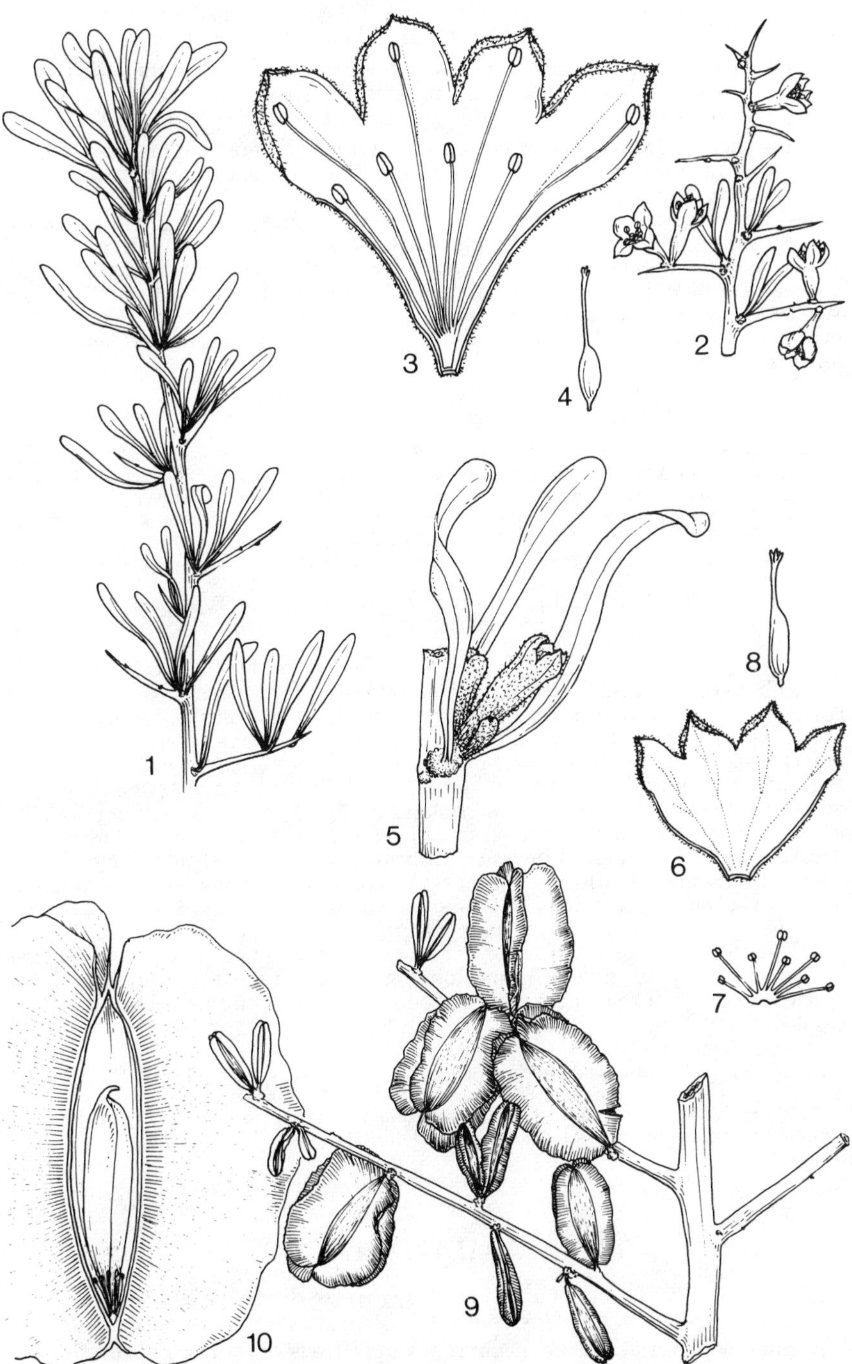

Tab. 6. PHAEOPTILUM SPINOSUM. 1, habit (×$\frac{2}{3}$); 2, hermaphrodite flowering shoot (×$\frac{2}{3}$); 3, perianth and stamens of hermaphrodite flower (×4); 4, gynoecium of hermaphrodite flower (×4), 1–4 from *Giess jun.* 133; 5, female flowers (×4); 6, perianth of female flower (×4), 5–6 from *Bryant* 531; 7, reduced stamens of female flower (×4); 8, gynoecium of female flower (×4), 7–8 from *Bryant* 531; 9, fruiting shoot (×$\frac{2}{3}$); 10, longitudinal section of fruit (×2), 9–10 from *Seydel* 1679g.

5. PHAEOPTILUM Radlk.

Phaeoptilum Radlk. in Abh. Naturwiss. Ver. Bremen **8**: 435 (1883).
Nachtigalia Schinz ex Engl., Bot. Jahrb. **19**: 133 (1894).
Amphoranthus S. Moore in Journ. Bot., Lond. **4**: 305, 408 (1902).

Spiny shrubs, densely branched, spreading. Leaves usually in fascicles, sessile to subsessile more or less fleshy, entire. Flowers solitary or in fascicles, polygamous, usually on leafless branches. Perianth infundibuliform, 4 (5)lobed; lobes spreading (less so in female flowers), more or less $\frac{1}{3}-\frac{1}{2}$ length of perianth. Stamens 8 (9), of two different lengths, the longer exserted, much reduced in female flowers; filaments connate into short more or less fleshy cup at base; anthers oblong-elliptic. Gynoecium reduced in functionally male flowers. Ovary stipitate, 1-ovulate; style asymmetrically inserted at apex of ovary, stout. Fruit enclosed in hardened, elongated, longitudinally 4-winged perianth-tube (anthocarp).

One variable species frequent in southern Africa.

Phaeoptilum spinosum Radlk. in Abh. Naturwiss. Ver. Bremen 8: 436 (1883).—Heimerl in Engl. & Prantl, Pflanzenfam. 3, **1b**: 28 (1889); ed. 2, **16 C**: 123—124, fig. 59 (1934).—Heimerl in Bull. Herb. Boiss. 5, append. **3**: 68 (1897).—Baker & C.H. Wright in F.T.A. **6**, 1: 9 (1909).—Cooke in Fl. Cap. **5**, 1: 397 (1910).—Pohnert & A. Schreiber in Merxm., Prodr. Fl. SW. Afr. **25**: 6—7 (1969). TAB. **6**. Type from S. Africa.
Nachtigalia protectoratus Schinz ex Engl., Bot. Jahrb. **19**: 133 (1894) nom. nud. in syn.
Phaeoptilon heimerlii Engl., Bot. Jahrb. **19**: 133 (1894). Type from S. Africa.
Amphoranthus spinosus S. Moore in Journ. Bot. **40**: 305, 408, fig. 441A (1902). Type from Namibia.

Shrub up to 3 m. tall. Branches terminating in spines with crowded, alternate, spine-like lateral branches. Bark greyish yellow to greyish brown. Leaves greyish green, 5—35 × 1—4 mm., linear-cuneiform, coriaceous, apex rounded or emarginate, glabrous to crisped pubescent. Flowers creamy yellow, scented. Perianth 6—8 mm. long, densely crisped pubescent to tomentose on outside; lobes rounded to (in female flowers) triangular, slightly cucullate, curling inwards as anthocarp develops. Stamens usually 4 longer, 6—12.5 mm. long and 4 shorter, 4—10 mm. long; filaments filiform; anthers 0.5—2 mm. long. Ovary 1.5—3.5 mm. long, fusiform-elliptic; style 1.5—3.5 mm. long; stigma multifid; ovule attached to one side of base of ovary. Anthocarp 15—25 × 12—20 mm. (incl. wings), pubescent to glabrescent, yellowish-green turning pink, red or purple with age; wings more or less semi-circular, parchment-like; central portion fusiform; fruit 7—8 × 2.5—3 mm., oblong-elliptic in outline, square in cross section, remains of reduced stamens around base; seed more or less filling fruit cavity.

Botswana. SW: Mamuno Airstrip, 1067 m., fr. 13.ii.1970, *Brown* 21 (SRGH). SE: near Mahalapye Village, fl. & fr. 9.ii.64, *Yalala* 459 (K; PRE; SRGH).
Southern Africa. On rocky, gravelly and sandy soils, often near dry river courses, reported as being valued as a fodder plant; 840—1300 m.

132. AMARANTHACEAE

By C. C. Townsend

Annual or perennial herbs or shrubs, rarely trees or lianes. Leaves simple, alternate or opposite, exstipulate, entire or nearly so. Inflorescence a dense head, loose or dense and spike-like thyrse, spike, raceme or panicle, basically cymose, bracteate; bracts hyaline to membranous, stramineous to white, subtending one or more flowers. Flowers hermaphrodite or unisexual (plants dioecious or monoecious), mostly actinomorphic, usually bibracteolate, frequently in ultimate 3-flowered cymules; lateral flowers of such cymules sometimes sterile, modified into scales, spines, hooks or hairs. Perianth uniseriate, membranous to firm and finally indurate, usually falling with the ripe fruit included, tepals free or more or less fused below, frequently more or less pilose or lanate, green to white or variously coloured. Stamens isomerous with and opposite the tepals, rarely fewer; filaments free or frequently more or less fused below, sometimes almost completely fused and 5-toothed at the apex with entire or deeply lobed teeth, occasionally some anantherous, alternating with variously shaped pseudostaminodes or not; anthers unilocular or bilocular. Ovary superior, unilocular; ovules commonly solitary, sometimes more numerous, erect to pendulous, placentation basal; style obsolete to long and slender; stigmas capitate to long and filiform. Fruit an irregularly rupturing or circumcissile capsule, rarely a berry or crustaceous, usually with rather thin, membranous walls; seeds round to lenticular or ovoid, embryo curved or circular, surrounding the more or less copious endosperm.

A large and predominantly tropical family of some 65 genera and over 1000 species, including several cosmopolitan weeds and a large number of xerophytic plants; some are locally important vegetable or grain crops, and some are grown as decoratives.

Key to the genera

1. Bracteoles with a projecting dorsal keel along at least the upper part of the midrib - - - - - - - - - - - - - - - - - - 23. **Gomphrena**
– Bracteoles with no dorsal keel along midrib - - - - - - - - - 2
2. Plants monoecious, the male flowers situated towards the apex of the inflorescences or scattered among the females - - - - - - - - 3. **Amaranthus**
– Plants with hermaphrodite flowers or dioecious - - - - - - - - - 3
3. Modified sterile flowers consisting of hairs, spines, bristles or scales present, frequently on one side of a sterile flower, sometimes concealed within the bracteoles in the flowering stage - - - - - - - - - - - - - - - 4
– Modified sterile flowers absent, all flowers fertile - - - - - - 11
4. Sterile flowers of greatly elongated plumose hairs; tall climber attaining c. 6 m. or more - - - - - - - - - - - - - - 4. **Sericostachys**
– Sterile flowers not of long plumose hairs; plant not a tall climber - - - 5
5. Stamens without intermediate pseudostaminodes - - - - - - - 6
– Stamens with intermediate pseudostaminodes - - - - - - - - - 9
6. Modified sterile flowers consisting of numerous uncinately hooked spines - - - - - - - - - - - - - - - - - - 12. **Pupalia**
– Modified sterile flowers without hooked spines - - - - - - - - - 7
7. Inflorescence capitate; leaves narrowly or widely elliptic to lanceolate or ovate, densely canescent - - - - - - - - - - - - - - - 9. **Leucosphaera**
– Inflorescence spicate; leaves linear-acicular to narrowly oblong, glabrous - - 8
8. Leaves linear-acicular, alternate or occasionally opposite; sterile flowers of strap-shaped processes furnished with long unicellular hairs, forming a silky-haired ball in fruit - - - - - - - - - - - - - - - 7. **Sericorema**
– Leaves broadly linear to narrowly oblong, opposite; sterile flowers of single rigid spines - - - - - - - - - - - - - - - - 8. **Centema**
9. Leaves narrowly linear; ovary asymmetrically horned at the apex - - - - - - - - - - - - - - - - - - 5. **Kyphocarpa**
– Leaves not narrowly linear; ovary without an apical horn - - - - - 10
10. Sterile flowers of long plumose-pilose processes - - - - - - 6. **Nelsia**
– Sterile flowers of spines or bracteoliform processes, never plumose-pilose - - - - - - - - - - - - - - - - - - 11. **Cyathula**
11. Stamens 1–2; flowers minute, c. 1.25 mm.; leaves, branches and flower clusters varying alternate to opposite - - - - - - - - - - 14. **Nothosaerva**
– Stamens 4–5; flowers minute to larger; leaves and branches constantly alternate or opposite, or else flowers 4.5–6 mm. - - - - - - - - - - 12
12. Leaves alternate - - - - - - - - - - - - - - - - 13
– Leaves opposit - - - - - - - - - - - - - - - - 16
13. Stamens fused for at least half their length in a tube, either with large bifid, intermediate pseudostaminodes, or with a sterile tooth on each side of the antheriferous tooth, or each filament widely expanded - 2. **Hermbstaedtia**
– Stamens not fused to half their length, pseudostaminodes absent or small, filaments never widely expanded or with a sterile tooth on each side of the anther - 14
14. Inflorescences white-woolly, tepals dorsally densely lanate - - 13. **Aerva**
– Inflorescences not white-woolly, tepals glabrous or pilose about the base only - - - - - - - - - - - - - - - - - - 15

15. Inflorescence capitate, crimson or flushed with bright pink; ovules solitary - - - - - - - - - - - - - - - - - 15. **Mechowia**
— Inflorescence spicate or spiciform-thrysoid, never crimson though occasionally pinkish; ovules mostly numerous - - - - - - - - - - 1. **Celosia**
16. Inflorescence sessile and axillary - - - - - - - - - - - - -17
— Inflorescence not sessile and axillary - - - - - - - - - - - -18
17. Tepals densely woolly-lanuginose, 1-nerved, fused to about halfway - - - - - - - - - - - - - - - - - 21. **Guilleminea**
— Tepals glabrous and 1-nerved or barbellate-pilose and 3-nerved, free - - - - - - - - - - - - - - - - - 22. **Alternanthera**
18. Inflorescence capitate - - - - - - - - - - - - - - -19
— Inflorescence spiciform or cymose-fastigiate- - - - - - - - - -20
19. Inflorescence subtended by mostly 4—5 unequal leaves, whitish - - - - - - - - - - - - - - - - - 23. **Gomphrena**
— Inflorescences not involucrate, pink to red - - - - - - 15. **Mechowia**
20. No Pseudostaminodes present between the filaments - - - 16. **Psilotrichum**
— Pseudostaminodes present between the filaments - - - - - - - -21
21. Robust aquatic or marsh perennial; upper tepal slightly longer than the remainder, with a sharper and often slightly recurved apex - - 18. **Centrostachys**
— Plant not truly aquatic or paludal; upper tepal not thus differentiated - - 22
22. Fruiting perianth strongly deflexed, closely appressed to the inflorescence axis; bracteoles spinous-aristate, the arista longer than the basal wings - - - - - - - - - - - - - - - - - 17. **Achyranthes**
— Fruiting perianth not or less deflexed, not closely appressed to the inflorescence axis; bracteole frequently with the midrib excurrent, but not long and spinous - 23
23. Inflorescence repeatedly branched and cymose-fastigiate - - 10. **Centemopsis**
— Inflorescence spiciform - - - - - - - - - - - - - - -24
24. Outer tepals densely pilose on the dorsal surface - - - - 20. **Pandiaka**
— Outer tepals glabrous or sparingly floccose - - - - - - - - - -25
25. Outer tepals with a single nerve (midrib - - - - - - 19. **Achyropsis**
— Outer tepals 3—5-nerved - - - - - - - - - - - - - - -26
26. Tuberous-rooted perennial; outer tepals with very broad white margins each more or less as wide as the green, 3-nerved centre - - - - - 20. **Pandiaka**
— Annual or short-lived perennial with a slender rhizome; outer tepals with the pale margins much narrower than the centre - - - - - - 10. **Centemopsis**

1. CELOSIA L.

Celosia L., Sp. Pl. **1**: 205 (1753); in Gen. Pl. ed. 5: 96 (1754).—Townsend, The genus *Celosia* (Subgenus *Celosia*) in tropical Africa, Hook. Ic. Pl. **38**, 2: 1—123 (1975).

Annual or perennial herbs, occasionally rather woody at the base, occasionally scandent. Leaves alternate, simple, entire or somewhat lobed. Flowers bibracteolate, in spikes or more generally dense to lax terminal and axillary bracteate thyrses (the uppermost frequently forming a panicle), lateral cymes lax to dense and forming lateral clusters and the inflorescence then spiciform. Perianth segments 5, free, all more or less equal. Stamens 5, the filaments fused below into a short cup, the free portions deltoid below and narrow above or more or less swollen below; pseudostaminodes absent but minute blunt intermediate teeth occasionally found; anthers bilocular. Ovary with few to numerous ovules (one or two species occasionally with solitary ovules in some ovaries); style distinct and elongate to almost obsolete; stigmas 2—3, circular to filiform. Capsule circumcissile, sometimes thickened at the apex. Seeds black, usually strongly compressed and shining, subcircular, feebly to strongly reticulate, grooved, punctate or tuberculate; endosperm present.

About 50 species in the tropics and subtropics of both Old and New Worlds.

1. Inflorescence a simple spike - - - - - - - - - - - - - - 2
— Inflorescence a spiciform thyrse, the partial inflorescence dense to rather lax - - - - - - - - - - - - - - - - - - 4
2. Style 5—7 mm. long, tepals 6—10 mm. long- - - - - - 11. *argentea*
— Style less than 5 mm. long, tepals less than 6 mm. long - - - - - - - 3
3. Tepals 3—3.5 mm. long; style with stigmas 2 mm. long; ovary 4—6-ovulate - - - - - - - - - - - - - - - - - 10 *brevispicata*
— Tepals 1.5—2 mm. long; style very short, c. 0.25 mm.; ovary biovulate - - - - - - - - - - - - - - - - - 9. *vanderystii*

4. Testa of the seeds very shiny, almost smooth, faintly reticulate; style short - 5
– Testa of seeds deeply sulcate or reticulate with convex areolae, and often punctate (but unknown in *C. richardsiae*, which has however a slender style) - - - 6
5. Tepals with the midrib excurrent in a short mucro, drying silvery or stramineous, the margins erose-denticulate at least around the apex - - - 2. *trigyna*
– Tepals cucullate, the midrib not excurrent, drying chestnut-coloured or blackish, not erose-denticulate - - - - - - - - - - - 1. *schweinfurthiana*
6. Lower leaves hastate, provided with lower and broader or longer and narrower basal lobes - - - - - - - - - - - - - - - - - - - 7
– Lower leaves not hastate - - - - - - - - - - - - - - 8
7. Tepals 2.5–3 mm. long; inflorescences slender, at least the lower clusters separated; ovules 5–9 - - - - - - - - - - - - - - - 5. *nervosa*
– Tepals 4–5 mm. long; inflorescences stout, dense; ovules 14 or more - - - - - - - - - - - - - - - - - 8. *pandurata*
8. Tepals with a single midrib and no lateral veins - - - - 3. *chenopodiifolia*
– Tepals 3 or more nerved, the laterals usually reaching the middle of each tepal or more - - - - - - - - - - - - - - - - - - - 9
9. Tepals 2–2.25 mm. long; ovary 1–2-ovulate - - - - 6. *stuhlmanniana*
– Tepals 2.75–6 mm. long; ovary with 5 or more ovule - - - - - 10
10. Tepals (3) 3.5–5 mm. long; ovules mostly 13–25 - - - - - 7. *isertii*
– Tepals 2.75–3.25 mm. long; ovary 5–7-ovulate - - - - - - - - 11
11. Free portion of filaments half as long as the sheath; stigmas rather stout, erect, shorter than or subequalling the c. 0.75 mm. long style - - - - - - 5. *nervosa*
– Free portion of filaments longer than the sheath; stigmas rather slender, flexuous, longer than the c. 1 mm. long style - - - - - - - - 4. *richardsiae*

1. **Celosia schweinfurthiana** Schinz in Engl. Bot. Jahrb. **21**: 178 (1895).—Baker & Clarke in F.T.A. **6**, 1: 22 (1909).—Schinz in Engl. & Prantl Pflanzenfam. ed. 2, **16 C**: 29 (1934).—Hauman in F.C.B. **2**: 18 (1951).—Townsend in Hook. Ic. Pl. **38**, 2: 17, t. 3727 (1975). Type from Zaire.

Celosia schweinfurthiana var. *sansibariensis* Schinz in Bull. Herb. Boiss. Sér. 2, **3**: 9 (1903). Type from Dar es Salaam.

Perennial herb, frequently somewhat woody below, vary variable in habit, prostrate and rooting at the lower nodes to erect or scandent up to c. 5 m., moderately to considerably branched. Stem and branches usually obviously striate or sulcate, glabrous or sometimes (especially at the nodes) furnished with a few multicellular hairs. Leaves lanceolate to lanceolate or deltoid- ovate, acute or acuminate, glabrous or more frequently furnished on the inferior surface about the base with scattered, short, multicellular hairs; larger stem-leaves 2.2–10 × 1–7 cm., shortly cuneate to truncate or subcordate at the base, with a petiole 1–4.5 cm. long; upper and branch leaves progressively smaller and usually narrower. Inflorescences scattered or approximate, few or up to 10-flowered, dense and rounded or lax (with peduncles and pedicels up to c. 3 mm. long), 2–12 mm. in diam. Bracts and bracteoles 0.5–1.25 mm. long, lanceolate or deltoid, glabrous, pale-membranous with the brownish or blackish midrib not excurrent. Tepals 2 mm. long, glabrous, elliptic-oblong, obtuse and subcucullate, pale greenish or whitish (pale brown or blackish when dry), the midrib not excurrent, margins narrowly scarious. Filaments with the free apices slightly to distinctly longer than the cup; no intermediate teeth present. Ovary (5) 6–9-ovulate; stigmas 2, reflexed, much longer than the very short style. Capsule oblong, c. 3 × 1 mm., usually clearly exerted from the perianth when fully mature, brown, more or less truncate with the apex slightly to distinctly rugose-incrassate. Seeds lenticular, black, very shiny, c. 1 mm., feebly reticulate with flat areolae.

Malawi. S: Malanga, Shire Valley, x.1887, *Scott* s.n. (K). **Mozambique**. MS: Between Inhamitanga and Lacerdónia, 7.ii.1942, *Torre* 4082 (LISC).

Widespread in E. and C. Africa. Although in Angola it has not yet been found in Zambia. A forest plant in Mozambique, elsewhere it also occurs in a wider range of habitats as a weed of cultivation, in coastal scrub and on rocky hill slopes.

2. **Celosia trigyna** L., Mant. Pl. Alt.: 212 (1771).—Baker & Clarke in F.T.A. **6**, 1: 19.—Schinz in Engl. & Prantl Pflanzenf. ed. 2, **16 C**:29 (1934).—Hauman in F.C.B. **2**: 17 (1951).—Meeuse in Kirkia 2: 155 (1961).—Cavaco in Mém. Mus. Hist. Nat. Paris Sér. B, **13**: 44 (1962).—Podlech & Meeuse in Merxm., Prodr. Fl. SW. Afr. **33**: 13 (1966);

J.H. Ross Fl. Natal: 158 (1973).—Townsend in Hook. Ic. Pl. **38**, 2: 27, t. 3729 (1975). Type from Senegal.

Celosia melanocarpos Poir., Encycl. Méth. Bot. Suppl. **4**: 318 (1816). Type from Senegal.

Celosia laxa Schumach. & Thonn., Beskr. Guin. Pl.: 141 (1827). Type from Ghana.

Celosia triloba E. Mey. ex Meissn. in Hook., Lond. Journ. Bot. **2**: 548 (misprinted 448) (1843). Type from S. Africa (Natal).

Celosia semperflorens Baker in Kew Bull. 1897: 277 (1897). Type: Malawi, Blantyre, *Buchanan* 52 (K, holotype).

Celosia digyna Suesseng. Trans. Rhod. Sci. Assoc. **43**: 8 (1951); in Mitt. Bot. Staatss. München **3**: 73 (1951), excl. vars. Type: Zimbabwe, Rusape, *Dehn* 1121/52 (K, isotype; M, holotype).

Celosia trigyna var. *convexa* Suesseng. in Mitt. Bot. Staatss. München **1**: 75 (1951). Type: Mozambique, Namagoa, *Faulkner* K.46 (K, lectotype).

Celosia trigyna var. *longistyla* subvar. *triangularis* Suesseng. in Mitt. Bot. Staatss. München **1**: 75 (1951). Type: Mozambique, Quelimane, *Faulkner* 46/a (K, holotype).

Celosia loandensis sensu Suesseng. ex Brenan in Mem. N.Y. Bot. Gard. **9**: 56 (1954) non Baker.

Annual herb, erect, simple or branching from near the base upwards, (8) 30—120 (180) cm. Stem and branches green to reddish, sulcate or striate, glabrous or with short, few-celled hairs especially about the nodes. Leaves narrowly lanceolate to broadly ovate, occasionally (mostly in S. Africa) with a broad, obtuse lateral lobe on each side near the base, acute to acuminate, glabrous or with short, few-celled hairs on the inferior surface about the base; lamina of main stem leaves (1) 2—8.5 (10) × (0.4) 1—4 (5) cm., subcordate to truncate or attenuate below, the lower margins often scabrid, more or less decurrent along the slender, up to more or less 5 cm. long petiole; superior and branch leaves smaller and often narrower, more shortly petiolate; all leaves often fallen by the fruiting stage. Inflorescences axillary and terminal, simple or branched spike-like thyrses more or less 6.5—35 cm. long, formed of distant or (at least above) approximate, few-to many-flowered lax or congested and subglobose, white or pinkish clusters 2—20 (30) mm. in diam., in well-grown individuals the superior leaves much reduced so that a terminal panicle is formed; inflorescence axis glabrous or furnished with multicellular hairs. Bracts and bracteoles ovate to ovate-elliptic, c. 1.25—2 mm. long, scarious with a darker nerve, margins minutely (or more coarsely at the base) erose-denticulate, glabrous. Tepals ovate-elliptic, 1.75—2.75 mm. long, shortly mucronate with excurrent midrib, glabrous, scarious with a narrow less translucent band along the midrib, margins minutely denticulate at least above. Free portion of filaments subequalling the sheath, sinuses rounded with no intermediate teeth; anthers reddish. Stigmas 2—3, longer than the very short style; ovary 6—8-ovulate. Capsule ovoid, 1.75—2.25 mm. long, included or a little exserted, rounded and not thickened at the apex. Seeds c. 0.75 mm. in diam., compressed, black, shining, with a rather fine reticulate pattern, the areolae only slightly convex.

Botswana. N: Sandbank near Okavango R., 15.ii.1979, *Smith* 2652 (PRE; SRGH). **Zambia**. B: Siwelewele 9.viii.1952, *Codd* 7464 (K; PRE; SRGH). N: Mbala Distr., Chilongowelo, 1360 m., 11.iii.1952, *Richards* 1015 (K). W: Mwinilunga Distr., dambo at Chisenga R., Mwinilunga-Solwezi Rd., c. 1400 m., 26.ii.1975, *Hooper & Townsend* 413 (K). C: Kabwe Distr., 24 km. NW. of Kabwe, Chitakata R., 1170 m., 8.iv.1972, *Kornas* 1560 (K). E: Mpangwe Hills, 1150 m., 18.iii.1956, *Wright* 90 (K). S: Monze, 7.vi.1962, *Fanshawe* 6864. (NDO; SRGH). **Zimbabwe**. N: Centenary Distr., Musengezi R., 16.v.1955, *Watmough* 119 (K; LISC; SRGH). W: Bulalima-Mangwe Distr., Isi baba, 18.iv.1942, *Feiertag* 45454 (K; LISC; SRGH). C: Gweru Distr., Whitewaters Dam, 20.ii.1963, *Loveridge* 583 (K; SRGH). E: Chipinge Distr., Chibuya Project, 515 m., v.1958, *Davies* 2459 (K; SRGH). S: Mwenezi Distr., Malangwe R., SW. Mateke Hills, c. 622 m., 6.v.1958, *Drummond* 5649 (K; SRGH). **Malawi**. N: Karonga Distr., Ngala, 27 km. N. of Chilumba, 492 m., 8.vii.1973, *Pawek* 7173B (K; MAL; MO). C: Lilongwe Agricultural Research Station, 31.iii.1955, *Jackson* 1544 (K; SRGH). S: Chikwawa Distr., Kapachira Falls, W. bank of Shire R., c. 110 m., 21.vi.1970, *Brummitt* 9992 (K; EA; LISC; MAL; PRE; SRGH; UPS). **Mozambique**. N: Massangulo, 14.iv.1933, *Gomes e Sousa* 1359 (K). Z: Quelimane Distr., Namagoa, viii.1947, *Faulkner* K46 (K; SRGH). T: Rd. Tete-Songo, near R. Sanângoè a c. 16 km. from Marueira, 7.iv.1972, *Macêdo* 5154 (K; LMA; SRGH). MS: Beira, 1894, *Kuntze* s.n. (K). GI: Gaza, Xai-Xai, Inhamissa, 13.viii.1957, *Barbosa & Lemos* 7819 (K;

LISC; LMJ). M: Maputo, Marracuene Campo Experimental de Estudos Arboricolas, 1.viii.1957, *Barbosa & Lemos* 7868 (K; LISC; LMJ).

Throughout most of tropical Africa (including Madagascar); in Namibia and S. Africa; also in Arabia and Madeira. Recently recorded as naturalised in Florida. The commonest species in the Flora Zambesiaca region, as elsewhere in Africa. Most frequently a weed of cultivation or disturbed ground on sandy, loamy or clay soils, also along watercourses (even in standing water), along forest tracks and in open woodland, on rocky ground, in grassland; near sea level to c. 1960 m.

3. **Celosia chenopodiifolia** Baker in Kew Bull. **1897**: 276 (1897).—Townsend in Hook. Ic. Pl. **38**, 2: 33, t. 3730 (1975). Type from Angola.

Annual (? sometimes biennial) herb, with one to several stems from the base, stems generally with rather few, long and slender branches; stem and branches striate, glabrous. Leaves elliptic or rhomboid-lanceolate, the inferior leaves of the main stem to c. 4.5 × 1.7 cm. with a distinct petiole up to 1.25 cm. long, the superior and branch leaves reducing (usually rapidly so) and more shortly petiolate. Inflorescences terminal or terminal and axillary, narrowly spiciform, 3–25 × 0.4–1 cm., the partial inflorescence usually small and very compact, rarely looser with the lowest shortly pedunculate; the lowest clusters usually increasingly distant. Bracts and bracteoles deltoid, rather obtuse, 1–1.5 mm. long. hyaline, very shortly mucronate with the excurrent stramineous midrib, glabrous, the apex minutely denticulate. Tepals 1.75–2 mm. long, oblong, glabrous, rounded above or usually more or less cucullate, hyaline or pale-stramineous, minutely denticulate, very shortly mucronate with the excurrent costa. Filaments with the free apex distinctly shorter than the basal cup, no intermediate teeth present. Ovary 2–4 (6) ovulate; stigmas 2–3, erect, more or less recurved, slightly longer than the very short style. Capsule subglobose, 1–1.25 mm., not exserted beyond the perianth. Seeds lenticular, c. 1 mm. in diam., black and moderately shiny, deeply sulcate-reticulate centrally with broadly convex areolae, and these longer and narrower towards the edge.

Zambia. (Only once collected) W: Mwinilunga Distr., 96 km. S. of Mwinilunga on Kabompo Rd., 3.vi.1963, *Loveridge* 742 (K; SRGH).

Also in Angola. In Zambia found by a roadside; in Angola also occurs in abandoned cultivation on "dampish soil".

4. **Celosia richardsiae** Townsend in Hook. Ic. Pl. **38**, 2: 45, t. 3733 (1975). Type: Zambia, Mporokoso Distr., *Richards* 9128 (K, holotype).

Herb, duration and habit not known. Stem and branches slender, terete, rather finely striate, moderately furnished with pale multicellular hairs, branches diverging at 45 degrees. Inferior leaves not seen; superior stem leaves lanceolate-ovate, to c. 3.5 × 1.6 cm., shortly acuminate, acute or rather blunt, rapidly diminishing in size upwards (those of the branches very small), glabrous above, sparsely pilose along the nerves on the lower surface, cuneate to c. 5 mm. long petiole. Inflorescence creamy, terminal and (by branch reduction) axillary, unbranched, spiciform-thyrsoid, up to c. 20 × 1.6 cm.; partial inflorescences condensed cymes up to 1.5 cm. in diam. Bracts and bracteoles deltoid-ovate, stramineous when dry, 1.75–2.25 mm., shortly mucronate withe the excurrent, brownish midrib, glabrous or sparsely ciliate with multicellular hairs. Tepals 3–3.25 mm. long, oblong-elliptic, acute, glabrous, stramineous when dry, mucronate with the excurrent midrib, which is subtended on either side by one pair of long nerves extending well towards the apex and a pair of shorter nerves extending to c. halfway or less. Filaments with the free apices longer than the basal cup; no intermediate teeth present. Ovary 6–7-ovulate; stigmas 2–3, slender and flexuose, slightly longer than the c. 1 mm. long style. Capsule and seeds unknown.

Zambia. N: Mporokoso Distr., Mweru-Wantipa, steep path to lake near Kapinda, 318 m., 10.iv.1957, *Richards* 9128 (BR; K).

Not known from elsewhere.

5. **Celosia nervosa** Townsend in Hook. Ic. Pl. **38**, 2: 49, t. 3734 (1975). Type: Mozambique, S. of Maputo, 13.viii.1948, *Gomes e Sousa* 3779 (K, holotype).

Herb, probably perennial, habit and size unknown, perhaps trailing or scrambling. Stem and branches long and slender, terete, finely striate, more or less densely furnished with multicellular hairs. Leaves deltoid-lanceolate or feebly and obtusely hastate with low lateral lobes near the base, lamina to 4.5 × 1.5 cm., glabrous above, moderately pilose on the lower surface of the nerves, with a more or less densely pilose petiole up to 1.5 cm. long; superior leaves narrower and more shortly petiolate. Inflorescence white, terminal and axillary (that terminal on the main stem sometimes with one or more branches), spiciform, to c. 28 × 1.4 cm.; partial inflorescence compact, fasciculiform, up to c. 1.2 cm. in diam. Bracts and bracteoles deltoid, acute, membranous, stramineous when dry, 1.5–2 mm., shortly mucronate with the excurrent brownish midrib, glabrous or with slender multicellular hairs below. Tepals 2.75–3 m. long, oblong, acute, glabrous, white, stramineous when dry, the midrib excurrent in a short mucro and subtended on each side by two pairs of lateral nerves - one pair long and reaching almost to the apex of the tepals, the outer pair shorter. Filaments with the free apices only c. half the length of the basal cup; no intermediate teeth present. Ovary 5–7-ovulate; stigmas 2–3, equalling or exceeding the very short style. Capsule shortly ovoid, c. 2 mm., not exceeding the perianth. Seeds lenticular, black, shiny, c. 1.2 mm. in diam., densely sulcate-punctate.

Mozambique. N: Liupo, 26.ix.1948, *Pedro & Pedrógão* 4652 (EA). GI: Panda, 25.ii.1955, *Exell, Mendonça & Wild* 602 (BM; LISC). M: Maputo, vi.1914, Maputoland Exped. 45 (LISC).

Not known from elsewhere. Sandy clay soil, in open forest.

6. **Celosia stuhlmanniana** Schinz in Bull. Herb. Boiss. **4**: 419 (1896).—Baker & Clarke in F.T.A. **6**, 1: 21 (1909).—Hauman in F.C.B. **2**: 25 (1951).—Townsend in Hook. Ic. Pl. **38**, 2: 53, t. 3735 (1975). TAB. **7**. Type from Zaire.

Spreading or scandent branched perennial herb, up to 9 m. but not rarely only 1–2 m. in the Flora Zambesiaca area. Stem and branches striate, glabrous or (especially about the nodes) more or less furnished with brownish multicellular hairs. Leaves glabrous or sparsely furnished with scattered multicellular hairs along the veins of the inferior surface, those of the median part of the stem 3.5–15 × 1.5–8 cm., broadly ovate to ovate-lanceolate, lanceolate or elliptic, acutely acuminate, shortly cuneate or rounded into the 1–3 cm. long petiole, the superior leaves narrower and more shortly petiolate. Inflorescences terminal and axillary, the axillary simple or branched, c. 8–25 × 0.5 cm., the terminal usually a broad panicle up to 70 × 30 cm. with branches diverging more or less at right angles; indumentum of axes resembling that of the branches; partial inflorescences always condensed, 2–6 mm. in diam., usually distant at least below. Bracts and bracteoles c. 1 mm. long, deltoid-ovate, acute, very shortly mucronate with the excurrent midrib, with at least the margins and usually the base of the midrib furnished with long, multicellular hairs. Tepals 2–2.5 mm. long, oblong-elliptic, acute, mucronate with the excurrent midrib, which is subtended on each side by one long inner and one shorter outer nerve, scarious, with a central stramineous or buff-coloured area, moderately furnished towards the base with long brownish, flexuous, multicellular hairs. Filaments with the free apices shorter than the basal cup; short, blunt intermediate teeth often present, these evanescent at maturity. Ovary 1–2-ovulate; stigmas 2, divergent, longer than the 0.5 mm. long style. Capsule shortly ovoid, rounded into the style, 1.25–1.5 mm. long, included. Seeds 1–1.25 mm. in diam., lenticular, black, shiny, depressed-verruculose, the areolae of the reticulum strongly convex.

Zambia. N: Between Nkamba and R. Lufubu, near L. Tanganyika, 28.vi.1933, *Michelmore* 454 (K).

Also in Zaire, Uganda, Tanzania. Recorded mostly from shaded places, in forest and scrub, in dense vegetation on valley sides and in a gorge, but also in grassland; frequently but not invariably near water; 900–1500 m.

7. **Celosia isertii** Townsend in Hook. Ic. Pl. **38**, 2: 57, t. 3736 (1975). Type from Ghana.
Celosia laxa sensu Engl. in Pflanzenw. Ost-Afr. C: 172 (1895).—Baker & Clarke in F.T.A. **6**, 1: 18 (1909).—Schinz in Engl. & Prantl Pflanzenfam. ed. 2, **16 C**: 29

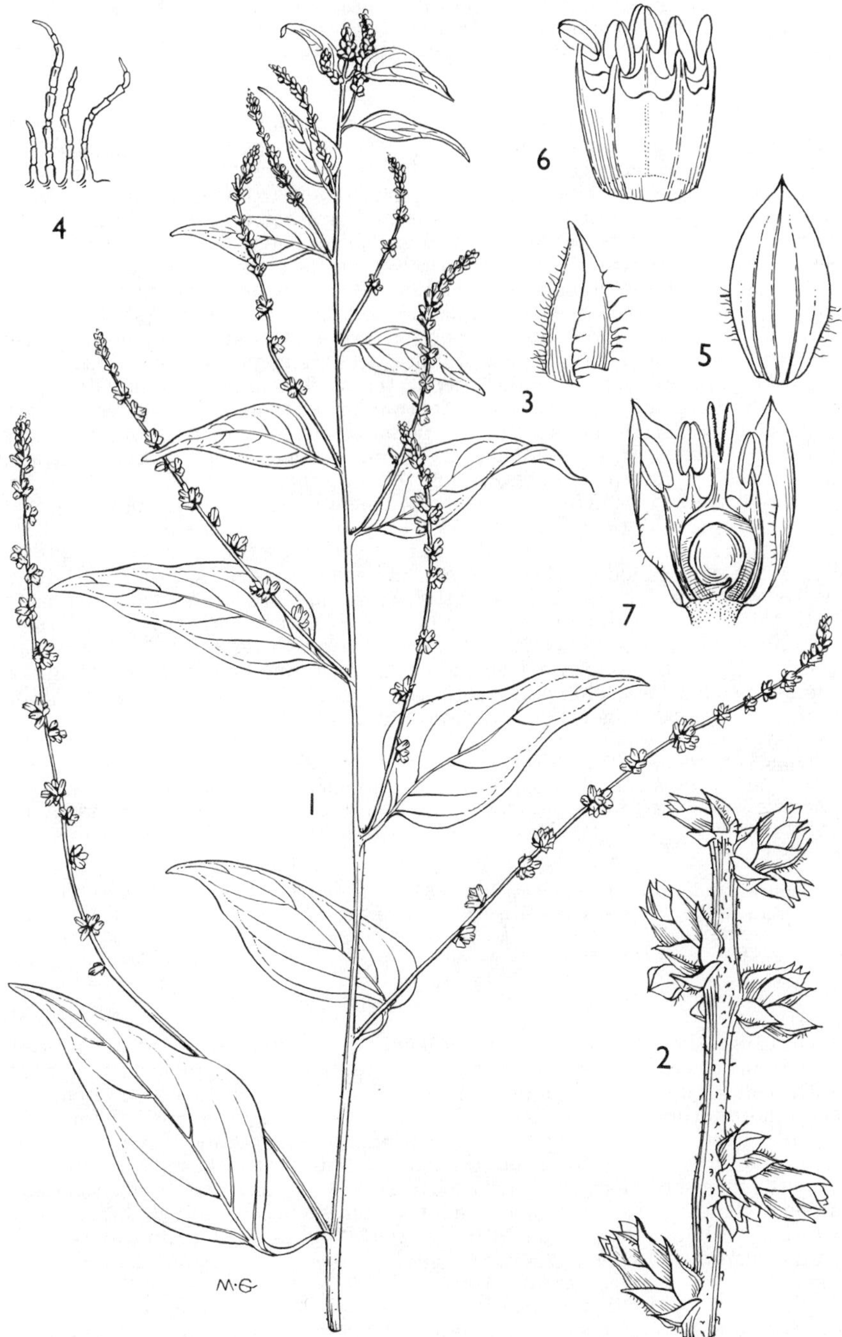

Tab. 7. CELOSIA STUHLMANNIANA. 1, flowering branch (×1) from *Stuhlmann* 3584; 2, part of inflorescence (×6); 3, bract (×14); 4, enlargement of hairs on inflorescence axis (×50); 5, tepal (×14); 6, androecium (×14); 7, longitudinal section of flower (×14), 2–7 from *Haarer* 2218.

(1834).—Hauman in F.C.B. 2: 20 (1951).—Cavaco in Mém. Mus. Hist. Nat. Paris Sér. B, **13**: 43 (1962) non Schum. & Thonn.

Celosia laxa var. *pilosa* Schinz in Bull. Herb. Boiss. Sér. 2, **3**: 9 (1903). Type from Cameroon.

A weak-stemmed perennial herb, often woody below, variable in habit from lax and prostrate, supported by other plants or scrambling to 4 m. (or probably more). Stem and branches glabrous or thinly to rather densely furnished with brownish multicellular hairs. Leaves lanceolate-ovate to broadly or deltoid-ovate, acuminate, glabrous or with multicellular hairs below (especially along the nerves), or rarely also the superior surface; larger leaves 3.5—10 × 2.5—5 cm., shortly cuneate to truncate or subcordate at the base, attenuate along a slender petiole; superior and branch leaves smaller, often narrower and more attenuate at the base. Inflorescences white, axillary and terminal, densely or more rarely laxly thyrsoid, the upper usually more or less congested to become spiciform or lobed and subpaniculate; partial inflorescences condensed or with unequal branches, approximate or slightly separated occasionally more distant), up to c. 1.5 cm. long; inflorescence axis glabrous or with multicellular hairs, often concealed by the densely set flowers; peduncle up to 9 cm. long but generally shorter. Bracts and bracteoles 2—3 mm. long, broadly deltoid-ovate, membranous, glabrous or marginally ciliate, mucronate with the shortly excurrent midrib, which is furnished with multicellular hairs on its dorsal surface. Tepals elliptic-oblong, (3) 3.5—5 mm. long, obtuse or subacute, whitish-membranous with (at least in the dry state) a darker central vitta; midrib shortly excurrent, subtended on each side at the base by one or two pairs of lateral nerves which are evanescent at or a little above the centre of the tepal. Filaments with the free apices longer than the basal cup. Ovary with numerous ovules (usually c. 13—25); stigmas (2) 3 (4), more or less reflexed, slightly longer than the c. 1—1.5 mm. long style. Capsule ovoid, 2—4 mm. long, not exceeding the tepals. Seeds black, shiny, lenticular, c. 1 mm., sulcate-punctate with the marginal areolae longer, narrower and more concentric.

Zambia. N: Samfya, 5.v.1958, *Fanshawe* 4399 (K; LISC; NDO).

Widespread in W. and C. tropical Africa to Uganda and bordering NW. Tanzania. Characteristically scrambling at forest edges, along water-courses and roadsides with thick undergrowth; also as a weed of plantations or abandoned cultivation and other disturbed ground; 1490—1800 m.

8. **Celosia pandurata** Baker in Kew Bull. **1897**: 276 (1897).—Baker & Clarke in F.T.A. **6**, 1: 19 (1909). Pro Parte.—Townsend in Hook. Ic. Pl. **38**, 2: 63, t. 3737 (1975). Type: Mozambique, iii/iv. 1860, *Kirk* s.n. (K; holotype).

Celosia cuofolia Baker in Kew Bull. **1897**: 276 (1897).—Baker & Clarke in F.T.A. **6**, 1, 1: 21 (1909).—Meeuse in Kirkia 2: 156 (1961). Type: Mozambique, near the foot of Morrumbala Hill, x.1858, *Kirk* s.n. (K, lectotype, selected here).

Herb, duration and habit not known, 0.6—1 m., sparsely branched above. Stem and branches slender, terete, striate, more or less densely furnished throughout with whitish or yellowish multicellular hairs. Leaves hastate with a shortly acuminate terminal lobe up to c. 5 × 2 cm. in length and breadth and usually obtuse basal lateral lobes up to c. 3 × 1.2 cm., (in one specimen lamina without lobes and up to c. 5 × 2.5 cm., but this has the appearance of atypical secondary growth), furnished along the veins with multicellular hairs, the lamina decurrent along a slender petiole 1.5—4 cm. long. Inflorescences terminal and axillary, whitish or pale mauve, spiciform-thyrsoid, the axillary pedunculate with peduncles up to 3 cm. long, the terminal solitary or paniculate by reduction of the upper leaves, c. 1.5—10 × 1.2—1.5 cm.; partial inflorescences fasciculiform, approximate or the lower separated. Bracts and bracteoles deltoid-lanceolate or deltoid-ovate, c. 2.5—3 mm. long, acute, whitish or stramineous, mucronate with the excurrent prominent midrib, the margins dentate-erose to lacerate, sometimes plicate-undulate. Tepals 3.5—4 mm. long, oblong, acute, glabrous, with a pair of long nerves on either side of the midrib and a further outer pair of shorter nerves (the long nerves occasionally forked), the midrib excurrent in a short mucro, margins minutely lacerate-dentate above. Filaments with the free apices equalling or shorter than the basal cup; no intermediate teeth present. Ovary 14–17-ovulate; stigmas 2, erect or flexuose, c. 0.75 mm. long, shorter than the distinct c. 1.25 mm. long style. Capsule shortly ovoid or subglobose, c. 2–2.5 mm., not exceeding the perianth. Seeds rotund-quadrate, strongly compressed, black, shiny, c. 1 mm. wide, densely sulcate-punctate.

Mozambique. Z: Massingire, slope of Serra da Morrumbala, near. Rd., 13.v.1943, *Torre* 5307 (LISC). T: Lower valley of Shire R., v.1861, *Meller* s.n. (K). MS: Below Chigogo, Lower Zambezi, iii.iv.1860, *Kirk* s.n. (K).

Not known from elsewhere. The only habitat recorded is for forests, in two of the three cases described as "dense".

One specimen named as *C. pandurata* but not cited above is *Tinley* 2594, from Gorongosa National Park. This has the leaves without lateral lobes at the base and had quite the appearance of *C. isertii* but with the very long inner pair of lateral nerves as in *C. pandurata*; it appears however to be secondary growth and may thus be aberrant.

9. **Celosia vanderystii** Schinz in Viert. Nat. Ges. Zürich **76**: 135 (1931).—Hauman in F.C.B. 2: 25 (1951).—Townsend in Hook. Ic. Pl. **38**, 2: 107, t. 3748 (1975). Type from Zaire.

Erect annual herb, 10–40 cm. tall, simple or with a few to numerous long, ascending branches; stem and branches slender, striate, glabrous or with short, asperulous hairs especially towards the base. Leaves elliptic or sometimes narrowly oblanceolate, 6–27 × 2–7 mm., glabrous or sometimes asperulous along the 2–5 mm. long petiole. Inflorescence creamy, stramineous or pale yellowish tinged with pink, densely spicate, 5–20 × 4–5 mm., solitary, geminate or clustered at the apices of the stem and branches, usually subtended at the base by one or more reduced leaves. Bracts and bracteoles persistent after fruit-fall, 1.25–1.5 mm. long, lanceolate, acute or shortly acuminate, membranaceous, more or less serrulate at the apex, very shortly mucronate with the excurrent midrib. Tepals 1.75–2.25 mm. long, glabrous, oblong, acute, distinctly denticulate in the apical third with the teeth commonly divergent or even recurved, less distinctly so below, very shortly mucronate with the excurrent midrib. Filaments with the free apex subequalling the basal cup; no intermediate teeth present. Ovary biovulate; stigmas 2, erect, very short, shorter than the c. 0.25 mm. long style. Capsule subglobose, c. 1.5 mm. long, not exceeding the perianth. Seeds lenticular, black, shiny, c. 1.25 mm. in diam., reticulate centrally with convex areolae, almost smooth marginally.

Zambia. B: 40 km. W. of Kaoma, 5.iv.1966, *Robinson* 6911 (K). N: Kasama-Mpika Rd., near Chambeshi pontoon, 1260 m., 29.iv.1962, *Richards* 16384 (K). W: Kalene Hill, 23.ii.1975, *Hooper & Townsend* 342 (K). C: Serenje, 18.ii.1955, *Fanshawe* 2090 (K; NDO).

Also in Zaire and Angola. Apparently always in light sandy or loamy soil, along roads and tracks, in open Miombo woodland, near streams. The author and Miss S. S. Hooper found it to be a common species in the Mwinilunga region from Kalene Hill southward in 1975; c. 1350–1400 m.

10. **Celosia brevispicata** Townsend in Hook. Ic. Pl. **38**, 2: 111, t. 3749 (1975). Type: Zambia, *Robinson* 6856 (K, holotype).

Annual herb, erect, 15–30 (52) cm. tall, glabrous throughout, simple or branched from the base and/or upwards with long, widely divaricate branches; stem and branches slender and wiry, striate-sulcate. Leaves linear-filiform, 8–45 × 0.3–1 mm., rarely oblanceolate and up to 34 × 8 mm., subacute, the midrib prominent on the inferior surface. Inflorescences silvery-stramineous, of dense short spikes up to 4 cm. long and 6–9 mm. broad, terminal on the stem and branches. Bracts and bracteoles similar, deltoid-ovate, acute, c. 1.25–1.5 mm. long, shortly mucronate with the thick, yellowish, excurrent midrib. Tepals oblong-elliptic, glabrous, shortly mucronate with the excurrent midrib, basally with a narrow green vitta and a short lateral nerve on each side of the midrib, otherwise hyaline, 3–3.5 × 1.5 mm., minutely denticulate at the apical margins. Filaments with the free apex shorter than the basal cup; no intermediate teeth present. Ovary 4–6-ovulate; style c. 2 mm. long; stigma solitary, linear or very shortly bilobed. Capsule subglobose, not exserted beyond the perianth. Seeds compressed-oval, black, shiny with a prominent hilum, the testa reticulate

Tab. 8. CELOSIA ARGENTEA. 1, flowering stem (×$\frac{2}{3}$); 2, flower with 2 tepals and part of androecium removed (×6); 3, tepal (×6); 4, androecium (×6); 5, longitudinal section of ovary (×6). All from *Chandler* 164.

centrally and punctate in the corners of the areolae, smoother and epunctate along the margins.

Zambia. W: Mongu, 5.iii.1966, *Robinson* 6856 (EA; K; LISC; SRGH).
Also in Zaire and Angola. In Zambia in well-drained woodland, elsewhere also in plantations, on well-drained, fine, sandy soils.

11. **Celosia argentea** L., Sp. Pl. **1**: 205 (1753).—Baker & Clarke in F.T.A. **6**, 1: 17 (1909).—Schinz in Engl. & Prantl Pflanzenfam. ed. 2, **16 C**: 29 (1934).—Hauman in F.C.B. **2**: 16 (1951).—Cavaco in Mém. Mus. Hist. Nat. Paris Sér. B, **13**: 41 (1962).—Townsend in Hook. Ic. Pl. **38**, 2: 115, t. 3750 (1975). TAB. **8**. Type, Linnean specimen 288.1 (LINN, lectotype).

Erect annual herb, 0.4—2 m. tall, simple or much branched with the branches ascending. Stem and branches obviously striate and often sulcate, quite glabrous. Leaves oblong-lanceolate to narrowly linear, acute or obtuse, glabrous, shortly mucronate with the excurrent midrib; leaves from the centre of the stem 2—15 × 0.1—3.2 cm., attenuate below into a slender indistinct petiole; superior and branch leaves smaller, obviously reduced; small-leaved sterile shoots often present in the leaf axils. Inflorescences dense, (sometimes laxer below), spicate, many-flowered, 2.5—20 × 1.5—2.2 cm., silvery or pink, at first conical but later cylindrical, on long peduncles up to 20 m. or more at the ends of the stem and branches. Bracts and bracteoles lanceolate (or the lower deltoid), 3—5 mm. long, hyaline, more or less aristate with the excurrent midrib, persistent after fruit-fall. Tepals 6—10 mm. long, narrowly oblong-elliptic, acute to rather obtuse, mucronate with the excurrent midrib, with one or two pairs of lateral nerves of which the inner reach about halfway up the tepal or more, centrally greenish or yellowish, with hyaline margins. Filaments delicate, the free part equalling or exceeding the basal cup; intermediate teeth none or very rarely minute; both filaments and anthers creamy to magenta. Ovary 4—8-ovulate; style slender, 5—7 mm. long, with 2—3 very short stigmas. Capsule ovoid or subglobose, 3—4 mm. long. Seeds lenticular, black, shiny, c. 1.25—1.5 mm. in diam., faintly reticulate.

Zambia. N: Kawambwa Distr., Kafulwe, Lake Mweru, 960 m., 24.iv.1957, *Richards* 9432 (K). W: Mwinilunga Distr., Matonchi Farm, 1936, *Paterson* 22 (K). C: Mfuwe, Mpika Distr., 610 m., 17.vii.1969, *Astle* 5710 (K; SRGH). E: Chipata Distr., left bank of Luangwa R., Kakumbi, 12.v.1965, *Mitchell* 2935 (SRGH). **Malawi**. S: Tandanko Valley, Zomba, 3.ii.1955, *Jackson* 1447 (K).
Throughout tropical Africa. Practically a pantropical weed, occurring in some warm temperate regions. In the Flora Zambesiaca area recorded as growing on sandy or clay soil, apparently as a weed of gardens or disturbed ground and also on a flood-plain, but elsewhere on a much wider range of soils and habitats; 46—960 m.

A note on *Richards* 9432 observes that it is grown as a relish in villages, and the leaves eaten with fish. The Cockscomb, f. *cristata* (L.) Schinz (*C. cristata* L.), a form of *C. argentea* with variously monstrous inflorescences, is grown in Mozambique (Maputo, *Barbosa & Domingues* 5304, BM; LMJ) and no doubt throughout the Flora Zambesiaca area.

2. HERMBSTAEDTIA Reichenb.

Hermbstaedtia Reichenb., Consp. Regn. Veg.: 164 (1828).—Townsend in Kew Bull. **37**: 82—90 (1982).
Berzelia Mart., Nova Acta Acad. Caesar. Leop. Carol. **13**, 1: 292 (1826) non Brongn. in Ann. Sci. Nat. Bot. Sér. **1**, 8: 370 (1826).
Hyparte Raf., Fl. Tellur. **3**: 43 (1836).

Annual or perennial herbs, sometimes somewhat woody at the base, but never scandent. Leaves alternate, simple, entire. Flowers bracteate and bibracteolate in spikes or heads terminal on the stem and branches; bracts and bracteoles persistent on fruit-fall. Perianth segments 5, equal or slightly dissimilar. Stamens 5, the filaments fused for at least half their length, the alternating pseudostaminodes either free and bifid, or each segment adnate to a variable degree to the adjoining filament; antheriferous apex thus appearing as a short tooth overtopped by a free pseudostaminode on each side, or with one short or elongate tooth on each side, or appearing as apical on a broadly flattened, abruptly narrowed or bluntly shouldered "filament"; anthers bilocular. Ovary with (2) numerous (over 20) ovules; style distinct and elongate to almost obsolete; stigmas 2–3 (5), filiform. Capsule circumcissile. Seeds black, strongly compressed and shiny, feebly reticulate, endosperm present.

A genus of 14 species, exclusively in tropical and southern Africa.

1. Anthers set on a short tooth projecting from the apex of what appears as a widely compressed, round-tipped or bluntly shouldered filament - - - - - - 2
— Anthers set on a short tooth subtended by a short or longer tooth on either side, or with a free bifid pseudostaminode on each side of the filaments - - - 4
2. Ovary long-tapering above, with colourless papillose hairs in upper part - 3. *scabra*
— Ovary not or less tapering above, smooth or mammillate-verruculose but never with colourless papillose hairs - - - - - - - - - - - - - - 3
3. Inflorescence very stout and dense-flowered, soon rounded at the apex, rarely attaining 7 cm. in length; tepals usually exceeding 5 mm.; ovules more than 10 and up to 18 - - - - - - - - - - - - - - - 1. *angolensis*
— Inflorescence slender, elongating, tapering at the apex; tepals not usually exceeding 5 mm. in length, and then the inflorescence finally much exceeding 7 cm.; ovules not more than 8 - - - - - - - - - - - - - - 2. *fleckii*
4. Style very short or obsolete, very rarely to 1 mm. long, shorter than the stigmas, only exceptionally exceeding the perianth in fruit; pseudostaminodes long - - - - - - - - - - - - - - - - - - - 5. *odorata*
— Style distinct, slender, rarely as short as 1 mm., longer than the stigmas which usually equal or project from the perianth in fruit; pseudostaminodes very short, or a short tooth present on each side of the anther - - - - - - - 4. *linearis*

1. **Hermbstaedtia angolensis** C.B. Clarke in F.T.A. **6**, 1: 29 (1909).
Celosia welwitschii Schinz in Engl. Bot. Jahrb. **21**: 189 (1895); in Engl. & Prantl Pflanzenfam. ed. 2, **16 C**: 30 (1934) non *Hermbstaedtia welwitschii* Baker in Kew Bull. **1897**: 278 (1897) quid est *H. argenteiformis* Schinz. Type from Angola.

Erect annual herb, c. 0.2–0.75 m. tall, simple to considerably branched from the base or in the upper half only, branches at c. 30 degrees from the stem, stem and branches more or less angled, striate, moderately furnished with multicellular hairs. Inflorescences terminal on the stem and branches, whitish, very densely spicate and remaining so in fruit, 1.5–7 × 1.25–1.5 cm., the axis (not visible until fruit-fall) more or less densely furnished with whitish, multicellular hairs. Bracts narrowly lanceolate, 4–5 mm. long, hyaline with a narrow white central band, glabrous or ciliate, often slightly erose-denticulate, reaching over halfway along the perianth; bracteoles almost identical; bracts and bracteoles clothing the rhachis with a dense scaly mass after fruit fall. Tepals narrowly elliptic, 5–7 mm. long, glabrous, all more or less similar, with narrow hyaline margins, a shortly excurrent midrib, a pair of long lateral veins and 1–2 shorter pairs. Stamens fused into a tube for about half their length, less than half as long as the perianth, with no free pseudostaminodes, the free portions of the tube rounded into the antheriferous apex or the latter longer than the short blunt teeth on each side of it. Ovary 10–18-ovulate, rounded above; stigmas 3, spreading or recurved and shorter than the finally exserted, 1.5–2 mm. long style. Capsule c. 5–6 mm. long, circumcissile, the apex rather shortly narrowed to the style and sometimes with a slight shoulder, glabrous or rarely (not in the Flora Zambesiaca area?) with a few colourless papillose hairs. Seeds c. 1 mm. in diam.

Zambia. W: Masese, 10.v.1961, *Fanshawe* 6551 (K; LISC; SRGH). **Zimbabwe**. W: Hwange Distr., Impofu Area, c. 6 km. NW. of Main Camp, Hwange Nat. Park, 18.iii.1969, *Rushworth* 1708 (K; SRGH). C: Nkai Distr., Gwampa Forest Reserve, 910 m., v.1956, *Goldsmith* 111/56 (K; LISC; SRGH).
Also in Angola and Namibia. On sandy soil in woodland or seasonally damp grassland.

Only one specimen has been seen with the papillate hairs on the ovary, as mentioned in the description. This is *Dinter* 7324, from Karakowisa in Namibia (K).

2. **Hermbstaedtia fleckii** (Schinz) Baker & Clarke in F.T.A. **6**, 1: 28 (1909). TAB. **9** fig. B. Type from Namibia.

Celosia fleckii Schinz in Bull. Herb. Boiss. Sér. **2**, 3: 8 (1903).—Podlech & Meeuse in Merxm. Prodr. Fl. SW. Afr. **33**: 12 (1966).

Celosia namoensis Schinz in Viert. Nat. Ges. Zürich **76**: 134 (1931). Type from Namibia.

Hermbstaedtia rogersii Burtt Davy in F.P.F.T. **1**: 43 (1926). Type from Transvaal.

Annual herb, erect, simple or branched from the base upwards, or occasionally branched from the base and decumbent, c. 15—60 cm. tall. Stem and branches terete and sulcate to more or less quadrangular, striate, thinly to moderately furnished with pale to yellowish multicellular hairs. Leaves linear to linear-oblanceolate, 15—60 (80) × 1—6 (8) mm., acute or subacute, attenuate to the base, subglabrous or the midrib and primary venation of the inferior surface together with the margins more or less furnished with divergent scabrid hairs, surfaces with softer multicellular hairs; superior and branch leaves smaller and narrower, small-leaved axillary shoots commonly present. Inflorescences terminal on the stem and branches, 1.5—15 (25) × (0.75) 1—1.5 cm., spicate, pinkish-white to bright magenta, the axis densely furnished with multicellular hairs. Bracts narrowly lanceolate, 2—4 mm. long, hyaline with a whitish central band, very acute with a shortly excurrent midrib, glabrous or ciliate, sometimes erose-denticulate; bracteoles similar but somewhat shorter and broader. Tepals elliptic-oblong, 3.5—6 mm. long, glabrous, the outer with a shortly excurrent midrib and 1—2 progressively shorter pairs of laterals (one or more of which is occasionally branched), narrowly hyaline-bordered; inner tepals similar but slightly narrower, more narrowed below and often only 3-nerved. Stamens fused into a tube for usually two thirds to three quarters of their length, 1—1.5 mm. shorter than the tepals; free apical portion more or less rounded, truncate or incised on each side of the antheriferous apex. Ovary c. 3—7-ovulate, rounded or slightly tapering above; stigmas (2) 3, spreading or recurved and much shorter than the 1.5—3 mm. long, finally more or less exserted style. Capsule c. (2) 3—4 mm. long, oblong-ovoid, circumcissile, the lid campanulate and rather shortly narrowed into the style, almost smooth or mammilose with low to prominently rounded colourless verrucae. Seeds c. 1 mm. in diam.

Botswana. N: S. of Mandunyane, Bokalaka area, 15.ii.1967, *McClintock* K62 (K). SW: Mahulitlhaki Pan, c. 340 km. W. of Kanye, 22—27.v.1967 *Cox* 313 (K). SW: Bohelabatho Pan, 15.i.1976, *Skarpe* S-6 (PRE; SRGH). SE: Shashi Siding, 28.ii.1967, *McClintock* K34 (K). **Zimbabwe** S: Beitbridge Distr., Limpop Ranches Rd., between turn off from Bulawayo Rd. and Umzingwane R. 25.iii.1959, *Drummond* 6022 (K; LISC; SRGH).

Also in S. Africa and Namibia (Transvaal and Cape Prov.). Usually recorded from sand (red or grey-brown Kalahari) on basalt or paragneiss, in "pans", valley bottom, alluvium or river banks, along irrigation canals or in open woodland; 452—1000 m.

Close to *H. scabra* but the ovary and capsule lid campanulate, not long-conical, and either smooth or rounded-verruculose and without the blunt, elongate unicellular papillose hairs of that species.

The types of *H. fleckii* and *H. rogersii* apparently differ in no respect save the strongly verrucose ovary of the latter. Not surprisingly, the Zimbabwean material of the affinity resembles *H. rogersii*, since most of it is from Beitbridge Distr. - only just across the border from the type locality (Messina) of *H. rogersii*. Namibian material of *H. fleckii* has almost smooth ovaries, as has material from N. and W. Botswana, while gatherings from SE. Botswana have feebly (*Coleman* 102) to quite obviously (*Cox* 313) verruculose ovaries on boiling out. The variation, as far as one can judge without field studies, appears to be clinal and thus scarcely recognisable by formal taxonomic rank.

I have examined the holotype of *H. longistyla* C.B. Clarke (Damara-land, *T.G. Een* s.n. (1879), BM) and find it conspecific with *H. argenteiformis* Schinz, not the present species as opined by some authorities.

3. **Hermbstaedtia scabra** Schinz in Verh. Bot. Ver. Prov. Brand. **31**: 209 (1809).—Baker & Clarke in F.T.A. **6**, 1: 28 (1909). TAB. **9** fig. C. Type from Namibia.

Celosia scabra (Schinz) Schinz in Engl. & Prantl Pflanzenfam. 3 **1a**: 100 (1893); in ed. 2, **16 C**: 30 (1934).—Meeuse in Kirkia **2**: 157 (1961).—Podlech & Meeuse in Merxm. Prodr. Fl. SW. Afr. **33**: 13 (1966).

Erect annual herb, 0.5—1.1 m. tall, simple to considerably branched with long ascending branches. Stem and branches terete to bluntly quadrangular, striate, moderately to densely furnished with pale to yellowish multicellular hairs. Leaves lanceolate or narrowly elliptic, those of the main stem 1.4–9 × 0.4–1.5 (2.5) cm., attenuate below or in broader-leaved forms with a petiole up to 1.2 cm. long, obtuse to subacute, the midrib and primary venation of the inferior surface together with the margins more or less furnished (usually more or less densely) with divergent scabrid hairs, surfaces with softer multicellular hairs; superior and branch leaves rapidly smaller and narrower, small-leaved axillary shoots commonly present. Inflorescences terminal on the stem and branches, spicate, c. (1.5) 4–30 × 1–1.25 cm., bright rose-pink, the axis densely furnished with multicellular hairs. Bracts lanceolate, 3–5 mm. long, hyaline with a whitish central band, very acute with a shortly excurrent midrib, glabrous or ciliate, sometimes erose-denticulate; bracteoles similar but somewhat shorter and broader. Tepals elliptic-oblong, 4.5–6.5 mm. long, glabrous, the outer with a very shortly excurrent midrib and usually 2 progressively shorter pairs of lateral (one or more of which is sometimes branched), hyaline-bordered; inner tepals similar but slightly narrower and often only 3-nerved. Stamens fused into a tube for c. three quarters or more of their length, c. 1 mm. shorter than the tepals; free apical portions more or less rounded, truncate or slightly emarginate on each side of the antheriferous apex. Ovary c. 6–8-ovulate, conical above; stigmas (2) 3, spreading or recurved and shorter than the slender, 2–4 mm. long, finally more or less exserted style. Capsule c. 5 mm. long, circumcissile with the apex longly and gradually narrowed into the finally exserted style, furnished with more or less colourless unicellular papillose hairs. Seeds c. 1 mm. in diam.

Botswana. N: W. bank of Okavango R., 28.iv.1975, *Biegel, Müller & Gibbs-Russell* 5033A (K; SRGH). **Zambia.** W: Nangweshi (Cult. ex) *Codd* 8507 (K; PRE). S: Kazungula, s.d., *Gairdner* 419 (K). **Zimbabwe.** W: Hwange Distr., near Sesheke Pan, c. 11 km. S. of Main Camp, Hwange Nat. Park, 1030 m., 28.ii.1971, *Rushworth* 2539 (K; SRGH).

Also in Namibia. On sandy soil in open savanna, in open Mopane woodland, on river banks and in abandoned cultivation; 900–1030 m. recorded.

4. **Hermbstaedtia linearis** Schinz in Verh. Bot. Ver. Prov. Brand. **31**: 210 (1890).—Baker & Clarke in F.T.A. **6**, 1: 27 (1909). TAB. **9** fig. D. Type from Namibia.

Celosia linearis (Schinz) Schinz in Engl. & Prantl Pflanzenfam. ed. 2, **16 C**: 30 (1934).—Meeuse in Kirkia 2: 159 (1961).—Podlech & Meeuse in Merxm. Prodr. Fl. SW. Afr. **33**: 13 (1966).

Hermbstaedtia schinzii C.B. Clarke in F.T.A. **6**, 1: 27 (1909). Type from Namibia.

Celosia schinzii (C.B. Clarke) Suesseng. in Mitt. Bot. Staatss. München **1**: 42 (1952). Type as above.

Erect annual herb, 8–35 cm. tall, with several branches from the base and frequently branched above. Stem and branches terete to bluntly quadrangular, striate, glabrous or sparingly furnished in the upper parts with fine multicellular hairs. Leaves narrowly linear, those of the main stem 16–45 × 1–2 (3) mm., obtuse to acute, glabrous or very finely and sparingly pilose; superior and branch leaves diminishing, sterile axillary fascicles sometimes present. Inflorescence terminal on the stem and branches, rounded-capitate to spicate, c. 0.7–15 × 0.75–1 cm., pink-tinged to deep rose-pink, the axis more or less densely furnished with white multicellular hairs. Bracts lanceolate-ovate, 2–3 mm. long, hyaline, acuminate with an aristate long-excurrent midrib, glabrous or sparingly ciliate; bracteoles more or less similar. Tepals narrowly oblong-elliptic, (3) 4–5 mm. long, glabrous, narrowly hyaline-bordered, with a very shortly excurrent midrib and a single pair of (sometimes sparingly branched) lateral nerves about attaining the middle of the tepal or longer. Stamens fused into a tube for c. three quarters or more of their length, c. 1 mm. shorter than the tepals; usually with the free apices short, with a tooth shorter than or slightly exceeding it on each side of the anther, rarely the staminal tube fused even higher so that there appears to be a short free pseudostaminode on each side of each fertile tooth. Ovary (2) 4–8 (10)-ovulate, rounded or shortly narrowed above, somewhat verruculose; stigmas (2) 3, spreading or recurved, usually shorter than the (1) 1.25–2 mm. slender style, more or less equalling the tepals or frequently exserted in fruit. Capsule c. 1.75–3 mm. long, circumcissile with the apex

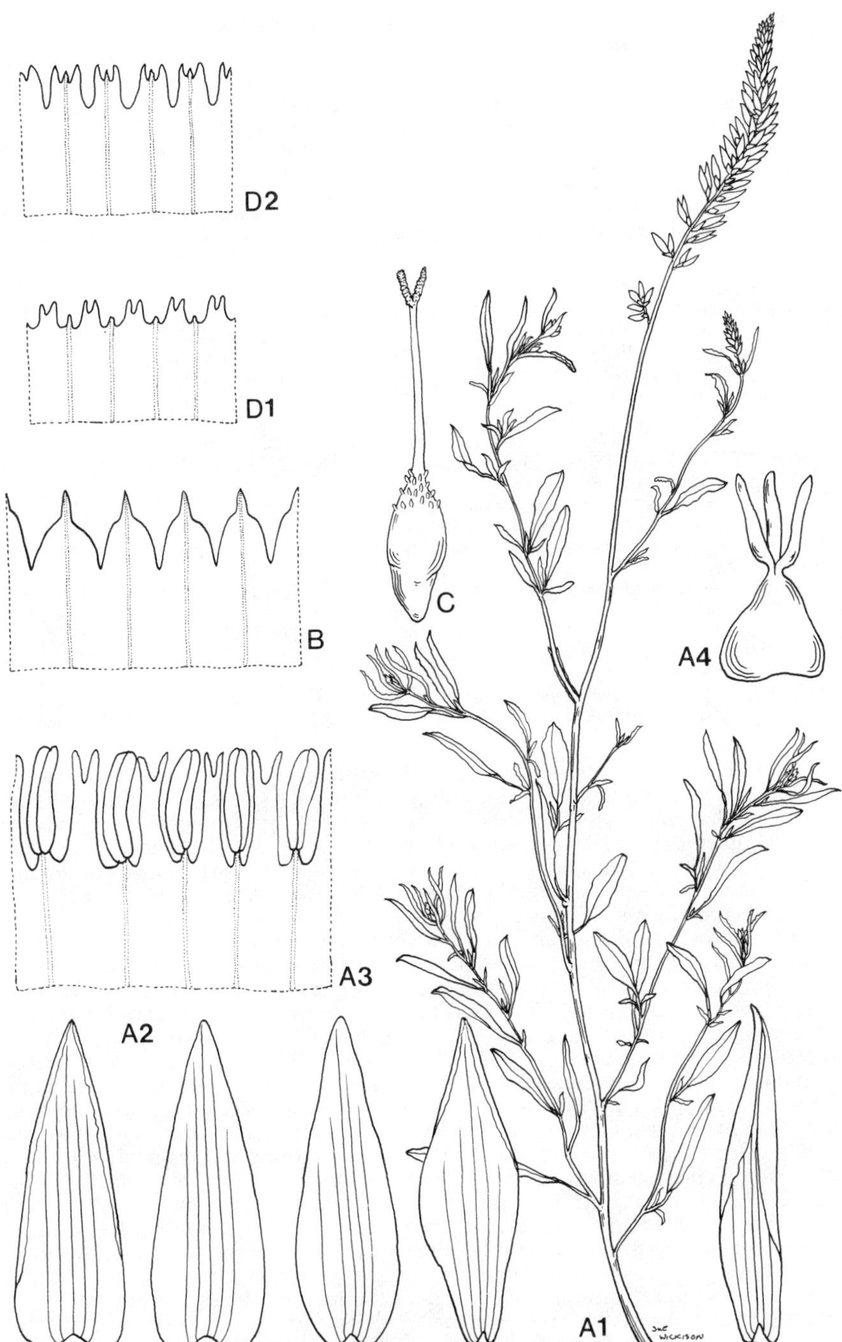

Tab. 9. A.—HERMBSTAEDTIA ODORATA. A1, flowering stem (×⅔); A2, tepals, inner surface, outer to innermost from left to right (×8); A3, androecium (×8); A4, gynoecium (×14), A1—4 from *McClintock* K66; B.—HERMBSTAEDTIA FLECKII. B1, androecium (×8) *Clarke* 310. C.—HERMBSTAEDTIA SCABRA. C1, gynoecium (×6) *Codd* 8507. D.—HERMBSTAEDTIA LINEARIS. D1—2 androecium (two forms) (×8), D1 from *Richards* 14653; D2 from *Wild* 4785.

narrowed into the style, oblong-ovoid, often constricted centrally, glabrous. Seeds c. 1 mm. in diam.

Botswana. N: Ngamiland, Makarikari Pan, 930 m., 10.iii.1961, *Richards* 14653 (K). SE: Boteli delta area N.E. of Mopipi, 900 m., 16.iv.1973, *Galloway* 2 (K; SRGH). SW: 9 km. W. of Ghanzi along Rd. to Mamuno, 15.ii.1978, *Skarpe* S258 (PRE). **Zimbabwe**. W: Hwange Game Reserve, 22.ii.1956, *Wild* 4785 (K; SRGH).

Also in Namibia and S. Africa (Transvaal). Apparently one of the rarer species, but others often misidentified as it. In short grassland, cracks in limestone pavement, open Mopane woodland and Acacia scrub - frequently in sandy ground in the vicinity of lakes or pans.

5. **Hermbstaedtia odorata** (Burchell) T. Cooke in Harv. & Sond. F.C. **6**, 1: 407 (1910).—Schinz in Engl. & Prantl Pflanzenfam. ed. 2, **16** C: 31 (1934).—Meeuse in Kirkia 2: 166 (1961).—Podlech & Meeuse in Merxm. Prodr. Fl. SW. Afr. **33**: 16 (1966).—J.H. Ross Fl. Natal: 158 (1973). TAB. **9** fig. A. Type from S. Africa.

Celosia odorata Burchell, Trav. S. Afr. **1**: 389 (1822).

Celosia recurva Burchell, Trav. S. Afr. **2**: 226 (1824). Type from S. Africa.

Hermbstaedtia elegans Moq. in DC., Prodr. **13**, 2: 247 (1849).—Baker & Clarke in F.T.A. **6**, 1: 26 (1909).—Cavaco in Mém. Mus. Hist. Nat. Paris Sér. B, **13**: 47 (1962). Type from S. Africa.

Hermbstaedtia laxiflora Lopr. in Engl. Bot. Jahrb. **30**: 105 (1901). Type: Mozambique, Ressano Garcia, *Schlechter* 11876 (BM; K, isotype).

Hermbstaedtia transvaalensis Lopr. in Engl. Bot. Jahrb. **30**: 105 (1901). Type from S. Africa.

Hermbstaedtia dammarensis C.B. Clarke in F.T.A. **6**, 1: 26 (1909). Type from Namibia.

Hermbstaedtia recurva (Burchell) C.B. Clarke in F.T.A. **6**, 1: 25 (1909).

Hermbstaedtia rubro-marginata C.H. Wright in Kew Bull. **1910**: 228 (1910). Type from Transvaal.

Hermbstaedtia tetrastigma Suesseng. in Mitt. Bot. Staatss. München **1**: 6 (1950). Type: Zimbabwe, Birchenough Bridge, *Hopkins* 1026 (M, holotype; SRGH, isotype).

Erect or ascending perennial herb, 15—60 cm., with a tuberous rootstock; stems several, simple to considerably branched, sometimes quite bushy, branches divaricate at 30 degrees or more. Stem and branches subterete or somewhat angled, striate, moderately to rather densely furnished with fine, flexuose, whitish hairs, rarely glabrous. Leaves narrowly linear to linear-oblanceolate to obovate-spathulate, more or less obtuse (or rather acute in very narrow-leaved forms), narrowed below, or distinctly petiolate and lower leaves with a narrowly elliptic lamina in broader-leaved forms, 10—50 × 0.8—10 mm. at the centre of the stem and branches but decreasing upwards, more or less glabrous to somewhat pilose with rather stiff divergent hairs along the venation of the lower surface; small-leaved axillary shoots frequently but not invariably present in the axils of the upper or all leaves. Inflorescences terminal on both the stem and branches, spicate, c. (1) 5—30 × 0.6—1.25 cm., considerably elongating, the axis moderately to densely furnished with whitish, floccose hairs. Bracts c. (1.5) 2.5—3.5 mm. long, lanceolate to lanceolate-ovate, hyaline with a darker midrib which is excurrent in a short, fine arista, glabrous or finely ciliate, sometimes erose-denticulate; bracteoles shorter and broader, 2—3 (3.5) mm. long. Tepals narrowly to broadly elliptic-oblong or oblong-obovate, the outer (2.5) 4—6 (7) mm. long, white or pink-tinged to orange or carmine, 3—7 (9)-nerved with the midrib excurrent in a very short mucro and 1—3 pairs of (sometimes branched) lateral nerves reaching well above the middle, and usually other shorter nerves; inner tepals similar but not infrequently fewer-nerved and slightly narrower. Stamens fused into a tube for two thirds to three quarters of their length, clearly shorter than the tepals; antheriferous tooth much shorter than the alternating deeply bifid pseudo-staminodes, which more or less equal or overtop the anthers. Ovary (2) 4—8-ovulate, rather squat, rounded or very shortly narrowed above; stigmas 2—3 (5), spreading or recurved, much longer than the very short (rarely to 1 mm. long) or almost obsolete style, rarely exceeding the androecium. Capsule oblong-ovoid, c. 2.5—3.5 mm. long, circumcissile, the lid campanulate and shortly tapering above, smooth. Seeds c. 1 mm.

Var. **albi-rosea** Suesseng. in Mitt. Bot. Staatss. München **1**: 192 (1953) nom. provis. No type cited, merely a remark that all white or pink-flowered forms belong here.

Flowers usually white or pink-tinged, the tepals rather rarely with more than 1 pair of lateral nerves (which may however be branched) attaining well above the middle, though often with shorter lateral nerves. Leaves variable in form but usually linear to linear-oblanceolate.

Botswana. N: c. 9 km. N. of Toteng, 17.ix.1964, *Story* 4705 (K; PRE). SW: c. 16 km. W. of Lephepe village, .iv.1969, *Kelaole & Chiwita* 535 (K; SRGH). SE: 8 km. N. of Gaberone, 1090 m., 19.i.1960, *Leach & Noel* 225 (K; SRGH). **Zimbabwe**. W: Bulawayo, 18.ii.1912, *Rogers* 5928 (K; SRGH). E: roadside 52 km. S. of Mutare, 910 m., 8.xii.1951, *Chase* 4231 (BM; SRGH). S: Beitbridge, 13 km. SE. of Tuli on road to Shashi Irrigation Scheme, 22.iii.1959, *Drummond* 5914 (K, SRGH). **Mozambique**. GI: Gaza, Mabalane, 4.vi.1959, *Barbosa & Lemos* 8606 (COI; K; LISC; LMJ).

Also in S. Africa and Namibia. In grassland, savanna, riverine vegetation, scrub mopane etc. and as a weed of roadsides and cultivation; in loose red sand, gritty loam, gravel (hard limestone) and on serpentine and basalt.

Var. **aurantiaca** (Suesseng.) Townsend in Kew Bull. **37**: 82 (1982). Type: Mozambique, Polana flats, *Hornby* 2070 (K, lectotype).

Hermbstaedtia quintasii Gandoger in Bull. Soc. Bot. France **66**: 222 (1919). Type: Mozambique, Maputo, *Quintas* 65 (COI, isotype).

Hermbstaedtia elegans var. *aurantiaca* Suesseng. in Mitt. Bot. Staatss. München **1**: 192 (1953), incl. forma *bilobata* Suesseng. and forma *irregularis* Suesseng.

Flowers carmine to vermilion, the tepals with usually 2–3 pairs of lateral nerves attaining well above the middle, with additional shorter nerves. Leaves narrowly oblanceolate to obovate-spathulate.

Mozambique. M: Maputo, between Catembe and Mogazine, 5 km. to Catembe, 18.ii.1952, *Barbosa & Lemos* 8292 (K; LMJ).

Also in S. Africa (Natal, Transvaal). Always recorded from sandy soil, in grassland, as a weed of cultivations or roadsides, or in "bushveld".

In the type of *H. laxiflora* the capsule is exceptionally long, almost equalling the perianth, so that the style and stigmas are exserted. This is exceedingly unusual for *H. odorata*. Even in such plants, however, the species is readily recognisable from forms of *H. linearis* by the short rather stout style and longer pseudostaminodes.

3. AMARANTHUS L.

Amaranthus L., Sp. Pl. 2: 989 (1753).—Brenan, "The genus *Amaranthus* in Southern Africa", Journ. S. Afr. Bot. **47**: 451–492 (1981).

Annual or more rarely perennial herbs, glabrous or furnished with short and gland-like or multicellular hairs. Leaves alternate, long-petiolate, simple and entire or sinuate. Inflorescence basically cymose, bracteate, consisting entirely of dense to lax axillary clusters or the upper clusters leafless and more or less approximate to form a lax or dense "spike" or panicle. Flowers monoecious or (not in Africa) dioecious, bibracteolate. Perianth segments (2) 3–5, free or connate at the base, membranous, those of the female flowers sometimes slightly accrescent in fruit. Stamens free, usually similar in number to the perianth segments; anthers bilocular. Stigmas 2–3. Ovule solitary, erect. Fruit a dry capsule, indehiscent, irregularly rupturing or commonly dehiscing by a circumcissile lid. Seeds usually black and shining, testa thin; embryo annular, endosperm present.

A genus of about 60 species in the tropical and warmer temperate regions of both Old and New Worlds, impermanent and sporadic as weeds in cooler temperate regions.

Amaranthus must be considered at least in part as a difficult genus. Its taxonomy has been complicated by the cultivation of certain species (e.g. *A. lividus* and *A. tricolor*) as green vegetables, and others (particularly *A. caudatus* and *A. hybridus* agg.) as grain crops. A good modern account is that by Aellen in the second edition of Hegi's "Illustrierte Flora von Mitteleuropa" Vol. 3 2, Lfg. 1: 465–516 (1959). Sauer's revision, "The grain amaranths and their relatives: a revised taxonomic and geographic study", is to be found in Ann. Missouri Bot. Gard. 54: 103–137 (1967). This employs a narrower species concept than the present writer is prepared to allow - though the case for allowing *A. cruentus* L. specific distinction from *A. hybridus* L. and *A. hypochondriacus* L. certainly seems stronger than that for separating the latter two - but it is a very careful and useful piece of work.

Several species of *Amaranthus* have been noted by collectors as being used as "a vegetable" in the Flora Zambesiaca region - viz., *A. spinosus*, *A. dubius*, *A. graecizans* subsp. *silvestris*, *A. hybridus* subsp. *hybridus* and subsp. *cruentus*, *A. thunbergii*.

In the preparation of this account of the genus *Amaranthus* the author has benefited considerably from the long experience and advice of Professor J.P.M. Brenan, whose revision of the genus *Amaranthus* in southern Africa was proceeding simultaneously.

1. Leaf axils with paired spines - - - - - - - - - - 4. *spinosus*
— Leaf axils without paired spines - - - - - - - - - - - - 2
2. Inflorescence consisting entirely of axillary, cymose clusters, no terminal leafless spike or panicle present - - - - - - - - - - - - - 3
— Inflorescence not entirely of axillary, cymose clusters - a terminal, leafless spike or panicle present - - - - - - - - - - - - - - - 8
3. Female flowers with 4–5 perianth segments; leaves narrowly oblanceolate to narrowly elliptic-oblong; female perianth segments long-aristate - - 8. *praetermissus*
— Female flowers with 3 perianth segments; if leaves so narrow, then female perianth-segments very shortly mucronate - - - - - - - - - - - - - 4
4. Female perianth-segments distinctly shorter than the fruit - - - - - 5
— Female perianth-segments equalling or exceeding the fruit - - - - - - 6
5. Leaves generally broadly and conspicuously emarginate; fruit strongly compressed, indehiscent - - - - - - - - - - - - - - - 10. *lividus*
— Leaves not, or narrowly and inconspicuously, emarginate; fruit not strongly compressed, circumcissile or more rarely indehiscent - - - - 9. *graecizans*
6. Leaves of the main stem broadest distinctly below the middle - - 5. *tricolor*
— Leaves of the main stem broadest at or above the middle - - - - - - 7
7. Stems furnished above with long, crisped hairs; female perianth-segments with a long (0.75–1.75), slender, flexuose or divergent, generally colourless arista - - - - - - - - - - - - - - - - - - - 6. *thunbergii*
— Stems glabrous (sometimes puberulous above especially about the nodes), without elongate crisped hairs, frequently papillose-scabrid; female perianth-segments with a short, 0.1–0.5 (0.75) erect or divergent mucro - - - - - - 7. *dinteri*
8. Capsule indehiscent - - - - - - - - - - - - - - - 9
— Capsule globular and extremely muricate, not or scarcely exceeding the perianth - - - - - - - - - - - - - - - - - - 11. *viridis*
9. Capsule compressed, not muricate, exceeding the perianth - - - 10. *lividus*
— Capsule circumcissile - - - - - - - - - - - - - - - 10
10. Female perianth-segments spathulate to narrowly oblong-spathulate, greenish along the midrib, which ceases below the apex or becomes colourless and faint; plant densely clothed with floccose, multicellular hairs - - - - - - - 1. *retroflexus*
— Female perianth-segments lanceolate to oblong, shortly aristate to mucronate with the excurrent midrib; plant glabrous to moderately pilose, usually with shorter hairs - 11
11. Male flowers confined to a normally quite short length at the apex of each spike, rarely mixed with the females; female perianth-segments obtuse, narrowly spathulate to narrowly oblong; lid of capsule strongly wrinkled, with a swollen neck formed by the inflated style-bases - - - - - - - - - - - - 3. *dubius*
— Male flowers mixed with the females; female perianth-segments acute, or if obtuse then the capsule not strongly wrinkled, with a short, smooth, firm beak, and the style-bases not inflated - - - - - - - - - - - - - 2. *hybridus*
12. Capsule circumcissile; bracteoles and perianth segments with a long, fine, colourless awn - - - - - - - - - - - - - - - - - - 5. *tricolor*
— Capsule indehiscent; bracteoles and perianth segments mucronate or shortly aristate only - 13
13. Capsule more or less globular, extremely muricate, not or scarcely exceeding the perianth segments; seeds with shallow, scurfy verrucae on the reticulate pattern of the testa - - - - - - - - - - - - - - - - - - 11. *viridis*
— Capsule ellipsoid or distinctly compressed, not muricate, distinctly exceeding the perianth segments; seeds with no shallow verrucae - - - - - - - 14
14. Capsule ellipsoid, scarcely compressed, seed also ellipsoid; leaves rarely (feebly) retuse - - - - - - - - - - - - - - - - - - - 12. *deflexus*
— Capsule compressed, lenticular or shortly pyriform, seed lenticular; leaves usually broadly and distinctly emarginate, rarely broadly truncate - - 10. *lividus*

1. **Amaranthus retroflexus** L., Sp. Pl.: 991 (1753).—Schinz in Engl. & Prantl Pflanzenfam.

ed. 2, **16** C: 38 (1934).—Aellen in Hegi, Illustr. Fl. Mitteleurop. ed. 2, **3**, 2: 485, f.5 (1959).—Brenan in Journ. S. Afr. Bot. **47**: 463 (1981). Type, cultivated material from Uppsala Botanic Garden, Linnean specimen 1117/22 (LINN, lectotype).

Annual herb, erect or with ascending branches, (6) 15—80 (100)cm., simple or branched (especially from the base to about the middle of the stem). Stem stout, subterete to angled, densely furnished with multicellular hairs. Leaves furnished with multicellular hairs along the lower surface of the primary venation and often the lower margins, long-petiolate (petioles up to c. 6 cm., in robust plants not rarely equalling the lamina), lamina ovate to rhomboid - or oblong-ovate, (1) 5—11 × (0.6) 3—6 cm., obtuse to subacute at the mucronulate apex, shortly cuneate to attenuate into the petiole. Flowers in greenish or rarely somewhat pink-suffused, stout, axillary and terminal spikes, which are usually shortly branched to give a lobed appearance, more rarely with longer branches, the terminal inflorescence more or less paniculate, very variable in size, male and female flowers intermixed, the latter generally much more plentiful except sometimes at the apex of the spikes. Bracts and bracteoles lanceolate-subulate, pale-membranous with a prominent green midrib excurrent into a stiff, colourless arista, longer bracteoles subequalling to twice as long as the perianth. Perianth segments 5, those of the male flowers 1.75—2.25 mm., lanceolate-oblong, blunt to subacute, those of the female flowers 2—3 mm., narrowly oblong-spathulate to spathulate, obtuse or emarginate, more or less green-vittate along the midrib, which ceases below the apex or is excurrent in a short mucro. Stigmas 2—3, patent-flexuose or erect, c. 1 mm. Capsule subglobose, c. 2 mm., usually shorter than the perianth, circumcissile, with an indistinct neck, rugose below the lid. Seed black and shining, compressed, c. 1 mm., almost smooth centrally, faintly reticulate around the margins.

Zambia. S: 22 km. W. of Livingstone, 23.iii.1982, *Drummond & Vernon* 10938 (K; LISC; MAL; MO; NDO; PRE; SRGH). **Zimbabwe**. E: Sabi Valley Experimental Station, on irrigated land, xii.1959, *Soane* 207 (K; LISC; SRGH).

A native of N. America south to Mexico, introduced into the Old world as a weed, but mostly in more temperate regions than many of its allies, occurring in S. and C. Europe, Mediterranean N. Africa and temperate Asia from Cyprus and Turkey to China and Japan; also in Australia, S. Africa, S. America (Bolivia etc.) and probably elsewhere. Only the above single gathering seen from the Flora Zambesiaca region.

2. **Amaranthus hybridus** L., Sp. Pl. 2: 990 (1753).—Schinz in Engl. & Prantl Pflanzenfam. ed. 2, **16** C: 37 (1934).—Podlech & Meeuse in Merxm. Prodr. Fl. SW. Afr. **33**: 8 (1966).—J.H. Ross Fl. Natal: 158 (1973).—Brenan in Journ. S. Afr. Bot. **47**: 456 (1981). Type, Linnean specimen 1117/19 (LINN, lectotype).

Annual herb, erect or less commonly ascending, up to 2 (3) m. in cultivated forms but less in wild populations, not infrequently reddish-tinted along the leaf venation or throughout. Stems mostly stout, simple or almost so to considerably branched, angular, glabrous to thinly or moderately furnished with short or occasionally longer multicellular hairs (increasingly so above, especially in the inflorescences). Leaves glabrous, or thinly pilose on the lower margins and underside of the primary venation, long-petiolate (petioles up to 15 cm. long but even then scarcely exceeding the lamina), lamina broadly lanceolate to rhomboid or ovate, 3—19 (30) × 1.5—8 (12) cm., gradually narrowed to the blunt to subacute mucronulate tip, attenuate to shortly cuneate into the petiole below. Flowers in yellowish, green, reddish or purple axillary and terminal spikes formed of cymose clusters which are increasingly closely approximate upwards, the terminal inflorescence varying from a single spike to a broad, much-branched panicle up to c.45 × 25 cm., in length and breadth, the ultimate spike not infrequently more or less nodding; male and female flowers intermixed throughout the spikes. Bracts and bracteoles deltoid-ovate to deltoid-lanceolate, pale-membranous, acuminate, with a long, pale to reddish-tipped straight arista formed by the stout, excurrent, yellow to greenish midrib; bracteoles subequalling to considerably exceeding the perianth. Perianth segments 5, 1.5—2.5 mm. long and paler in the male, 1.5—3.5 mm. long in the female, lanceolate to oblong, shortly aristate or mucronate, acute or the inner segments of the female sometimes blunt, usually only the midrib greenish. Stigmas (2) 3, erect

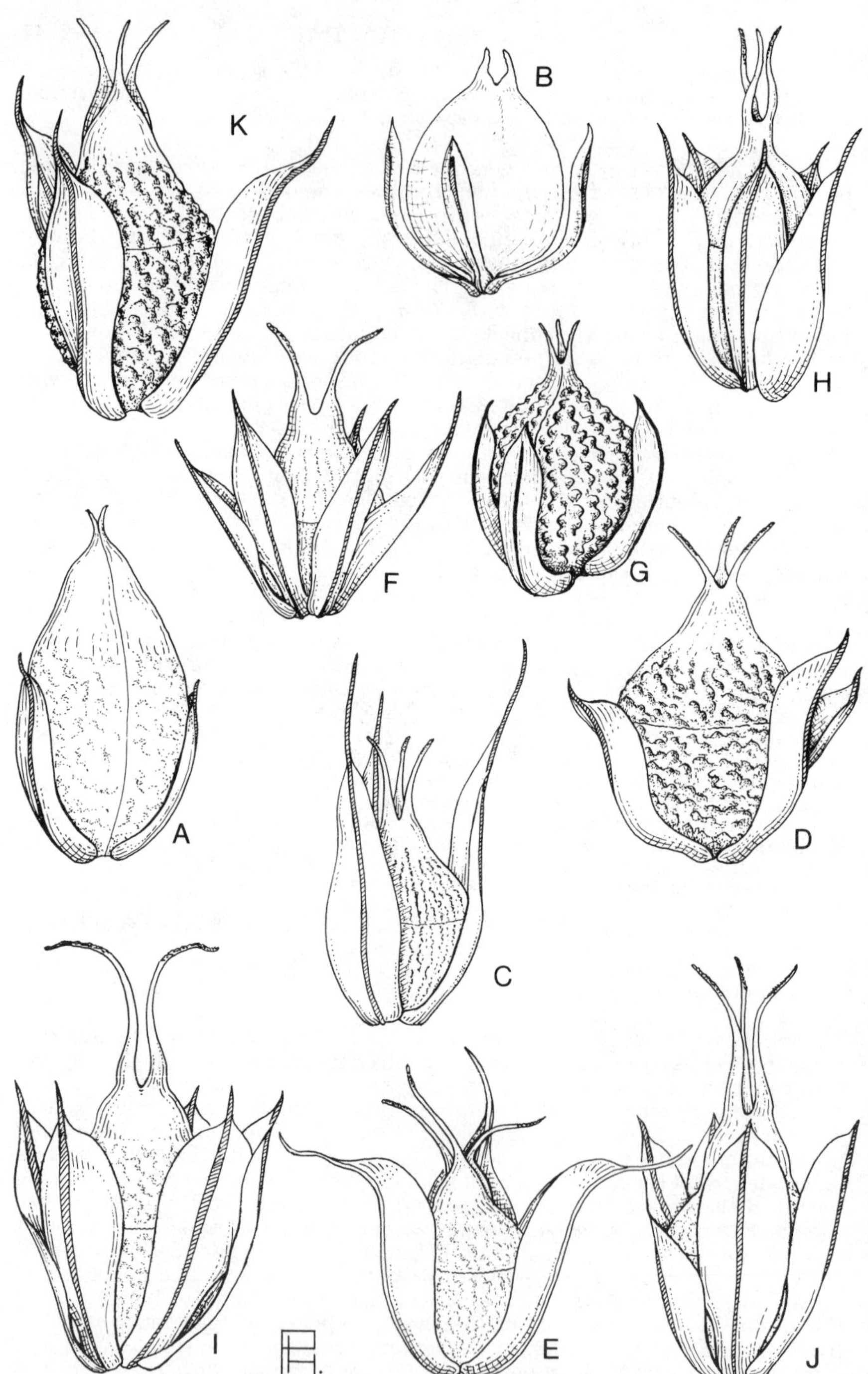

Tab. 10. CAPSULES OF AMARANTHUS. A.—AMARANTHUS DEFLEXUS (×20) *Bally* 8067. B.—AMARANTHUS LIVIDUS SUBSP. POLYGONOIDES (×20) *Archbold* 643. C.—AMARANTHUS THUNBERGII (×10) *Wilson* 621. D.—AMARANTHUS GRAECIZANS SUBS SYLVESTRIS (×20) *Dümmer* 471. E.—AMARANTHUS TRICOLOR (×6) *Greenway* 980. F.—AMARANTHUS HYBRIDUS SUBSP. HYBRIDUS (×10) *Lugard & Lugard* 638. G.—AMARANTHUS VIRIDIS (×20) *Richards* 10900. H.—AMARANTHUS HYBRIDUS SUBSP. CRUENTUS (×20) *Culwick* 4. I.—AMARANTHUS SPINOSUS (×20) *Bally* 7828. J.—AMARANTHUS DUBIUS (×20) *Conrads* 13288. K.—AMARANTHUS GRAECIZANS SUBSP. THELLUNGIANUS (×20) *Bally* 914.

to flexuose or recurved, 0.5–1.25 mm. long. Capsule subglobose to ovoid or ovoid-urceolate, c. 2–3 mm. long, circumcissile, with a moderately distinct to obsolete beak, the style-bases not to considerably dilated, the lid smooth to longitudinally striate or rugulose. Seed black and shining (occasionally pale and dull), compressed, roundish, 0.75–1.25 mm. in diam., almost smooth centrally and faintly reticulate around the margins.

Subsp. **hybridus** TAB. **10** fig. F.

Amaranthus hypochondriacus L., Sp. Pl. 2: 991 (1753). Type, Linnean specimen 1117/24 (LINN, lectotype).

Amaranthus chlorostachys Willd. Hist. Amaranth.: **34**, t. x f. 19 (1790).—Aellen in Hegi Illustr. Fl. Mitteleurop. ed. 2, **3**, 2: 480 (1959). Type, *Willdenow* 17521 (B, holotype; IDC. microfiche neg. 1265 No. 10!).

Amaranthus patulus Bertol., Comment. It. Neap. 19, t. 2 (1837).—Aellen in Hegi Illustr. Fl. Mitteleurop. ed. 2, **3**, 2: 483 (1959). Type from Italy.

Amaranthus incurvatus Tim. ex Gren. & Godr., Fl. France Prosp.: 8 (1846) Type from France.

Amaranthus powellii S. Wats. in Proc. Am. Acad. **10**: 347 (1875). Type a cultivated specimen grown from Arizona, seed leg. Col. *Powell* s.n. (US, holotype).

Amaranthus hybridus subsp. *hypochondriacus* (L.) Thell., Fl. Adv. Montpellier: 204 (1912).—Hauman in F.C.B. 2: 29 (1951). Type as for *A. hypochondriacus*.

Amaranthus hybridus subsp. *hypochondriacus* var. *chlorostachys* (Willd. & Thell., ibid.: 205 (1912). Type as for *A. chlorostachys*.

Amaranthus hypochondriacus subsp. *chlorostachys* var. *verticillatus* Suesseng. in Mitt. Bot. Staats. München **1**: 104 (1952). Type from S. Africa (Natal).

Stigma-bases and upper part of lid of fruit more or less swollen, so that the fruit has a more or less inflated beak. Inner perianth segments of female flowers commonly (but not always!) acute. Longer bracteoles of female flowers mostly about twice as long as the perianth.

Botswana. SE: Gaberone University Campus, 965 m., 22.iii.1974, *Mott* 179 (K; UBLS). **Zambia**. N: Mbala Distr., Nkali Dambo, Rd. to Kawimbi Mission, 1535 m., 5.i.1952, *Richards* 312 (K). W: Ndola, 22.v.1953, *Fanshawe* 28 (K; NDO; SRGH). C: Lusaka, 24.vii.1956, *Simwanda* 16 (K; SRGH). E: Chipata, 1030 m., 6.i.1936, *Winterbottom* 62 (K). S: Mazabuka, vi.1963, *van Rensburg* 1601 (K; SRGH). **Zimbabwe**. W: Bulalima, Mangwe Distr., Dombodema Mission, 395 m., 27.iv.1972, *Norrgrann* 141 (K; SRGH). C: Chikwingwizha Seminary grounds 14 km. from Gweru, 1450 m. 30.v.1966, *Biegel* 1203 (K; SRGH). E: Inyanga, ad villam Cheshire, 300 m., 15.i.1931, *Norlindh & Weimarck* 4379 (K; SRGH) S: Chiredzi Distr., Mabagwashe Nyajena T.T.L., 18.v.1971, *Taylor* 177 (K; SRGH). **Malawi**. C: Dedza Distr., Chongoni Forest Reserve, 23.ii.1968, *Salubeni* 982 (K; SRGH). S: Blantyre Distr., Matenje Rd. 1–2 Km. N. of Limbe, 1180 m., 12.iv.1970, *Brummitt* 9812 (K; MAL; SRGH).

This subspecies occurs throughout the tropical and subtropical regions of the world as a wild or spontaneous plant, also occurring as a casual in temperate regions, e.g. as a common wool adventive in Europe; it is of American origin. In the Flora Zambesiaca region a weed of cultivated and disturbed ground along roadsides, in trampled grassland etc., 300–1520 m.

Forms of this subspecies with red inflorescences can be called var. *erythrostachys* Moq., the colour, however, fades and herbarium material eventually becomes green.

An abnormal form occurs with the lower partial inflorescences globular and their final branches somewhat elongated and pectinate. This is Suessenguth's "var. *verticillatus*". It has been received as grown in Harare from seed found as a contaminant of "Red Manna" (*Setaria* cf. *italica*) seed (Seed Services VS/110 & VS/118). An even more extreme abnormality of this type was collected in an old maize field in the Chilanga District of Zambia by Mrs. C I Sandwith (No. 143, K) in 1924. The extreme variability in size of this plant is shown by the note on the herbarium sheet (*Best* 96 from Lusaka, K) that it was "7.5 cm. to 2.1 m. or more in height" at this one locality.

Subsp. **cruentus** (L.) Thell. in Fl. Adv. Montpellier: 205 (1912).—Hauman in F.C.B. **2**: 27 (1951).—Cavaco in Mém. Mus. Hist. Nat. Paris Sér. B, **13**: 55 (1962).—Brenan in Journ. S. Afr. Bot. **47**: 462 (1981). TAB. **10** fig. H. Type a Linnean specimen 1117/25 (LINN, lectotype).

Amaranthus cruentus L., Syst. Nat. ed. 10, **2**: 1269 (1759).—Podlech & Meeuse in Merxm. Prodr. Fl. SW. Afr. **33**: 7 (1966).

Amaranthus paniculatus L., Sp. Pl. **2**: 1406 (1753).—Aellen in Hegi, Illustr. Fl. Mitteleurop. ed. 2, **3**, 2: 484 (1959). Type a Linnean specimen 1117/20 (LINN, lectotype)

Amaranthus paniculatus var. *cruentus* (L.) Moq. in DC., Prodr. **13**, 2: 257 (1949). Type as for *C. cruentus*.

Amaranthus tricolor sensu Baker & Clarke in F.T.A. **6**, 1: 32 (1909) non L.

Amaranthus caudatus sensu Baker & Clarke in F.T.A. **6**, 1: 31 (1909).—Suesseng. & Beyerle in Bot. Arch. **41**: 79 (1940) non L.

Amaranthus hybridus L. subsp. *incurvatus* sensu Brenan in Watsonia **4**: 268 (1961); in J.H. Ross Fl. Natal: 158 (1973) non *Amaranthus incurvatus* Tim. ex Gren. & Godr.

Stigma bases and upper part of lid of fruit scarcely swollen, fruit with a short, firm, smooth beak. Inner perianth segments of female flowers obtuse. Longer bracteoles of female flowers mostly 1—1.5 times as long as the perianth.

Botswana. N: Mutsoi, NE. of Nokaneng, 23.iii.1967, *Lambrecht* 110 (LISC, SRGH). **Zambia**. N: Mporokoso Distr., Mweru-Wantipa, Mwawe R., 1050 m., 6.iv.1957, *Heany* Teacher Training College 43 (K, SRGH). C: Harare Distr., Greendale, vi.1939, *Leach* 8929 (K; SRGH), young but probably this. E: Nyamkwarara Valley, below 1210 m., ii.1935, *Gilliland* K1382 (K) **Malawi**. N: Nyika Plateau, ii—iii.1903, *Mc Clounie* 157 (K). S: Blantyre-Naperi R., 1000 m., 9.viii.1937, *Lawrence* 454 (K). **Mozambique**. "Mozambique" no further details, *Forbes* s.n. (K). N: Lichinga, 17.vi.1934, *Torre* 179 (LISC). T: Songo, cultivated, 4.iii.1972, *Macêdo* 4996 (LISC; SRGH). M: Maputo, iv.1945, *Pimenta* 4304 (LISC).

This subspecies is also presumed to be of (Central) American origin, but is now widespread in the tropical and subtropical regions of the world; the red form (which may well be of cultivated derivation) is cultivated as an ornamental in temperate regions also, and occurs occasionally as an escape. In the Flora Zambesiaca area a plant of disturbed ground and cultivation, frequently in damper ground; noted from 1000—1120 m.

As observed in the introductory remarks to the genus, on a world basis the case for separating *A. cruentus* specific rank is stronger than that for separating *A. hybridus* and *A. hypochondriacus*. (unfortunately, this does not apply in the Flora Zambesiaca region, where I have found more difficulty in naming material than from any other area. Such intermediates as Sauer saw he named as *A. hybridus* sensu stricto, but other authorities have named such plants as subsp. *cruentus*. Gatherings which seem to me to be intermediate between the two subspecies have been seen from Botswana (3) and Zambia (2).

With reference to the synonomy, the only Mozambique specimen collected by *Forbes* which has been seen, and is presumably the source of the F.T.A. record of *A. tricolor*, was nevertheless named by Clarke as *A. paniculatus* - used by him in F.T.A. as a synonym of *A. caudatus*. There is evidently a slip of the pen here somewhere.

As with subsp. *hybridus*, subsp. *cruentus* has forms with red and green inflorescences, the red being "typical". I have not traced a varietal name for the green plant. A plant which I collected in Malawi (Glyn Jones Street, Blantyre, waste ground, 7.v.1980, *Townsend* 2143) had an extremely long, more or less pendulous terminal spike and was collected as possible green-flowered *A. caudatus*. But the female perianth segments are too narrow for that species and the style-bases slender; it is probably a very robust individual of *A. hybridus* subsp. *cruentus*.

3. **Amaranthus dubius** Mart., Pl. Hort. Acad. Erlang.: 197 (1814) nomen nudum, ex Thell., Fl. Adv. Montpellier: 203 (1912).—Schinz in Engl. & Prantl Pflanzenfam. ed. 2, **16** C: 38 (1934).—Hauman in F.C.B. 2: 30 (1951).—Aellen in Hegi, Illustr. Fl. Mitteleurop. ed. 2, **3**, 2: 476 (1959).—J.H. Ross Fl. Natal: 158 (1973).—Brenan in Journ. S. Afr. Bot. **47**: 464 (1981). TAB. **10** fig. J. Type, cultivated specimen from Erlangen Botanic Garden.

Amaranthus tristis sensu Moq. in DC., Prodr. **13**, 2: 260 (1849) non L.

Amaranthus patulus sensu Baker & Clarke in F.T.A. **6**, 1: 33 (1909) non Bertol.

Amaranthus dubius var. *crassespicatus* Suesseng. in Mitt. Bot. Staatss. München **1**: 73 (1951). Type from Tanzania.

Erect annual herb, mostly up to c. 90 cm. (rarely to 1.5 m.) tall. Stem rather slender to stout, usually branched, angular, glabrous or increasingly furnished upwards (especially in the inflorescence) with short to rather long, multicellular hairs. Leaves glabrous, or thinly and shortly pilose on the inferior surface of the primary venation, long-petiolate (petioles up to c. 8.5 cm. long, sometimes longer than the lamina), lamina ovate or rhomboid-ovate, 1.5—8 (12) × 0.7—5 (8) cm., blunt or retuse at the apex with a distinct, fine mucro formed by the percurrent nerve, cuneate (usually shortly so) at the base; leaf axils without spines. Flowers green, in the lower part of the plant in axillary clusters 4—10 cm. in diam., towards the ends of the stem and branches the leafless clusters approximated to form simple or (the terminal at least) branched spikes c. 3—15 (25) cm. long and 6—8 (10) mm. wide. Lower clusters of flowers entirely female, the spikes generally showing a few male flowers at the apices only (rarely in more than the apical 1 cm.), occasionally with male flowers also scattered among the lower female flowers. Bracts and bracteoles deltoid-ovate, pale-membranous with an erect reddish awn formed by the excurrent green midrib, bracteoles somewhat shorter than or subequalling the perianth, rarely slightly exceeding it. Perianth segments (4) 5, those of the female flowers c. 1.5–2.75 mm. long, narrowly oblong or spathulate oblong, obtuse or sometimes (particularly those approaching the male flowers) acute, mucronulate, frequently with a greenish dorsal vitta above; those of the male flowers broadly lanceolate or lanceolate-oblong, generally acuminate, only the thin midrib green. Stigmas 3, flexuose or reflexed, c. 0.75–1 mm. long. Capsule subequalling the perianth, ovoid-urceolate, with a short inflated beak below the style base, c. 1.5–1.75 mm., circumcissile, the lid strongly rugulose below the neck. Seed 1–1.25 mm., compressed, black, shining, faintly reticulate.

Zambia. N: Mbala Distr., Mpulungu, Lake Tanganyika, 880 m., 29.xii.1951, *Richards* 180 (K). E: Luangwa Valley, Mulila Tundwe, Munkanya, 21.xii.1967, *Phiri* 18 (K). **Zimbabwe**. N: Kariba, Gorge just below the Dam wall, 456 m., v.1960, *Goldsmith* 76/60. C: Harare, Forbes Ave, 11.x.1971, *Biegel* 3613 (K; LISC; MAL; MO; PRE; SRGH). **Malawi**. N: "N. Nyassa", 1896, *Whyte* s.n. (K). S: Blantyre, Glyn Jones Street, 1040 m., 7.v.1980, *Townsend* 2144 (K). **Mozambique**. N: Mandimba, 7.iii.1942, *Hornby* 3721 (PRE). Z: Quelimane Distr., Namagoa, s.d., *Faulkner* K5 (K). MS: Sabi R., 27.vi.1950, *Chase* 2456 (BM; SRGH). GI: Inhambane, dans les jardins potagers, 10 m., iv.1936, *Gomes e Sousa* 1721 (K). M: Delagoa Bay, viii, *Wilms* 1250 (K). T: Tete, ii.1932, *Guerra* 41 (COI).

This species is of tropical American origin but now occurs almost throughout the tropics of the world; it is scarcer as an adventive in temperate regions than *A. spinosus*.

A. dubius is the only known polyploid *Amaranthus*, and it is postulated by *Grant* (Canadian Journ. Bot. **37**: 1063–70 (1959)) that it arose as an allotetraploid with *A. spinosus* as one parent and possibly *A. quitensis* as the other - a conclusion disputed by *Pal & Khoshoo* (Curr. Sci. **34**: 370–371 (1965)). Hybrids between *A. dubius* and *A. spinosus* appear to occur freely where these two species are associated, and Srivasta, Pal & Nair (Rev. Palaeobot. Palynol. **23**: 287–291 (1977)) claim that these may be distinguished by the presence of micrograins among the pollen. Such hybrids will certainly occur in tropical Africa, and the *Faulkner* specimen cited above may be such. No apparent hybrid was seen when I found the two species growing together in Blantyre. In spite of various characters used in the literature, I have been able to find no infallible means by which *A. spinosus* can be separated from *A. dubius* other than by the presence or absence of spines, though the generally considerably greater number of terminal male flowers in the spikes of *A. spinosus* seems a reasonably reliable character where only "tops" are collected. According to Srivasta, Pal & Nair (loc. cit.) the pollen of *A. dubius* has larger pores than those of *A. spinosus*. The leaves of *A. dubius* are commonly broader than those of *A. spinosus*. but this is a highly comparative character.

4. **Amaranthus spinosus** L., Sp. Pl. 2: 991 (1753).—Baker & Clarke in F.T.A. **6**, 1: 32 (1909).—Schinz in Engl. & Prantl Pflanzenfam. ed. 2, **16** C: 38 (1934).—Hauman in F.C.B. 2: 31 (1951).—Aellen in Hegi, Illustr. Fl. Mitteleurop. ed. 2, **3**, 2: 477 (1959).—Cavaco in Mém. Mus. Hist. Nat. Paris Sér. B, **13**: 53 (1962).—J.H. Ross Fl. Natal: 158 (1973).—Brenan in Journ. S. Afr. Bot. **47**: 465 (1981). TAB. **10** fig. I. Type, Linnean specimen 1117/27 (LINN, lectotype).

Annual herb, erect or slightly decumbent, up to c. 1.5 m. in height. Stem stout, sometimes reddish, usually branched, angular, glabrous or increasingly furnished above (especially in the inflorescence) with short or longer, multicellular, flocculent hairs. Leaves glabrous, or thinly pilose on the lower surface of the primary venation, long-petiolate (petioles up to c. 9 cm. long, sometimes longer than the lamina), lamina lanceolate-ovate to rhomboid-ovate, elliptic, lanceolate-oblong or lanceolate, c. 1.5–12 × 0.8–6 cm., subacute or more commonly blunt or retuse at the apex with a distinct, fine colourless mucro, cuneate or attenuate at the base; each leaf-axil bearing a pair of fine and slender to stout and compressed spines up to c. 2.5 cm. long. Flowers green, in the lower parts of the plant in axillary clusters 6–15 mm. in diam., towards the ends of the stem and branches the clusters leafless and approximated to form simple or (the terminal at least) branched spikes up to c. 15 cm. long and 1 cm. wide. Lower flower clusters entirely female, as are the lower flowers of the spikes; upper flowers of spikes male, mostly for the apical 1/4–2/3 of each spike. Bracts and bracteoles deltoid-ovate, pale-membranous, with an erect, commonly reddish awn formed by the excurrent green midrib; bracteoles shorter than to a little exceeding the perianth, commonly smaller than the bracts. Perianth segments 5, those of the female flowers c. 1.5–2.5 mm. long, narrowly oblong or spathulate-oblong, obtuse or acute, mucronulate, frequently with a greenish dorsal vitta; those of the male flowers broadly lanceolate or lanceolate-oblong, acute or acuminate, only the midrib green. Stigmas (2) 3, flexuose or reflexed, 1–1.5 mm. long. Capsule ovoid-urceolate with a short, inflated beak below the style-base, c. 1.5 mm. long, regularly or irregularly circumcissile or more rarely indehiscent, the lid rugulose below the beak. Seed 0.75–1 mm. across, black, compressed, shining, very faintly reticulate.

Botswana. N: Maun, vii.1967, *Lambrecht* 238 (K; SRGH). SE: The Mall, Gaberone, 1050 m., 30.vii. 1978, *Hansen* 3421 (C; GAB; K; PRE; SRGH). **Zambia**. B: 3 km. N. of Kabompo, 31.iii.1982, *Drummond & Vernon* 11081 (K; MO; SRGH). N: Nchelenge Distr., Kafulwe, Lake Mweru, 960 m., 24.iv.1957, *Richards* 9424 (K). W: Mwinilunga District, just S. of Matonchi Farm, 6.xii.1937, *Milne-Redhead* 3510 (BM; K). C: 8 km. E. of Lusaka, 1275 m., 20.v.1955, *King* 23 (K). E: Lutembwe R. gorge, E. of Machinje Hills, 900 m., 13.x.1958, *Robson* 99 (BM; K). S: Livingstone, 970 m., 29.x.1964, *Bainbridge* 1016 (K; SRGH). **Zimbabwe**. N: Rushinga Distr., Chimanda Common Land Winde Pools, 515 m., 5.ix.1958, *Phipps* 1315 (K; SRGH). W: Sawmills, 80 km. NW. of Bulawayo, 8.viii.1929, *Rendle* 350 (BM). C: Shurugwi Distr., Gwetshetshe R. bridge on Gwenoro Dam, Zvishavane Rd., 21.iv.1968, *Biegel* 2609 (K; LISC; SRGH). E: Near Lusitu/Haroni Rivers confluence, Ngorima Common Land 23.xi.1967, *Ngoni* 46 (K; LISC; SRGH). S: Chiredzi Distr., Hippo Valley Estate, 27.xii.1971, *Taylor* 198 (K; LISC; SRGH). **Malawi**. N: Mzimba Distr., Mzuzu, Marymount, 1370 m., 10.iv.1974, *Pawek* 8320 (K; MO; SRGH; UC). C: Dedza Distr., Chongoni Forest Reserve, 23.ii.1968, *Salubeni* 981 (K; SRGH). S: Nsanje Distr., between Muona and Shire R., c. 79 m., 20.iii.1960, *Phipps* 2572 (K; SRGH). One sheet, *Lawrence* 98 (K), describes this plant as "very common all over the Shire plain". **Mozambique**. N: Lichinga, ii.1934, *Torre* 37 (LISC). T: Boroma, 29.viii.1931, *Guerra* 40 (COI). MS: Chimoio, Gondola, prox. da serraçâo de Braunstein, 2.ii.1948, *Garcia* 11 (K; LISC). GI: Dumela, Limpopo R., 30.iv.1961, *Drummond & Rutherford-Smith* 7610 (K; SRGH). M: Maputo, iv.1945, s.d., *Pimenta* 177 (LISC; SRGH).

This species occurs throughout the tropical and subtropical regions of the world, also occurring sporadically as a weed in some temperate regions. In the Flora Zambesiaca area chiefly as a weed of disturbed or heavily grazed ground, along roadsides, in current or abandoned cultivation, often irrigated or near water, soils from sand to heavy black loam, recorded altitudes 79–1370 m., but certainly occurs at lower altitudes.

5. **Amaranthus tricolor** L., Sp. Pl.: 989 (1753).—Baker & Clarke in F.T.A. **6**, 1: 32 (1909) pro parte, excl. pl. Forbes.—Schinz in Engl. & Prantl Pflanzenfam. ed. 2, **16C**: 38 (1934).—Aellen in Hegi, Illustr. Fl. Mitteleurop. ed. 2, **3**, 2: 477 (1959). TAB. **10** fig. E. Type, Linnean specimen 1117/7 (LINN, lectotype).

Annual herb, ascending or erect, attaining c. 1.25 m. in cultivation in Africa (taller in Asia). Stem stout and usually branched, both it and the branches angular, glabrous or furnished in the upper parts with sparse (or denser in the inflorescence), more or less crisped hairs. Leaves glabrous, or thinly pilose on the inferior surface of the primary venation, green or purplish-suffused, long-petiolate (petioles to c. 8 cm. long), the lamina broadly ovate, rhomboid-ovate or broadly elliptic to lanceolate-oblong, acute to obtuse or emarginate at the apex, shortly cuneate to attenuate at the base. Flowers green to crimson, in more or less globose clusters c. 4–25 mm. in diam., all or only the lower axillary and more or less distant, the upper sometimes without subtending leaves and increasingly approximate to form a thick terminal spike of variable length; male and female flowers intermixed. Bracts and bracteoles broadly or deltoid-ovate; bracteoles subequalling or shorter than the perianth, pale-membranous, broadest near the base and narrowed upwards to the green midrib, which is excurrent to form a long, pale-tipped awn which is usually at least half as long as the basal part and not rarely equalling it. Perianth segments 3, 3-5 mm. long, elliptic or elliptic-oblong, narrowed above, pale-membranous, the green midrib excurrent into a long, pale-tipped awn; segments of the female flowers slightly accrescent in fruit. Stigmas 3, erect or recurved, c. 2 mm. long. Capsule ovoid-urceolate with a short beak below the style-bases, 2.25–2.75 mm., membranous, obscurely wrinkled, circumcissile. Seeds 1–1.5 mm. wide, black or brown, shining, faintly reticulate.

Mozambique. M: ornamental in the municipal garden of Maputo, 1945, *Pimenta* 4309 (LISC).

Apparently a native of tropical Asia, this species is well-known through much of the tropics both as an ornamental and a vegetable. Though only the above specimen has been seen, it must surely occur elsewhere in the Flora Zambesiaca area.

6. **Amaranthus thunbergii** Moq. in DC., Prodr. **13**, 2: 262 (1849).—Schinz in Engl. & Prantl. Pflanzenfam. ed. 2, **16 C**: 38 (1934).—Hauman in F.C.B. **2**: 32 (1951).—Aellen in Hegi, Illustr. Fl. Mitteleurop. ed. 2, **3**, 2: 496 (1959).—Podlech & Meeuse in Merxm. Prodr. Fl. SW. Afr. **33**: 8 (1966).—J.H. Ross Fl. Natal: 158 (1973).—Brenan in Journ. S. Afr. Bot. **47**: 467 (1981). TAB. **10** fig. C. Type from S. Africa.★

Amaranthus graecizans sensu Baker & Clarke, F.T.A. **6**, 1: 34 (1909) non L.

Annual herb, ascending or erect, simple or branched from the base and frequently also above, reaching 15—55 cm. Stem and branches stout, angular, glabrous or thinly hairy below, upwards increasingly furnished with long, crisped, multicellular, rather flocculent hairs. Leaves glabrous, or thinly pilose on the inferior surface of the primary venation, sometimes with a dark purple blotch, long-petiolate (petioles up to c. 4 cm. long, sometimes longer than the lamina), the lamina narrowly or broadly elliptic to rhomboid or spathulate, c. (5) 15—45 (60) × (4) 10—30 (40) mm., blunt or retuse at the apex with the midrib excurrent in a short mucro, at the base cuneate to attenuate, more or less decurrent along the petiole. Flowers green, males most frequent at the top of the upper clusters, all in axillary clusters 6—15 mm. across, approximate above, usually increasingly distant towards the base of the stem and branches with male and female flowers intermixed. Bracts and bracteoles deltoid-lanceolate, bracteoles subequalling or shorter than the perianth, pale-membranous, often greenish centrally above, the midrib often excurrent in a long, fine awn often as long as the basal portion, bracteoles shorter (to 2 mm. long), the awn colourless and often more or less reflexed above. Perianth segments 3, similar in male and female flowers, lanceolate to oblong, or in the females rarely narrowly spathulate, 3—6 mm., pale-membranous or (especially in the female flowers) somewhat greenish above, gradually or more abruptly narrowed into the long (0.75—1.5 mm.) awn formed by the excurrent midrib, the latter green but the divergent or flexuose awn colourless above; fruiting female perianth segments slightly accrescent, wider than those of the male flowers. Stigmas 3, flexuose or often reflexed, pale, 1.5—2 mm. long. Capsule ovoid-ellipsoid to pyriform, c. 2.5—3.5 mm. long, with a short beak, circumcissile, membranous, obscurely wrinkled, shorter than the perianth (attaining the base of the aristate apices). Seed 1—1.5 mm. across, black or chestnut, shining, feebly reticulate.

★ Moquin cited Thunberg specimens from Herbs. Krauss and Hooker. The Krauss specimen has not been traced and may be B†. No Thunberg specimen from Herb. Hooker exists at Kew. See Brenan, loc. cit.: 470 (1981).

Botswana. N: Boteli R., Samedupe Bridge, 13.xii.1976, *Smith* 1840 (K; SRGH). SW: Kgalagadi, Ditatso Pan, c. 40 km. NW. of Tshabong, 940 m., 24.ii.1963, *Leistner* 3061 (K; PRE). SE: Gaberone University Campus, 1000 m., 1.xii.1975, *Mott* 827 (K; SRGH). **Zambia**. B: near Senanga, 1030 m., 4.ix.1952, *Codd* 7388 (BM; K; PRE). N: Kanakashi School, 9.iii.1981, *Drummond & Vernon* 10753 (K; PRE; SRGH). W: Solwezi-Mwinilunga Rd., Kabompo R., c. 1400 m., 15.ii.1975, *Hooper & Townsend* 58 (K; NDO; SRGH). C: Lukanga R. near Shamguta, 80 km. NW. of Kabwe, 1120 m., 9.iv.1972, *Kornas* 1578 (K). S: near Mumbwa, 1911, *Macaulay* 688 (K). **Zimbabwe**. W: Bulawayo, Norfolk Rd., 1350 m., 15.v.1964, *Best* 396 (K; SRGH). C: 29 km. SSE. of Kwe Kwe, Mlezu School Farm, 1275 m., 7.i.1966, *Biegel* 769 (K; LISC; SRGH). E: Lower Sabi, Chibuwe, 485 m., 28.i.1948, *Wild* 2322 (K; SRGH). **Malawi**. N: Chitipa, s.d., *Whyte* s.n. (K). S: Mpasu, Shire Valley, Oct. 1887, *Scott* s.n. (K). **Mozambique**. MS: Gaza Distr., Dumela, 30.iv.1961, *Drummond & Rutherford-Smith* 7609 (K; SRGH). GI: Guijá, Missâo de S. Vicente de Paula, 23.vi.1947, *Pedro & Pedrógâo* 1197 (K; LMJ). M: Inhaca Is., sea level—152 m., 6.x.1957, *Mogg* 27711 (K).

Widespread in tropical Africa from Ethiopia and Somalia to Zaire and Angola, south to Namibia and S. Africa. Introduced into Australia and then to Europe as a frequent casual "wool adventive". In the Flora Zambesiaca area on waste and disturbed ground as a weed of cultivation, in overgrazed areas, in open woodland, frequently where irrigated or in seasonally wet areas along river banks, lake shores, in pans; apparently always on light soils (sand, loam); sea level—1400 m.

7. **Amaranthus dinteri** Schinz in Mém. Herb. Boiss. **20**: 15 (1900); in Engl. & Prantl Pflanzenfam. ed. 2, **16 C**: 39 (1934).—Podlech & Meeuse in Merxm. Prodr. Fl. SW. Afr. **33**: 7 (1966).—Brenan in Journ. S. Afr. Bot. **47**: 470 (1981). TAB. **11** fig. B. Type from Namibia.

Annual herb, decumbent or ascending with numerous stems from the base (the stems rather sparingly branched below), or erect and branched, chiefly in the lower half, glabrous or sometimes puberulous when young, mainly about the nodes; stem and branches more or less sulcate and angled, almost smooth to considerably papillose-scabrid. Main leaves of stem and branches glabrous, c. 10—45 × 3—13 (18) mm. including the slender petiole, which varies from slightly shorter to considerably longer than the broadly obovate or less commonly elliptic lamina, rounded-obtuse to slightly emarginate, shortly and feebly mucronate, with or without a purplish black blotch. Flowers green, in dense axillary clusters 4—8 mm. in diam., extending almost to the base of the plant, clusters becoming more approximate above but the superior leaves only gradually and never considerably reduced; male and female flowers intermixed, the males more numerous above. Bracts lanceolate, 1—1.5 mm. long, pale membranous, the arista somewhat shorter than the lamina; bracteoles c. 1.25—3 mm. long, lanceolate-ovate to lanceolate, green along the midrib above, usually more shortly aristate than the bracts, outwardly-curving, shorter than the perianth. Male flowers with 3 elliptic, shortly aristate tepals, c. 1.5—2 mm. long, pale-membranous with a greenish or brownish midrib. Female flowers with 3 tepals, tepals ovate to oblong, (1.5) 2—3 mm. long, frequently broadly greenish above with branched or anastomosing nervation towards the usually more or less outwardly curved apices, acute to somewhat obtuse, shortly but distinctly aristate. Stigmas 3, c. 0.5—0.75 mm. long, somewhat expanded below. Fruit ovoid to shortly obpyriform, shorter than or sometimes subequalling the perianth, circumcissile, strongly rugose, dark when ripe. Seeds compressed, lenticular, 1—1.5 mm. across, finely reticulate over the entire surface, less shining than in many species of the genus.

Botswana. N: Mumpswe Pan, 40 km. NNW. of mouth of Nata R., 896 m., 21.iv.1957, *Drummond & Seagrief* 5172 (K; SRGH). SW: Tshane Pan, 24.ii.1960, *Wild* 5126 (K; SRGH).

8. **Amaranthus praetermissus** Brenan in Journ. S. Afr. Bot. **47**: 478, f. 2 (1981). TAB. **11** fig. A. Type from S. Africa.

Amaranthus angustifolius sensu Adamson in Journ. S. Afr. Bot. 2, 4: 194 (1936) saltem pro parte, non Lam.

Amaranthus schinzianus sensu Suesseng. & Podlech in Merxm., Prodr. Fl. SW. Afr. **33**: 8 (1936) pro parte, non Thell.

Amaranthus thunbergii Suesseng. & Podlech in Merxm., loc. cit.: 8 (1936) pro parte, non Moq.

Erect annual herb, 15—75 (100) cm. in height, simple or branched from below and sometimes for some way up the stem, quite glabrous, stem and branches more or less sulcate and angled, smooth or minutely papillose when young, upper branches elongate and lax to short and very densely floriferous. Leaves glabrous, c.12—60 × 1—10 mm. including the slender petiole, which may be as long as the lamina; lamina linear to narrowly oblanceolate, narrowly elliptic or elliptic-oblong, long-attenuate below into the petiole, at the apex acute to obtuse with a very distinct, pale mucro up to 1.5 (2) mm. long formed by the excurrent nerve. Flowers green, in dense axillary clusters c. 3—6 mm. in diam., normally extending well down towards the base of the plant, the clusters approximate above, the superior leaves scarcely reducing, or sometimes so rapidly so that a few upper clusters are leafless; male and female flowers intermixed, the males more numerous in the upper clusters. Bracts and bracteoles pale-membranous with the lamina lanceolate, terminating in an arista (which may be up to as long as the lamina) formed by the percurrent nerve; bracteoles 2—3 mm., slightly more rigid and slightly longer than the bracts. Male flowers with 4 elliptic-ovate tepals 1.5—2 mm. long, pale with a brownish midrib, distinctly mucronate, the mucro very variable in length. Female flowers with 4—5 tepals, tepals narrowly oblong to narrowly oblong-lanceolate or more rarely oblanceolate, (1.5) 2—3 mm. long, tapering (rarely more abruptly narrowed) into the erect or spreading, pale

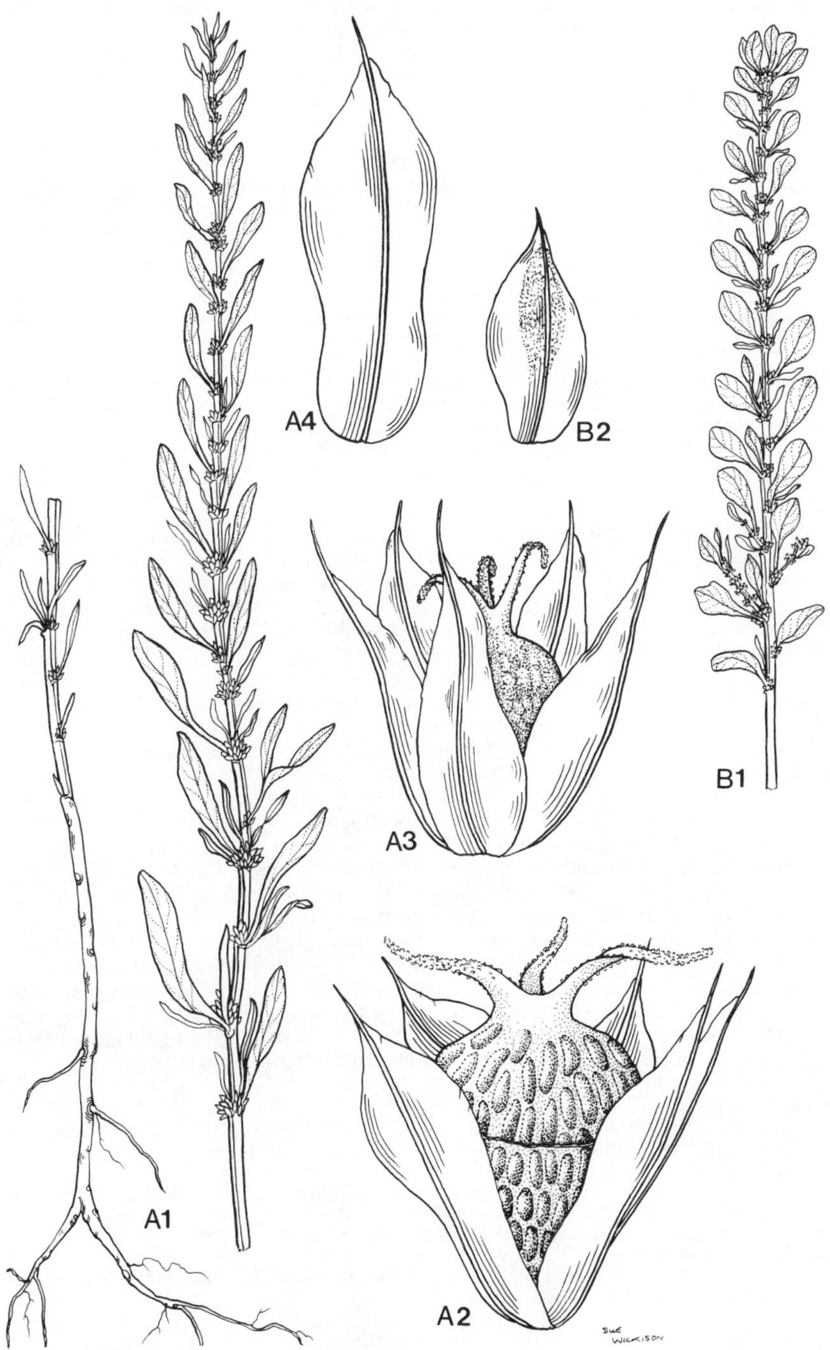

Tab. 11. A.—AMARANTHUS PRAETERMISSUS. A1, habit ($\times\frac{2}{3}$); A2, fruiting perianth ($\times$16); A3, younger perianth ($\times$16); A4, tepal of female flower ($\times$18), A1—4 from *Ngoni* 314. B.—AMARANTHUS DINTERI. B1, flowering stem ($\times\frac{2}{3}$); B2, tepal of female flower ($\times$18), B1—2 from *Wild* 5126.

or brownish, long mucro which is commonly c. half the length of the lamina. Styles 3, broad at the base and gradually narrowed above, c. 0.5−0.75 mm. long. Capsule ovoid to obpyriform, somewhat shorter than the perianth, circumcissile, convolute-rugose, commonly blackish when dry. Seeds shining, compressed, lenticular, c. 1 mm. in diam., only very faintly patterned centrally, more distinctly reticulate over a wide margin.

Botswana. N: Maun, above high-flood level of Thamalakane R., 910 m., 22.i.1972, *Biegel & Gibbs-Russell* 3723 (K; LISC; SRGH). SW: Tshane Police Station, 22.ii.1960, *de Winter* 7451 (K; PRE; SRGH). SE: Boteli delta area, N.E. of Mopipi, 850 m., 17.iv.1973, *Standish-White* 16 (K; SRGH). **Zimbabwe**. E: Chipinge Distr., Lower Sabi, Rupisi Hot Springs, 515 m., 28.i.1948, *Wild* 2308 (K; SRGH). S: Gwanda Distr., Tuli Breeding St. Offices, 16.ii.1965, *Norris-Rogers* 664 (K; SRGH).

Also in S. Africa, Namibia and Angola. In the Flora Zambesiaca area a species of Mopane woodlands, overgrazed areas, "pans", gardens, and vicinity of rivers; on Kalahari Sand, alluvium and fine basalt soils, 560−1020 m.

9. **Amaranthus graecizans** L., Sp. Pl. 2: 990 (1953).—Aellen in Hegi, Illustr. Fl. Mitteleurop. ed. 2, 3, 2: 500 (1959).—Cavaco in Mém. Mus. Hist. Nat. Paris Sér. B, **13**: 52 (1962). Type, from Uppsala Botanic Garden, Linnean specimen 1117/3 (LINN, lectotype).

Amaranthus blitum sensu Baker & Clarke, F.T.A. **6**, 1: 35 (1909) non L.
Amaranthus polygamus sensu Baker & Clarke, F.T.A. **6**, 1: 36 (1909) non L.

Annual herb, branched from the base and usually also above, erect, decumbent or prostrate, mostly up to c. 45 cm. (rarely to 70 cm. high). Stem slender to stout, angular, glabrous or thinly to moderately furnished with short to long, often crisped multicellular hairs which increase upwards, especially in the inflorescence. Leaves glabrous or sometimes sparingly furnished on the lower surface of the principal veins with very short, gland-like hairs, long-petiolate (petiole from 3−45 mm. long, sometimes longer than the lamina), lamina broadly ovate or rhomboid-ovate to narrowly linear-lanceolate, 4−55 × 2−30 mm., acute to obtuse or obscurely retuse at the mucronulate apex, cuneate to long-attenuate at the base. Flowers all in axillary cymose clusters, male and female intermixed, male commonest in the upper cymes. Bracts and bracteoles narrowly lanceolate-oblong, pale-membranous, acuminate and with a pale or reddish arista formed by the excurrent green midrib, bracteoles subequalling or usually shorter than the perianth. Perianth segments 3, all 1.5−2 mm. long; those of the male flowers lanceolate-oblong, cuspidate, pale membranous with a narrow green midrib excurrent in a short, pale arista; those of the female flowers lanceolate-oblong to linear-oblong, gradually to abruptly narrowed to the mucro, the midrib often bordered by a green vitta above and apparently thickened, the margins pale whitish to greenish. Stigmas 3, slender, usually pale, flexuose, c.0.5 mm. long. Capsule subglobose to shortly ovoid, 2−2.25 mm. usually strongly wrinkled throughout with a very short, smooth beak, exceeding the perianth, circumcissile or sometimes not, even on the same plant. Seeds shining, compressed, black, 1−1.25 mm. across, faintly reticulate especially towards the margin.

Subsp. **graecizans**

Amaranthus angustifolius Lam., Encycl. Méth. Bot. **1**: 115 (1783) nom. illegit.—Schinz in Engl. & Prantl Pflanzenfam. ed. 2, **16** C: 39 (1934).—Hauman in F.C.B. 2: 34 (1951).

Amaranthus blitum sensu Baker & Clarke, F.T.A. **6**, 1: 35 (1909) saltem quoad specimens *Whyte, Cameron & Scott-Elliot*.

Amaranthus oleraceus sensu Baker & Clarke, F.T.A. **6**, 1: 34 (1909) quoad specimens *Whyte & Johnson*.

Amaranthus polygamus sensu Baker & Clarke, F.T.A. **6**, 1: 36 (1909) saltem quoad specimens *Adamson & Buchanan*.

Amaranthus angustifolius subsp. *aschersonianus* Thell. in Asch. & Graebn., Syn. Mitteleur. Fl. 5, 1: 309 (1914). Type from Sudan.

Amaranthus angustifolius var. *graecizans* (L.) Thell., Rep. B.E.C. Brit. Is. **5**: 306 (1918). Type as for *A. graecizans*.

Leaf blade (in particular of the larger leaves of the main stem) at least 2.5 times as long as broad, oblong to linear-lanceolate. Perianth segments mucronate, mucro up to c. 0.25 mm. long, usually straight.

Zimbabwe. N: Kariba Distr., Sengwa, 10.viii.1965, *Jarman* 225 (K; SRGH).

This subspecies is scattered through most of Africa, except South Africa, otherwise distributed in the Old World from the warmer parts of Europe through subtropical and tropical Asia to India. Its habitat in Zimbabwe was not recorded, but in E. Africa it frequently occurs on seasonally flooded flats or by seasonal pools.

Subsp. **silvestris** (Vill.) Brenan in Watsonia 4: 273 (1961). TAB. **10** fig. D. Type, *Tournefort* specimen No. 1849 (P, holotype, IDC microfiche Neg. 90, No. 19!).

Amaranthus silvestris Vill., Cat. Pl. Jard. Strasb.: 111 (1807).—Cavaco in Mém. Mus. Hist. Nat. Paris Sér. B, **13**: 53 (1962).

Amaranthus graecizans var. *silvestris* (Vill.) Asch. in Schweinf., Beitr. Fl. Aethiop.: 176 (1867).—Aellen in Hegi, Illustr. Fl. Mitteleurop. ed. 2, **3**, 2: 500 (1959).

Amaranthus angustifolius var. *silvestris* (Vill.) Thell. in Schinz & Keller, Fl. Schweiz ed. 4, **1**: 222 (1923).

Amaranthus angustifolius subsp. *silvestris* (Vill.) Henkels, Geillustr. Schoolfl. voor Nederl. ed. **11**: 170 (1934).—Hauman in F.C.B. **2**: 35 (1951).

Leaf blade (particularly of the main stem leaves) broadly to rhomboid-ovate or elliptic-ovate, less than 2.5 times as long as broad. Perianth segments mucronate, the mucro very short and usually straight.

Zambia. N: Mbala Distr., Saisi R., 1500 m., 15.iv.1959, *Richards* 11220 (K). E: Luangwa Valley, Mulila Munkanya, 2.iii.1968, *Phiri* 42 (K). S: within 1.6 km. of the Zambesi c.112 km. upstream from Kariba Gorge, ii.1957, *Scudder* 30 (K; SRGH). **Zimbabwe**. E: Lower Sabi Chibuwe, 485 m., 28.i.1948, *Wild* 2320 (K; SRGH). **Malawi**. N: between M'pata and commencement of Tanganyika Plateau, 610—910 m., vii.1886, *Whyte* s.n. (K). S: Nsanje c. 80 m., 19.iii.1960, *Phipps* 2554 (K; SRGH). **Mozambique**. T: Machenga-Gomero (?), 8.viii.1907, *Johnson* 269 (K). MS: Chemba, Chiou, estacâo experimental do CICA, 23.v.1961, *Balsinhas & Marrime* 485 (K; LMJ).

This subspecies is distributed in the Old World from the warmer parts of Europe to the cooler regions of SW. Asia and NW. India, and also in most parts of Africa (not S. Africa). In the Flora Zambesiaca area a weed of cultivated or otherwise disturbed ground (heavily grazed or trodden, track edges) on light sandy soil to heavy clay alluvium; 80—1500m.

The above two subspecies are not as truly sympatric as might be judged from the literature, and are thus maintained at this rank rather than merely as varieties.

Subsp. **thellungianus** (Nevski) Gusev in Bot. Zhurn. **57**: 462 (1972). TAB. **10** fig. K. Type from Kugitang.

Amaranthus blitum var. *polygonoides* Moq. in DC., Prodr. **13**, 2: 263 (1849). Type from India.

Amaranthus angustifolius "proles" *polygonoides* (Moq.) Thell. in Aschers. & Graebn., Syn. Mitteleurop. Fl. **5**, 1: 308 (1914).

Amaranthus thellungianus Nevski in Act. Acad. Sc. URSS **1**, 4: 311 (1937).

Amaranthus thunbergii var. *grandifolius* Suesseng. in Mitt. bot. Staatss. München **3**: 73 (1951). Type from Tanzania.

Leaf-blade narrowly linear or lanceolate to rhomboid-spathulate; perianth segments narrow, more tapering, long-aristate, awns mostly 0.3—0.75 mm. in length and frequently somewhat divergent, bracteoles also long-aristate.

Mozambique. M: Inhaca Is., Monte Inhaca, 12.viii.1980, *Jansen & Koning* 7398.

This subspecies has occurred once in Kenya, but otherwise is confined to Middle Asia and the Indian continent; it may have been introduced into Africa from India. The plant described by Suessenguth as *A. thunbergii* var. *grandifolius* is a lush form of *A. graecizans* subsp. *thellungianus*, and was grown as a vegetable at Amani, Tanzania.

10. **Amaranthus lividus** L., Sp. Pl. **2**: 990 (1753).—Schinz in Engl. & Prantl Pflanzenfam. ed. 2, **16 C**: 38 (1934).—Aellen in Hegi, Illustr. Fl. Mitteleurop. ed. 2, **3**, 2: 505 (1959).—Brenan in Journ. S. Afr. Bot. **47**: 486 (1981). Type, early cultivated material (BM, neotype).

Annual herb, erect, ascending or prostrate, 6—60 (90) cm. Stem slender to stout, simple or considerably branched from the base or upwards, more or less angular, green to reddish or yellow, quite glabrous or more rarely with 1-few-celled, short hairs above and/or in the inflorescence. Leaves glabrous or more rarely with scattered few-celled hairs near the base on the lower surface of the primary venation, long-petiolate (petioles up to c. 10 cm., frequently longer than the lamina), lamina ovate to rhomboid-ovate, 1–8 × 0.6–6 cm., shortly cuneate below, the apex usually broad and almost always distinctly emarginate, mucronulate. Flowers green, in slender to stout terminal and axillary spikes (in small forms even the terminal sometimes indistinct) or rarely panicles, terminal spikes c. 0.6–11 cm. long and 0.3–2 cm. wide, or the lower axillary inflorescences of dense cymose clusters up to 2 cm. in diam.; male and female flowers intermixed. Bracts and bracteoles deltoid-ovate to lanceolate, whitish-membranous with a short yellow or reddish mucro formed by the excurrent midrib, bracteoles shorter than or rarely subequalling the perianth. Perianth segments 3 (occasionally 4 or even 5 in cultivated forms), membranous-margined, male and female both varying from lanceolate-oblong, subacute and mucronate to broadly spathulate and obtuse with the thick midrib ceasing below the summit but the female frequently blunter, 0.75–2 mm. long. Stigmas 2–3, short, erect or flexuose. Capsule subglobular to shortly pyriform, compressed, exceeding the perianth, 1.25–2.5 mm. long, usually rather smooth but sometimes wrinkled on drying, indehiscent or rupturing irregularly at maturity. Seeds 1–1.75 mm. in diam., circular, compressed, dark brown to black, the centre feebly reticulate and shining, the margin duller, minutely punctate-roughened over the reticulum.

Subsp. **lividus**

Amaranthus blitum L., Sp. Pl.: 990 (1753) nomen confusum. Type, Linnean specimen 1117/4 (LINN. lectotype).

Amaranthus oleraceus L., Sp. Pl. ed. 2: 1403 (1763).—Baker & Clarke in F.T.A. **6**, 1: 34 (1909) quoad sp. *Scott*. Type a Linnean specimen 1117/13 (LINN, lectotype).

Amaranthus ascendens Lois., Not. Pl. France: 141 (1810). Type, Bauhin, Hist. Plant. 2: 966, fig. *"Blitum pulchrum rectum magnum rubrum"* (lectotype).

Amaranthus oleraceus var. *maxima* C.B. Clarke in F.T.A. **6**, 1: 34 (1909). Type from Uganda.

Amaranthus lividus var. *ascendens* (Lois.) Hayward & Druce, Adventive Fl. Tweedside: 177 (1919).—Hauman in F.C.B. **2**: 33 (1951).—Aellen in Hegi, Illustr. Fl. Mitteleurop. ed. 2, **3**, 2: 506 (1959).

Amaranthus lividus subsp. *ascendens* (Lois.) Heukels, Geill. Schoolfl. voor Nederl. ed. **11**: 169 (1934). Type as above.

Plant robust to very robust, with large leaves, generally erect or ascending. Fruit 2 mm. or more in length.

Malawi. N: Mzimba Distr., Mzuzu, Marymount, 1365 m., 6.viii.1970, *Pawek* 3679 (K).
Widespread in the warmer parts of Europe, east to central Asia, China and Japan, N. Africa from Morocco to Egypt, tropical W. and E. Africa, N. America.

Subsp. **polygonoides** (Moq.) Probst, Wolladventivpfl. Mitteleur.: 74 (1949). TAB. **10** fig. B. Type from Brazil.

Euxolus viridis var. *polygonoides* Moq. in DC., Prodr. **13**, 2: 274 (1859). Type as above.

Amaranthus lividus var. *polygonoides* (Moq.) Thell. in Rep. B.E.C. 5: 574 (1920).—Aellen in Hegi, Illustr. Fl. Mitteleurop. ed. 2, **3**, 2: 506 (1959). Type as above.

Amaranthus adscendens subsp. *polygonoides* (Moq.) Priszter in Ann. Sect. Horti et Viticult. Univ. Sc. Agric. Budapest **1951**, 2: 221 (1953). Type as above.

Plant smaller and neater, with smaller leaves rarely as much as 4 cm. long and usually less, generally prostrate to decumbent. Fruit 1.25–1.75 mm. in length.

Zambia. N: Mbala, ii.1970, *Drummond & Williamson* 9982 (K; MO; NDO; SRGH). **Zimbabwe**. W: Bulawayo, 30.xii.1979, *Best* 1374 (K; MO; PRE;SRGH). C: Gweru Distr., Mlezu School Farm 29 km. SSE. of Kwe Kwe 1275 m., 7.i.1966, *Biegel* 768 (K; SRGH). **Mozambique**. M: Maputo, Jardin Vasco da Gama, 15.ix.1971, *Balsinhas* 2201 (LISC).
This subspecies is widespread in the warmer temperature regions and tropics of both Old and New Worlds; in Asia from the Caucasus through India to Malaya, Java and New Guinea; Australia; in tropical Africa mainly eastern; Macaronesia, Mascarene Is., S. America from Guyana S. to Argentina. Often casual in Europe and N. America.

These two taxa are maintained at subspecific rather than varietal rank in view of their differences in geographical range, subsp. *lividus* being chiefly found in the regions of the older civilisations of Europe and Asia and subsp. *polygonoides* chiefly in the tropics - being especially uniform in S. America. *Biegel* 768, cited above under subsp. *polygonoides*, is unusually narrow in the leaf apex with no broad emargination, the leaves in fact much resembling *A. viridis*, as which species it was received at Kew; but the fruits and seeds show it unequivocally to be *A. lividus*.

11. **Amaranthus viridis** L., Sp. Pl. ed. 2: 1405 (1763).—Baker & Clarke in F.T.A. **6**, 1: 34 (1909).—Brenan in Journ. S. Afr. Bot. **47**: 488 (1981). TAB. **10** fig. G. Type, Linnean specimen 1117/15 (LINN. lectotype).

Chenopodium caudatum Jacq., Collect. Bot. 2: 325 (1788); in Ic. Pl. Rar. 2: 12, t. 344 (1789). Type from "Guinea".

Amaranthus gracilis Desf., Table Ecole Bot.: 43 (1804) nomen nudum, ex Poir. in Lam. Encycl. Méth. Bot. Suppl. **1**: 312 (1810) nomen illegit.—Hauman in F.C.B. 2: 32 (1951).—Cavaco in Mém. Mus. Hist. Nat. Paris Sér. **B**, 13: 57 (1962); in J.H. Ross Fl. Natal: 158 (1973).—Aellen in Hegi, Illustr. Fl. Mitteleurop. ed. 2, **3**, 2: 503 (1959). Type, material from Botanical Garden, Paris (not seen).

Short-lived perennial herb (sometimes flowering in first year), erect or more rarely ascending, 10—75 (100) cm. Stem rather slender, sparingly to considerably branched, angular, glabrous or more frequently increasingly (but still rather sparsely) hairy upwards (especially in the inflorescence) with short or longer and rather floccose multicellular hairs. Leaves glabrous or shortly to fairly long-pilose on the lower surface of the primary or most of the venation, long-petiolate (petioles up to c. 10 cm. long and the longest commonly longer than the lamina), lamina deltoid-ovate to rhomboid-oblong, 2—7 × 1.5—5.5 cm., the margins occasionally obviously sinuate, shortly cuneate to subtruncate below, obtuse and narrowly to clearly emarginate at the apex, minutely mucronate. Flowers green, in slender, axillary or terminal, frequently paniculate spikes c. 2.5—12 cm. long and 2—5 mm. wide, or in the lower part of the stem in dense axillary clusters to c. 7 mm. in diam.; male and female flowers intermixed but the latter more numerous. Bracts and bracteoles deltoid-ovate to lanceolate-ovate, whitish-membranous with a very short pale or reddish awn formed by the excurrent green midrib, bracteoles shorter than the perianth (c. 1 mm. long). Perianth segments 3, very rarely 4, those of the male flowers oblong-elliptic, acute, concave, c. 1.5 mm. long, shortly mucronate; those of the female flowers narrowly oblong to narrowly spathulate, finally 1.25—1.75 mm. long, the borders white - membranous, minutely mucronate or not, midrib green and often thickened above. Stigmas 2—3, short, erect or almost so. Capsule subglobose, 1.25—1.75 mm. in diam., not or only slightly exceeding the perianth, indehiscent or rupturing irregularly, very strongly rugose throughout at maturity. Seed c. 1—1.25 mm. in diam., circular, only slightly compressed, more or less shining, reticulate and with shallow the verrucae on reticulum, the verrucae with shape of the areolae.

Zambia. B: Kaoma Tobacco Scheme, c. 64 km. E. of Kaoma Town, 3.iv.1982, *Drummond & Vernon* 11165 (K; MO; SRGH). N: Mununshi Banana Scheme, Luapula Valley, 11.ii.1970, *Anton-Smith* s.n. (K; SRGH). W: Chingola, 18.i.1956, *Fanshawe* 2748 (K; NDO). S: Livingstone, 900 m., iv.1909, *Rogers* 7066 (K). **Zimbabwe**. W: Bulawayo, Mbala St., 14.vi.1970, *Biegel* 3334 (K; LISC; SRGH). **Mozambique**. N: R.C. Mission, 10 km. N. of Angoche, 22.x.1965, *Mogg* 32497 (LISC). T: between Tete and Cahobra Bassa Rapids, xi.1858, *Kirk* s.n. (K). GI: Inhambane, i.1936, *Gomes e Sousa* 1710B (K). M: Maputo, iv.1945, *Pimenta* 4321 (LISC).

A practically cosmopolitan weed (possibly of Asian origin) in the tropical and pantropical regions of the world, penetrating more into the temperate regions than most of its allies (as, for example, in N. & S. America and Europe). In the Flora Zambesiaca area apparently always on disturbed or cultivated ground.

12. **Amaranthus deflexus** L., Mant. Alt.: 295 (1771).—Schinz in Engl. & Prantl Pflanzenfam. ed. 2, **16 C**: 39 (1934).—Aellen in Hegi, Illustr. Fl. Mitteleurop. ed. 2, **3**, 2: 504 (1959).—J.H. Ross Fl. Natal: 158 (1973).—Brenan in Journ. S. Afr. Bot. **47**: 484 (1981). TAB. **10** fig. A. Type, cultivated material from Uppsala Botanic Garden, Linnean specimen at Stockholm (S, lectotype, IDC microfiche Neg. 384 no. 19!). Linnean specimen No. 1117/18 (LINN), labelled as *A. scandens*, is also this species.

Perennial herb, prostrate to somewhat ascending, (8) 12–45 (80) cm. Stems several, slender to rather stout, usually much-branched from the base upwards, more or less angular, green to reddish, glabrous below but usually increasingly furnished from well below the middle upwards with yellowish, flexuose or crisped multicellular hairs. Leaves moderately to more or less densely furnished on the margins and inferior surface (especially of the principal veins) with similar multicellular hairs, long-petiolate (petioles 6–25 mm. long, but rarely, if ever, exceeding the lamina), lamina c. 1–4.5 × 0.5–2.5 cm, broadly ovate to lanceolate (most commonly rhomboid-ovate), subtruncate to shortly cuneate at the base, subacute to obtuse and sometimes shallowly retuse at the mucronulate apex. Flowers green, in slender and lax to stout and dense terminal and axillary spikes c. 2–10 cm long and 5–12 (20) mm. wide, the terminal spike not rarely with rather short, stout branches and the lowest inflorescences dense, subglobose clusters to c. 1 cm in diam.; male and female flowers intermixed, the males generally rather few. Bracts and bracteoles deltoid-ovate to ovate-lanceolate, pale-membranous with a shortly excurrent greenish midrib, bracteoles usually about half the length of the perianth. Perianth segments 2–3, 1.5–2 mm., linear- to oblong-spathulate, obtuse to subacute, male and female similar or the male shorter and somewhat more acute, pale-membranous with the thin to thick green midrib excurrent in a short, paler mucro. Stigmas 2–3, pale, slender, flexuose, c. 0.4 mm. long. Capsule ellipsoid, sometimes constricted above, 1.75–3 mm., obviously exceeding the perianth, scarcely compressed, smooth, indehiscent or irregularly rupturing at maturity. Seeds compressed-ellipsoid, c. 1–1.2 × 0.7–0.8 mm. in diam., black, shining and almost smooth, with a duller, slightly rugose border.

Zimbabwe. C: Harare, weed in Landscapes Nurseries, Highlands, 1650 m., 21.iii.1971, *Biegel* 3491 (K; SRGH).

A native of temperate S. America widely naturalised in the Mediterranean region (Europe, N. Africa, Turkey) as well as in Macaronesia, Polynesia, Kenya, S. Africa, N. America & c. In the Flora Zambesiaca area apparently a rare weed, but will certainly increase.

4. SERICOSTACHYS Gilg & Lopr.

Sericostachys Gilg & Lopr. in Engl. Bot. Jahrb. **27**: 50 (1899).

Scandent shrub with opposite branches and leaves; leaves entire; indument of jointed barbellate hairs. Inflorescence a broad panicle of "spikes", bracteate, bracts persistent; partial inflorescences sessile on the axis of the spike, consisting of one fertile bibracteolate flower and two modified sterile flowers, also bibracteolate. Sterile flowers consisting of a number of filiform, hairlike appendages which greatly elongate as fruit develops and are densely plumed with whitish hairs; bracteoles also greatly accrescent. Fertile flowers hermaphrodite, with 5 perianth segments which are all similar in form. Stamens 5, very shortly fused into a rather solid, disk-like rim at the extreme base, alternating with very small pseudostaminodes; anthers bilocular. Style filiform, stigma capitate; ovary with a single pendulous ovule, glabrous, obovoid-pyriform. Fruit a thin-walled indehiscent utricle enclosed by and falling with the persistent perianth and bracteoles; endosperm copious.

A monotypic genus.

Sericostachys scandens Gilg & Lopr. in Engl. Bot. Jahrb. **27**: 51 (1899); in Malpighia **14**: 27 (1901).—Baker & Clarke in F.T.A. **6**, 1: 71 (1909).—Schinz in Viert. Nat. Ges. Zürich **56**: 256 (1911); in Engl. & Prantl Pflanzenfam. ed. 2, **16 C**: 42 (1934).—Hauman in F.C.B. **2**: 70 (1951).—Cavaco in Mém. Mus. Hist. Nat. Paris Sér. B, **13**: 62 (1962). TAB. 12. Type from Cameroon.

Sericostachys tomentosa Lopr. in Engl. Bot. Jahrb. **27**: 51 (1899) et **30**: 26, t.1 f. P & Q (1901); in Malpighia **14**: 450 (1901).—Baker & Clarke in F.T.A. **6**, 1: 71 (1909).—Schinz in Viert. Nat. Ges. Zürich **56**: 256 (1911); in Engl. & Prantl Pflanzenfam. ed. 2, **16 C**: 42 (1934). Type from Uganda.

Sericostachys scandens var. *tomentosa* (Gilg & Lopr.) Cavaco in Mém. Mus. Hist. Nat. Paris Sér. B, **13**: 62 (1962). Type as for *S. tomentosa*.

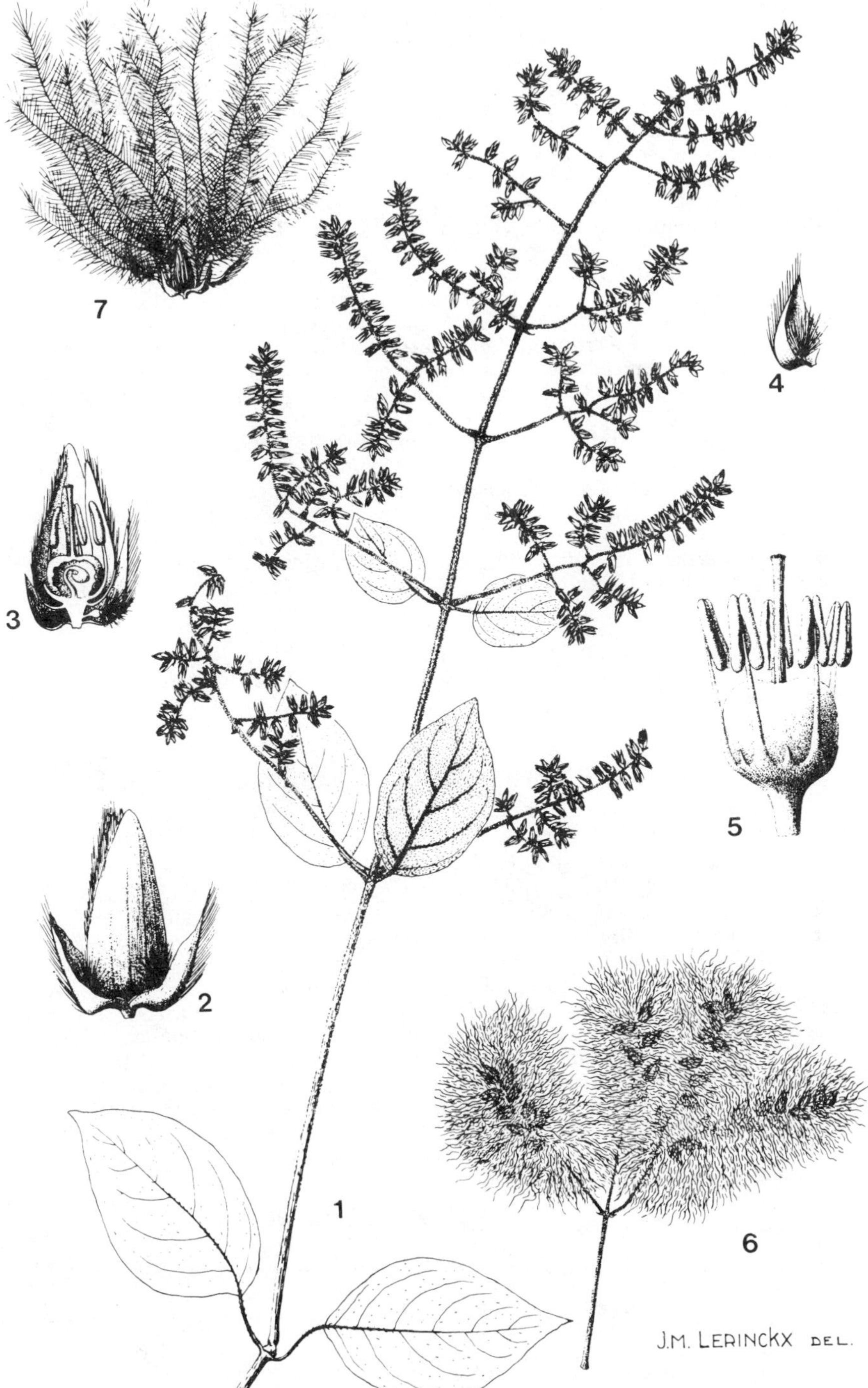

Tab. 12. SERICOSTACHYS SCANDENS. 1, flowering branch (×$\frac{1}{2}$); 2, cymule (×5); 3, cymule in longitudinal section (×5); 4, bracteole with sterile flower (×5); 5, androecium and gynoecium (×10); 6, part of infructescence (×$\frac{1}{2}$); 7, fruiting cymule, from *Ghesquire* 4411 and *Giorgi* 1624. Reproduced with permission from "Flore de Congo Belge".

Much-branched scandent perennial up to at least c. 20 m. high, the flowering branches hanging widely from trees and, with the plumose-hairy sterile flowers, having much the appearance of a *Clematis*. Branches terete or subtetragonous, finely striate, swollen at the nodes, glabrous or more usually increasingly tomentose towards the inflorescence. Leaves broadly ovate to lanceolate-ovate, acuminate, on the main branches below the inflorescence 5.5–14 × 3–7 cm., subglabrous to more or less densely tomentose on both surfaces, cuneate to attenuate at the base, shortly (c. 1–2 cm.) petiolate. Inflorescence large, broadly paniculate with divergent branches, the branches subtended by pairs of leaves which become small and bract-like upwards, the individual "spikes" up to c. 6 × 1.5 cm., peduncles shortening upwards. Partial inflorescences of the "spikes" sessile. Bracts deltoid-ovate, c. 2 mm. long, membranous with a pale midrib, glabrous or sparingly hairy; bracteoles of fertile flowers similar, 2–4 mm.; bracteoles of sterile flowers at first small and closed round the rudimentary appendages, finally accrescent, lanceolate-acuminate, to c. 4–6 mm. long. long-pilose. Tepals 5, lanceolate to lanceolate ovate, 3.5–8 mm. long, the outer 2 glabrous or floccose at the basal margin, the inner 3 more or less pilose on the upper dorsal surface, all pale-margined, greenish centrally with a scarcely excurrent midrib and numerous fine lateral nerves. Stamens 3.5–6 mm. long, very shortly connate into a rim-like ring below, alternating with very small subulate or linear-oblong and toothed pseudostaminodes. Style 1.5–2.5 (3) mm. long. Sterile flowers of up to c. 12 linear, much-accrescent appendages densely furnished with spreading, jointed, minutely barbellate hairs, much exceeding the fertile perianth in fruit. Capsule ovoid-cylindrical, c. 3–3.5 mm. long Seed brown, smooth and shining, ovoid, 2.5–3 mm. long.

Malawi. N: Chitipa Distr., Misuku Hills, Mughesse rain forest, 1600–1800 m., 21.iv.1975, *Pawek* 9434 (K; MA; MO; SRGH; UC).

E. W. and C. Africa from Fernando Po to Ethiopia and Angola to Tanzania. Scrambling over trees in forest.

5. KYPHOCARPA (Fenzl) Lopr.

Kyphocarpa (Fenzl) Lopr. in Engl. Bot. Jahrb. **27**: 42 (1899).
Sericocoma Sect. *Kyphocarpa* Fenzl in Linnaea **17**: 323 (1843).

Erect annual herbs. Leaves entire, opposite on simple or branched stems. Flowers in spike-like, densely pilose bracteate thyrses terminal on the stem and branches, shortly or long-pedunculate or subtended by 1–2 reduced leaves; each bract subtending a bibracteolate partial inflorescence of mostly 4 fertile flowers accompanied by 4 bracteolate sterile flowers consisting of slender finally spinous processes; partial inflorescences falling as a unit complete with bracteoles; bracts persistent on the inflorescence axis. Perianth segments 5, narrow, 3-nerved, mucronate, only a narrow central section coloured. Stamens 5; filaments delicate, fused below with very short and blunt pseudostaminodes between; anthers bilocular. Ovary scarcely compressed, uniovulate, considerably pilose above and corniculate to one side at the apex; style slender, stigma simple, capitate. Utricle scarcely compressed, irregularly ruptured below by the developing seed, the rounded apex considerably pilose, gibbous to one side of the style with a short horn-like process. Endosperm copious.

A genus of 3 or 4 species in south tropical and South Africa.

The generic name has been mis-spelt "*Cyphocarpa*" by numerous authors.

Kyphocarpa angustifolia (Moq.) Lopr. in Engl. Bot. Jahrb. **27**: 45 (1899).—Baker & Clarke in F.T.A. **6**, 1: 53 (1909).—Podlech & Meeuse in Merxm. Prodr. Fl. SW. Afr. **33**: 17 (1966); in J.H. Ross Fl. Natal: 158 (1973). TAB. **13**. Type from S. Africa.
Cyathula angustifolia Moq. in DC., Prodr. **13**, 2: 328 (1849).
Sericocoma angustifolia (Moq.) Hook.f. in Benth. & Hook. Gen. Pl. **3**, 1: 30 (1880). Type as above.
Kyphocarpa zeyheri sensu Lopr. in Engl. Bot. Jahrb. **27**: 45 (1899) et auctt. al. incl. Schinz in Engl. & Prantl Pflanzenfam. ed. 2, **16 C**: 43 (1934) non *Trichinium zeyheri* Moq. quid est *Sericocoma avolans* Fenzl.

Kyphocarpa zeyheri var. *petersii* (Lopr.) Schinz in Viert. Nat. Ges. Zürich **57**: 541 (1912) saltem quoad sp. *Kirk*.
Sericocoma hereroensis Suesseng. & Beyerle in Fedde, Repert. **44**: 47 (1938). Type from Namibia.

Annual herb with a slender taproot (sometimes woody at the base in grazed-off plants with secondary growth), (12) 20–70 cm. tall, simple or with several stems from about the base and sometimes branched above with branches diverging at 45 degrees or less. Stem and branches terete, striate, very slender, thinly pilose above especially at the nodes, rapidly glabrescent. Leaves narrowly linear, 15–75 × 0.5–4.5 mm., sparingly pilose when young, the upper folded inwardly, the lower frequently flat, all with a sharp, acute mucro up to 0.75 mm. long; axillary sterile, fasciculiform short shoots frequently present, occasionally with solitary abortive flowers. Inflorescence white or pinkish, considerably lengthening, 1.5–20 × 1.2–1.75 cm., the axis densely white-lanate, formed of numerous densely set partial inflorescences comprising generally four fertile flowers, two central and solitary, the others set on each side of one of these and subtended by a modified sterile flower on each side, partial inflorescences falling as a unit in fruit, the persistent bracts deflexed. Bracts ovate-acuminate, 3.5–5 mm. long, furnished with flexuose white hairs, deflexed after fruit-fall, brownish-aristate with the excurrent green or brownish midrib. Bracteoles of the partial inflorescences broadly deltoid-ovate, 4–6 mm., hyaline, dorsally long-pilose along the darker midrib, which is excurrent in a rigid mucro; bracteoles of secondary triads more narrowly deltoid-ovate, c. 5–8 mm. long, dorsally long-pilose and sharply aristate; bracteoles of sterile flowers similar but narrower and slightly more longly aristate; all bracteoles more rigid and accrescent (the longest to 10 mm. long or more) in fruit. Sterile flowers of mostly two broad-based spine-like processes finally c. 4–8 mm. long, some shorter processes sometimes found. Outer 2 tepals of fertile flowers narrowly oblong, c. 5–7 mm. long, hyaline except for a narrow central, 3-veined vitta, the 2 lateral veins joining the midrib below the apex, densely pilose over most of the central dorsal surface, midrib excurrent in a short, sharp mucro; inner 3 tepals somewhat shorter and narrower; all tepals somewhat sinuose in outline. Ovary turbinate, c. 1.5 mm. long; style slender, c. 3–4 mm. long, slender, glabrous, white or pinkish. Stamens c. 4–6 mm. long, filaments white or pinkish; anthers shortly oblong, c. 1 mm. Fruit ovoid, densely pilose over the dome-like apex, gibbous unilaterally with an acute horn-like process near the style, c. 2–2.5 mm. long; seed brown, smooth and shining, c. 1.75–2.25 mm. long, very feebly reticulate, only slightly compressed.

Botswana. N: Ngamiland, upper N. slopes of second highest of Tsodilo Hills, c. 1180 m., 2.v.1975, *Miller & Biegel* 2316 (K; SRGH). SW: Ghanzi, Farm 56, 29.iv.1969, *de Hoogh* 258 (K; SRGH). SE: Mahalapye Village, 16.xii.1963, *Yalala* 396 (K; SRGH). **Zambia**. S: c. 1.6 km. from Chirundu bridge on Lusaka Rd., 1.ii.1958, *Drummond* 5432 (K; LISC; SRGH). **Zimbabwe**. N: Makonde Distr., dyke near Rod Camp Mine, 22.ii.1961, *Rutherford-Smith* 568 (K; SRGH). W: Matobo Distr., Besna Kobila Farm, iii.1961, *Miller* 7761 (K; LISC; SRGH). C: Sebakwe R. near Kwe Kwe 1380 m., 6.iii.1961, *Richards* 14539 (K). E: Mutare Distr., 938 m., 13.i.1952, *Chase* 4315 (BM; SRGH). S: Masvingo Distr., Makholi Experiment St. 13.iv.1978, *Senderayi* 204 (K; SRGH). **Mozambique**. T: Moatize, andades 50 km. Zóbuè para Tete, 350 m., 12.iii.1964, *Torre & Paiva* 11170 (K; LISC). MS: near Sena, 8.iv.1860, *Kirk* s.n. (K). GI: Inhambane, entre Zavala e Inharrime, 3.iv.1959, *Barbosa & Lemos* 8485 (COI; K; LMJ; SRGH). M: Mangulane, 1931, *Gomes e Sousa* 510 (K).
Also in Angola, Namibia, S. Africa (all provinces). On sandy or sandy/clay soils, schist or paragneiss (one specimen records it from a copper outcrop), mostly in dry situations - open woodland, savanna and scrub, in grassland, along roadsides and overgrazed areas, even from cracks in rocky areas; 350–1460 m.

I have not located a type specimen of *Kyphocarpa petersii* Lopr. in Engl. Bot. Jahrb. 27: 43 (1899), but read nothing in the description to justify its specific separation from *K. angustifolia*. Schinz reduced it to varietal rank under the latter (as *K. zeyheri*) in Viert. Nat. Ges. Zürich **57**: 541 (1912), citing Kirk's Mozambique specimen gathered between Lupata and Tete. This is at Kew and I would not distinguish it from *K. angustifolia* at any rank. The same remarks apply to *K. wilmsii* Lopr., similarly reduced to varietal rank by Schinz, no specimens which I have seen under this name, including some from Mozambique, would I separate from *K. angustifolia*, but again, no type has been seen.

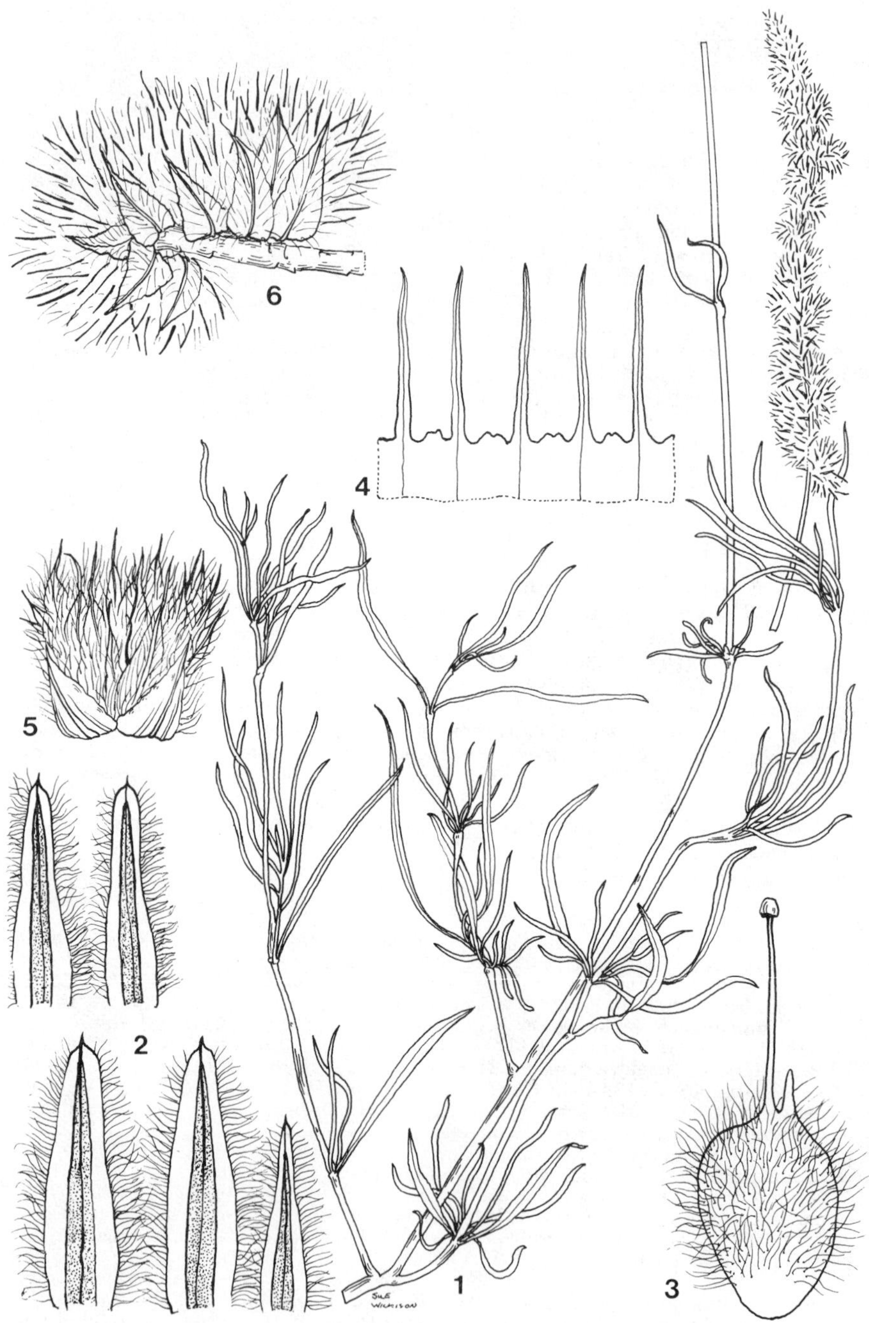

Tab. 13. KYPHOCARPA ANGUSTIFOLIA. 1, flowering branch (×⅔); 2, tepals, inner surface (×6); 3, gynoecium (×14); 4, androecium (×8); 5, flower with subtending sterile flowers (×4), 1–5 from *McClintock* K.30; 6, fruiting clusters (×2) *Müller & Biegel* 2316.

6. NELSIA Schinz

Nelsia Schinz in Viert. Nat. Ges. Zürich **56**: 247 (1911).

Erect annual or perennial herbs. Leaves entire, opposite on simple or branched stems. Flowers in sessile or pedunculate, spike-like, densely silky-hairy, terminal or axillary bracteate thyrses, each bract subtending a bibracteolate partial inflorescence. Partial inflorescences consisting of two solitary fertile flowers accompanied by one or two lateral bracteolate triads of a central fertile flower subtended by two lateral bracteolate sterile flowers; sterile flowers consisting of long, narrowly bracteoliform, plumose-pilose processes, some short scale-like processes (probably reduced perianths) frequently also present at fruit-fall; partial inflorescences (in the one species in which ripe fruit is known) very indurate at the base in fruit, falling as a burr-like unit complete with bracteoles. Bracts persistent on the inflorescence axis. Perianth segments 5, narrow, 3-nerved, mucronate, hyaline-margined. Stamens 5; filaments filiform, delicate, fused below with the interposed flabellate pseudostaminodes; anthers narrow, bilocular. Ovary uniovulate, somewhat compressed, delicate below; style slender, stigma solitary and capitate. Capsule scarcely compressed, the very delicate base more or less immersed in the indurate base of the partial inflorescence, the apex in the one species known in ripe fruit with a firm to rather narrow, raised apical rim; seed with copious endosperm.

A genus of two species, that described below and a second in Angola.

Nelsia quadrangula (Engl.) Schinz in Viert. Nat. Ges. Zürich **56**: 247 (1911); in Engl. & Prantl Pflanzenfam. ed. 2, **16 C**: 44 (1934).—Cavaco in Mém. Mus. Hist. Nat. Paris Sér. B, **13**: 73 (1962).—Podlech & Meeuse in Merxm. Prodr. Fl. SW. Afr. **33**: 20 (1966). TAB. **14**. Type from Namibia.

Sericocoma quadrangula Engl. in Engl. Bot. Jahrb. **10**: 7 (1889). Type as above.
Sericocoma nelsii Schinz in Pflanzenfam. ed. **1**, 3, 1a: 107 (1893). Type from Namibia.
Sericocoma welwitschii Baker in Kew Bull. **1897**: 278 (1897). Type from Angola.
Sericocomopsis welwitschii (Baker) Lopr. in Engl. Bot. Jahrb. **27**: 42 (189) in clav.
Sericocoma quadrangula (Engl.) Lopr., loc. cit.
Cyphocarpa welwitschii (Baker) C.B. Clarke in F.T.A. **6**, 1: 53 (1909). Type from Angola.
Cyphocarpa quadrangula (Engl.) C.B. Clarke, loc. cit.: 54 (1909). Type as for *Sericocoma quadrangula*.

Annual herb with a slender taproot, erect, 40–90 cm. tall, with few to numerous opposite branches diverging at c. 45 degrees, the uppermost pair of branches with a sessile to shortly (to c. 3 cm.) pedunculate inflorescence between. Stem and branches quadrangular, thinly to more or less densely upwardly-appressed pilose. Leaves narrowly lanceolate to broadly elliptic or ovate, subacute to acuminate, those of the main stem and large branches 2.5–10 × 0.8–4.5 cm., moderately white pilose on both surfaces or more densely so below, cuneate or attenuate at the base into a distinct, 3–15 mm. long petiole; superior leaves of branches smaller, generally sessile. Inflorescence considerably lengthening in fruit, (1.5) 2.5–16 × 1.5–2.5 (3) cm., axis more or less densely furnished with flexuose whitish hairs, formed of numerous densely set partial inflorescences comprising three or four fertile flowers, two central and solitary and the others subtended by a modified sterile flower on each side; partial inflorescences soon falling in fruit, the persistent bracts strongly deflexed. Bracts broadly ovate, 4.5–5.5 mm. long, furnished with appressed white hairs, rather abruptly acuminate, finely aristate with the long-excurrent midrib. Bracteoles of the partial inflorescences broadly cordate-ovate, 5–7 mm. long, ciliate or pilose, hyaline with the darker midrib excurrent in a fine arista up to 1.5 mm. long; bracteoles of secondary triads deltoid-ovate, c. 7–9 mm. long, pilose along the midrib with a rigid arista; bracteoles of the sterile flowers narrower, densely pilose along the midrib with multicellular hairs, the arista very long (almost as long as the lamina), more solid in fruit and finally up to c. 12 mm. long. Sterile flowers of 2 rapidly developing linear, bracteoliform processes densely clothed with long white hairs. Outer 2 tepals of fertile flowers linear-oblong, c. 6–7 mm. long, green centrally with 3 distinct ribs (the midrib distinctly excurrent in a fine, yellowish arista) broadly hyaline-margined,

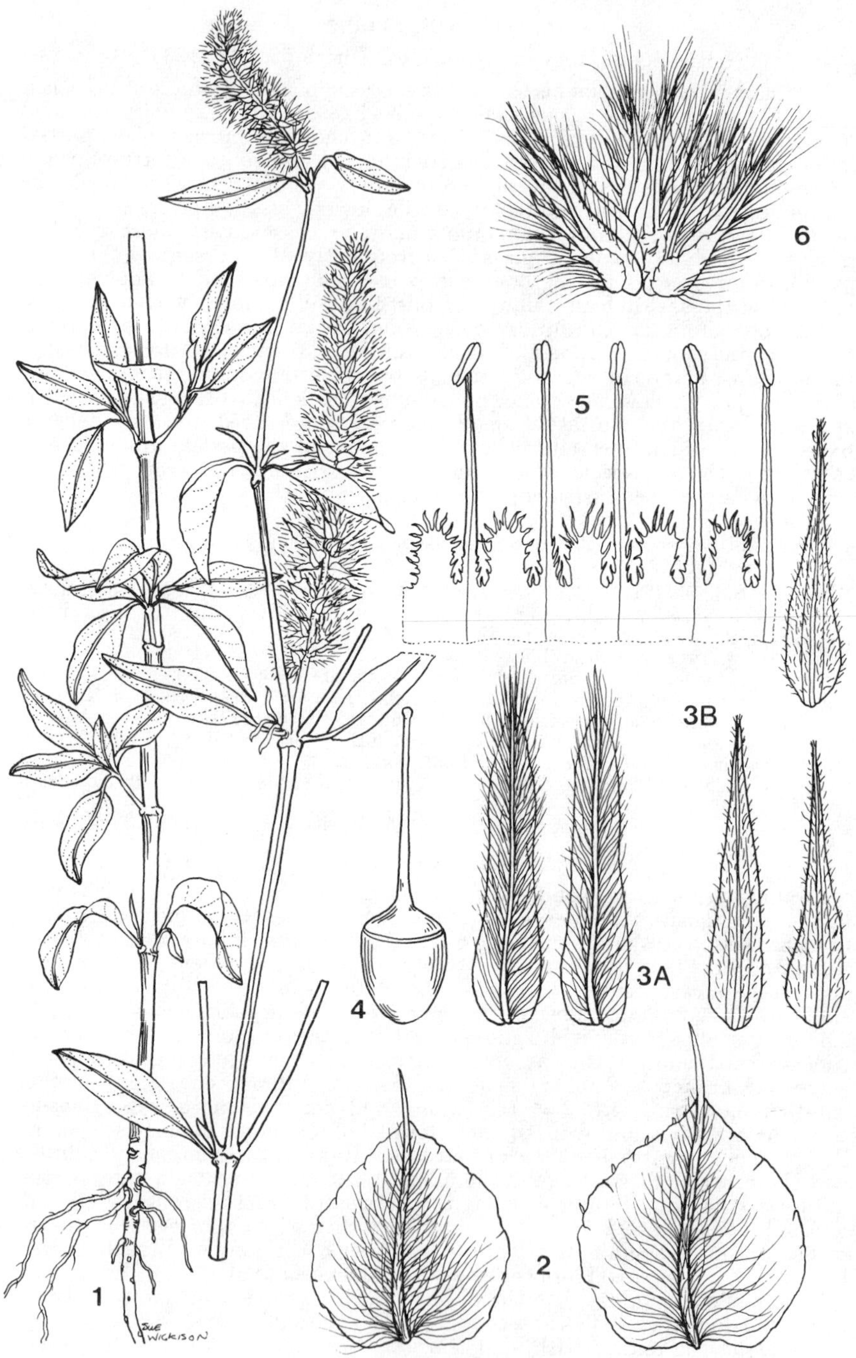

Tab. 14. NELSIA QUADRANGULA. 1, habit with some branches removed (×⅔); 2, bracteoles (×6); 3a, outer and 3b, inner tepals, outer surface (×6); 4, gynoecium (×12); 5, androecium (×10), 1–5 from *De Winter* 2403; 6, cluster of fertile and sterile flowers (×2) *Lugard* 221.

furnished with long, whitish multicellular hairs; inner 3 tepals similar but c. 1 mm. shorter, more narrowly hyaline-margined and somewhat widened at the base. Ovary c. 2 mm. long, obpyriform, delicate below with a flat, firm apex; style slender, c. 1.5–2 mm. long. Stamens c. 3.5–4.5 mm. long; filaments slender, fused for c. one quarter of their length to the pseudostaminodes; pseudostaminodes c. 1.5 mm. long, flabellate, dentate with a fimbriate dorsal scale; anthers narrowly oblong, c. 1 mm. long. Branches of partial inflorescence and bases of bracteoles incrassate and lignescent in fruit, the whole forming a soft, silky-hairy, burr-like unit. Fruit compressed-ovoid, c. 3 mm. long, apex concave with a distinct raised rim; seed brown, smooth and shining, c. 2 mm. long, finely reticulate.

Botswana. N: Maun, c. 90 m. from river, 8.iii.1967, *Lambrecht* 87 (K; SRGH). SW: Kuke fence 64 km. E. of gate, 27.iii.1970, *Brown* 8792 (K). SE: Boteli delta area, NE of Mopipi, 850 m., 17.iv.1973, *Root* 1 (K; SRGH).

Also in Angola, Namibia. On sandy or rocky soil sometimes with mica-schist and lime, in various habitats - open forest, roadsides, brackish flats, sandy places near rivers and boulder-strewn hill slopes, 850–c.1180 m.

7. SERICOREMA (Hook. f.) Lopr.

Sericorema (Hook. f.) Lopr. in Engl. Bot. Jahrb. **27**: 39 (1899).
Sericocoma Fenzl Sect. *Sericorema* Hook.f. in Benth. & Hook. Gen. Pl. **3**, 1: 30 (1880).

Much-branched erect, annual or perennial herbs. Leaves entire, alternate or occasionally opposite, semiterete from the more or less closely revolute margins, linear. Flowers in elongate, bracteate, spike-like thyrses formed of more or less distant, bibracteolate partial inflorescences consisting of a single fertile flower subtended by two modified sterile flowers to which a further fertile flower with a solitary sterile flower, and sometimes a further triad, may be added; sterile flowers consisting of a strap-shaped dentate process furnished on the abaxial surface with long, fine, smooth, unicellular hairs, these processes accrescent to become spinous-margined and indurate in fruit, fused with the indurate base of the partial inflorescence so that the latter falls as a unit complete with bracteoles. Bracts persistent on the inflorescence axis. Perianth segments narrow, 3-nerved, long- or short-aristate. Stamens 5; filaments filiform, delicate, fused at the base only, without intermediate pseudostaminodes; anthers narrow, bilocular, each theca with an awn-like tail at each end. Ovary uniovulate, somewhat compressed, delicate below; style very short, stigma with a tuft of penicillate hairs. Capsule scarcely compressed, with a firm apex, delicate below; seed with copious endosperm.

Two species in southern Africa, both occurring in the Flora Zambesiaca area.

All partial inflorescences with a solitary fertile flower; perianth segments with the shortly aristate apices erect; ovary and fruit lanate above. - - 1. *remotiflora*
At least some partial inflorescences with 2–3 fertile flowers; perianth segments with the long-aristate apices finally spreading; ovary and fruit quite glabrous. - - - - - - - - - - - - - - - - - - 2. *sericea*

1. **Sericorema remotiflora** (Hook.) Lopr. in Engl. Bot. Jahrb. **27**: 40 (1899).—Schinz in Viert. Nat. Ges. Zürich **56**: 245 (1911); in Engl. & Prantl Pflanzenfam. ed. 2, **16** C: 44 (1934).—Cavaco in Mém. Mus. Hist. Nat. Paris Sér. B, **13**: 75 (1962).—Podlech & Meeuse in Merxm. Prodr. Fl. SW. Afr. **33**: 22 (1966). TAB. **15** fig. B. Type from S. Africa.
Trichinium remotiflorum Hook., Ic. Pl. **6**: t. 596 (1843). Type as above.
Pupalia remotiflora (Hook.) Moq. in DC., Prodr. **13**, 2: 333 (1849). Type as above.
Sericocoma remotiflora (Hook.) Hook.f. in Benth. & Hook. Gen. Pl. **3**, 1: 30 (1880). Type as above.

Plant much-branched, rounded and rather bushy, probably a perennial flowering in the first year, 0.4–1 m. tall; stem and branches slender and wiry, terete, young parts green and somewhat floccose to sublanate at least when young (usually soon glabrescent), older cortex brownish-yellow and the base frequently purplish. Leaves linear-acicular, subterete, 6–32 × 0.5–1 mm., mucronate, glabrous or more rarely somewhat floccose, straight or somewhat recurved, usually with axillary short shoots or leaf-fascicles and tufts of floccose hairs. Inflorescence much elongating at maturity (up to c. 25 cm. but commonly less), the axis glabrous to floccose; partial inflorescences finally distant (to c. 1 cm. apart), of a single central fertile flower with a modified sterile flower on each side, the lateral brushes of hairs brown to whitish. Bracts ovate-lanceolate, 4−5 mm. long, scarious, glabrous and more or less floccose along the green, slightly excurrent midrib, more or less patent after fruit-fall. Bracteoles of partial inflorescence 4−5 mm. long, deltoid-ovate, finally deeply concave with the midrib bent at c. 90 degrees to accommodate the sterile modified flowers when mature, glabrous to slightly floccose, shortly mucronate with the excurrent midrib. Flowers white to dull yellow or greenish. Tepals all narrowly lanceolate, tapering from base to apex and straight-sided, of similar length (c. 8−10 mm. long) but the inner progressively narrower, hyaline-margined, the firm portion 3-nerved with the lateral pair of nerves evanescent but the midrib excurrent in a short but sharp and rigid, straight arista, all more or less floccose with frequently longer fine hairs intermixed. Ovary subglobular or obovoid, c. 2.25−2.5 mm. long, with a densely lanate apex. Stamens c. 6−7 mm. long at anthesis, finally subequalling the perianth, white or pinkish; anthers very long and narrow, c. 4 mm. long, exceeding the filaments at anthesis. Fruiting partial inflorescence forming a fluffy ball c. 1.5−2 cm. in diam. Fruit oblong-ovoid, c. 3.25−3.5 mm. long, densely lanate in the upper half or rather more; seed 3−3.25 mm. in diam., brown, smooth and shining.

Botswana. N: c. 6 km. SE. of Nata R. on Francistown Rd., 900 m., 24.iv.1957, *Seagrief* in CAH 2452 (K; LISC; SRGH). SW: Ghanzi, 550 m., 9.v.1969, *Brown* 6055 (K; SRGH). SE: Seleka Ranch, 830 m., 3.xi.1977, *Hansen* 3266 (C; GAB; K; PRE; SRGH). **Zimbabwe**. W: c. 4 km. upstream from Shashe & Shashane R. confluence, 4.v.1963, *Drummond* 8080 (K; LISC; MO; SRGH). E: Sabi Valley on road to Rupisi Hot Springs, 605 m., iv.1961, *Goldsmith* 26/61 (K; LISC; PRE; SRGH). S: Buhera District, near the Devuli R. on main road to Masvingo from the Sabi R., i.1969, *Goldsmith* 6/69 (K; LISC; SRGH).

Also in Namibia and S. Africa (Cape Prov., Transvaal, Orange Free State). Commonly in deep red sand, also recorded from black clay, mica-schist, lime, basalt and granite; in flats, bushland, eroded grassland, open Mopane woodland (*Grewia/Commiphora* and *Acacia/Terminalia/Combretum*), eroded areas and along roadsides; 550−1100 m.

2. **Sericorema sericea** (Schinz) Lopr. in Engl. Bot. Jahrb. **27**: 40 (1899).—Schinz in Viert. Nat. Ges. Zürich **56**: 244 (1911); in Engl. & Prantl Pflanzenfam. ed. 2, **16 C**: 44 (1934).—Podlech & Meeuse in Merxm. Prodr. Fl. SW. Afr. **33**: 23 (1966). TAB. **15** fig. A. Type, (three syntypes) from Namibia.

Sericocoma sericea Schinz in Engl. Bot. Jahrb. **21**: 181 (1895). Type as above.
Marcellia sericea (Schinz) C.B. Clarke in F.T.A. **6**, 1: 50 (1909). Type as above.

Much-branched, bushy annual herb with a slender taproot, c. 0.2−0.6 m. high; stem and branches slender and rather wiry, terete, young parts green and more or less floccose to sublanate at least when young, usually soon glabrescent. Leaves linear-acicular, subterete-canaliculate, (18)28−50(60) × 0.5−1 cm., mucronate, glabrous (floccose only when very young), straight or slightly flexuose, usually with axillary short shoots or leaf-fascicles and tufts of floccose hairs. Inflorescence elongating, (to c. 30 cm. or more but often less), the axis glabrous to more or less floccose; partial inflorescences finally distant (separated by 1.5−2.5(3) cm.), varying from 1−3 flowered as in the generic description, the lateral brushes of hair whitish. Bracts lanceolate, 4−5 mm. long, glabrous or somewhat floccose, the darker midrib excurrent in amucro, more or less patent after fruit-fall. Bracteoles of partial inflorescence, deltoid-lanceolate, 4−5 mm. long, rather similar to the bracts but more deeply concave basally to accommodate the sterile flowers, glabrous to more or less floccose, at anthesis. Fruiting partial inflorescence a fluffy ball c. 1.5−2 cm. in diam. Fruit oblong-ovoid, c. 3.5 mm. long, smooth and glabrous throughout; seed c. 3.25 mm. long, brown, smooth and shining.

Botswana. N: Xangwa, 27 km. N. of Aha Hills, 12.iii.1965, *Wild & Drummond* 6941 (K; LISC; SRGH). SW: Mobua, Shobea Pan, 117 km. S. of Tsane, *Wild* 5141 (BM; K; LISC; SRGH). E: Boteli delta area, NE. of Mopipi, 900 m., 16.iv.1973, *Tyers* 316 (K;

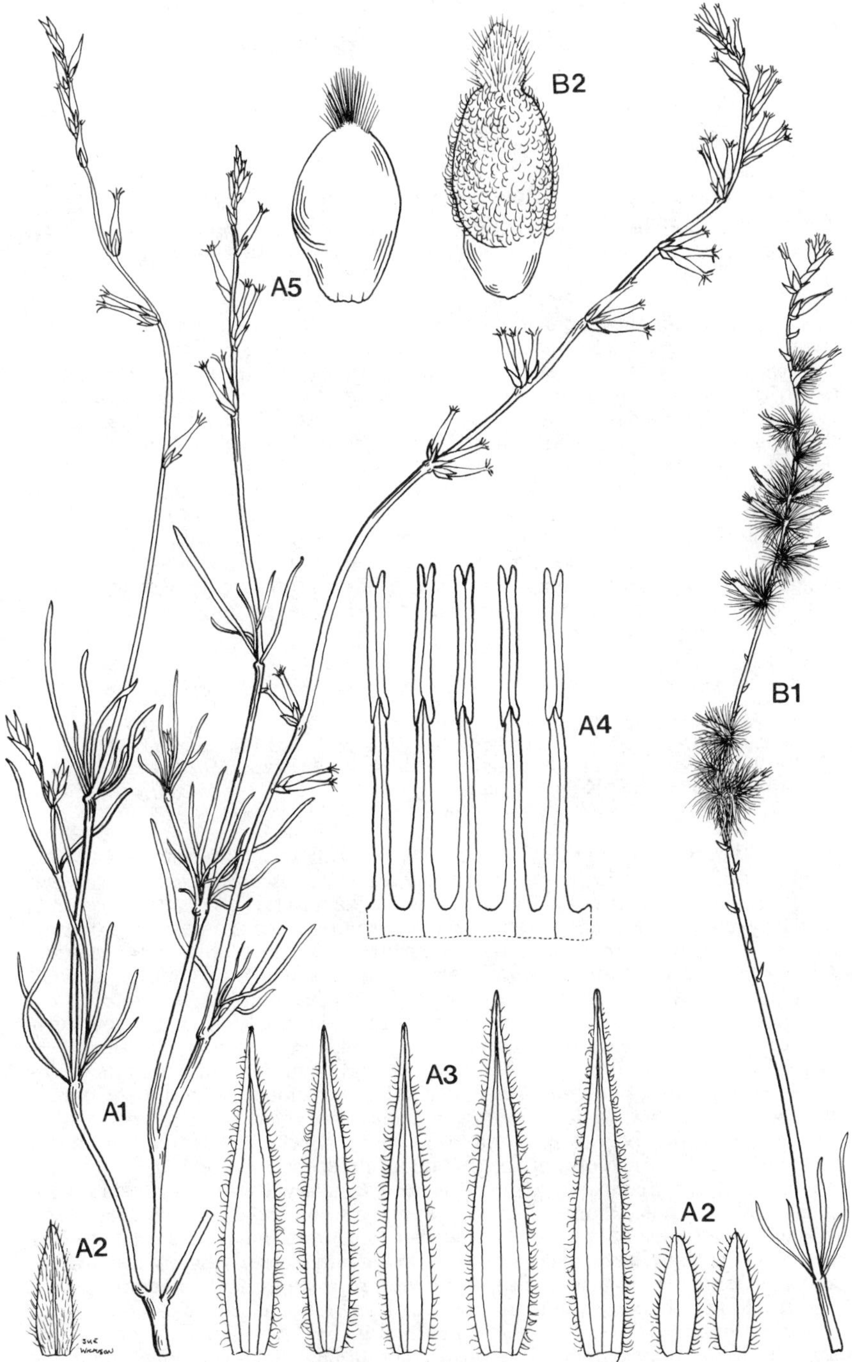

Tab. 15. A.—SERICOREMA SERICEA. A1, habit (×⅔); A2, bracteoles, dorsal and (right) ventral surfaces (×4); A3, tepals, ventral surface (×4); A4, androecium (×6); A5, gynoecium (×20), A1–5 from *Skarpe* 406; B.—SERICOREMA REMOTIFLORA. B1, flowering branch (×⅔); B2, gynoecium (×15), B1–2 from *Seagrief* 2452.

SRGH). **Zimbabwe**. S: Gwanda Distr., Tuli Exp. St., 1.ii.1965, *Norris-Rogers* 643 (K; SRGH).

Also in Angola, Namibia and S. Africa (N. Cape, Transvaal). On sandy flats or stony soil, sometimes in "black turf", in open *Acacia* scrub, mopane, sandveld, limestone pavement, among quartzite blocks.

8. CENTEMA Hook. f.

Centema Hook. f. in Benth. & Hook. Gen. Pl. **3**, 1: 31 (1880).

Erect perennial herbs or low shrubs. Leaves entire, opposite on mostly branched stems. Flowers in pedunculate (rarely some sessile) spike-like bracteate thyrses, each bract subtending a bibracteolate partial inflorescence. Partial inflorescences normally consisting of two (occasionally one in the upper part of the thyrse) fertile flowers; one of these is subtended dorsally by a single sterile flower (which is thus contained in a partial inflorescence bracteole), the other by a sterile flower on each side, each of which is contained in a bracteole of its own. The sterile flowers (which develop so rapidly as to be difficult to detect in young flowers) consist of a single finally hard and gibbous-based spine, the entire partial inflorescence thus falling as a 3-spined burr complete with bracteoles; bracts persistent on the inflorescence axis. Perianth segments 5, obviously narrower from the outer to the inner. Stamens 5; filaments filiform, flattened, fused only at the extreme base without intermediate pseudostaminodes; anthers narrow, bilocular. Ovary uniovulate, somewhat compressed, delicate; style slender, with two short or elongate stigmatal branches. Utricle scarcely compressed, delicate save for a small, firm area about the base of the style, irregularly ruptured by the developing seed; seed with abundant endosperm.

A genus of two species, the one described below and another in Angola.

Centema subfusca (Moq.) T. Cooke in Harv. & Sond., F.C. **5**, 1: 418 (1910).—Schinz in Viert. Nat. Ges. Zürich **57**: 547 (1912); in Engl. & Prantl Pflanzenfam. ed. 2, **16 C**: 45 (1934); in J.H. Ross Fl. Natal: 158 (1973). TAB. **16**. Type: Mozambique, Baia de Maputo, *Forbes* s.n. (K, holotype).

Pupalia subfusca Moq. in DC., Prodr. **13**, 2: 332 (1849). Type as above.

Perennial herb, woody at the base, 15–60 cm. tall with numerous stems from the base and usually a few opposite (rarely alternate by suppression) branches diverging at 30–45 degrees above. Stem and branches terete, glabrous or thinly pilose (especially about the nodes). Leaves broadly linear to narrowly oblong, 13-40 × 3–9 mm., somewhat crisped-margined, acute to obtuse, mucronate, glabrous or thinly pilose (especially on the lower surface of younger leaves), sessile or abruptly narrowed (and in broader leaves subauriculate) into a very short petiole, often darkening on drying and commonly with fasciculiform axillary short shoots. Inflorescence crimson, capitate to shortly cylindrical, 1–4 × 1.5–1.75 cm. but elongating to as much as 8.5 cm. in fruit, the axis thinly to moderately white-pilose, more or less sulcate, formed of closely set partial inflorescences. Bracts deltoid-ovate, scarious-margined with a firm centre, glabrous or thinly pilose, 3.5–4.5 mm. long, shortly mucronate with the excurrent midrib. Bracteoles of the partial inflorescences broadly deltoid-ovate, 4–5 mm. long, very concave and more or less carinate with the prominent darker midrib, which is not excurrent or very shortly so on the dorsal surface below the apex. Bracteoles of the sterile flowers c. 5–6 mm. long, ovate, the midrib ceasing below the blunt, hyaline apex or excurrent dorsally below the apex in a short or longer rigid arista. Sterile flowers of a single spine becoming to c. 4–5 mm. long in fruit, swollen and more or less lanate at the base. Outer tepal of fertile flowers c. 6.5–8 mm. long, oblong, 5–7-nerved, firm below and narrowly hyaline-margined above, slightly lanate dorsally near the base; remaining 4 tepals rapidly narrowing and more pilose, the innermost very narrowly lanceolate, densely lanate except about the apex, with only one faint pair of lateral nerves; all tepals with the midrib ceasing below the apex or only very shortly mucronate. Ovary oblong, c. 2 mm. long; style 4–5.5 mm. long, with short to very short stigmatal branches. Stamens finally 8–9 mm. long;

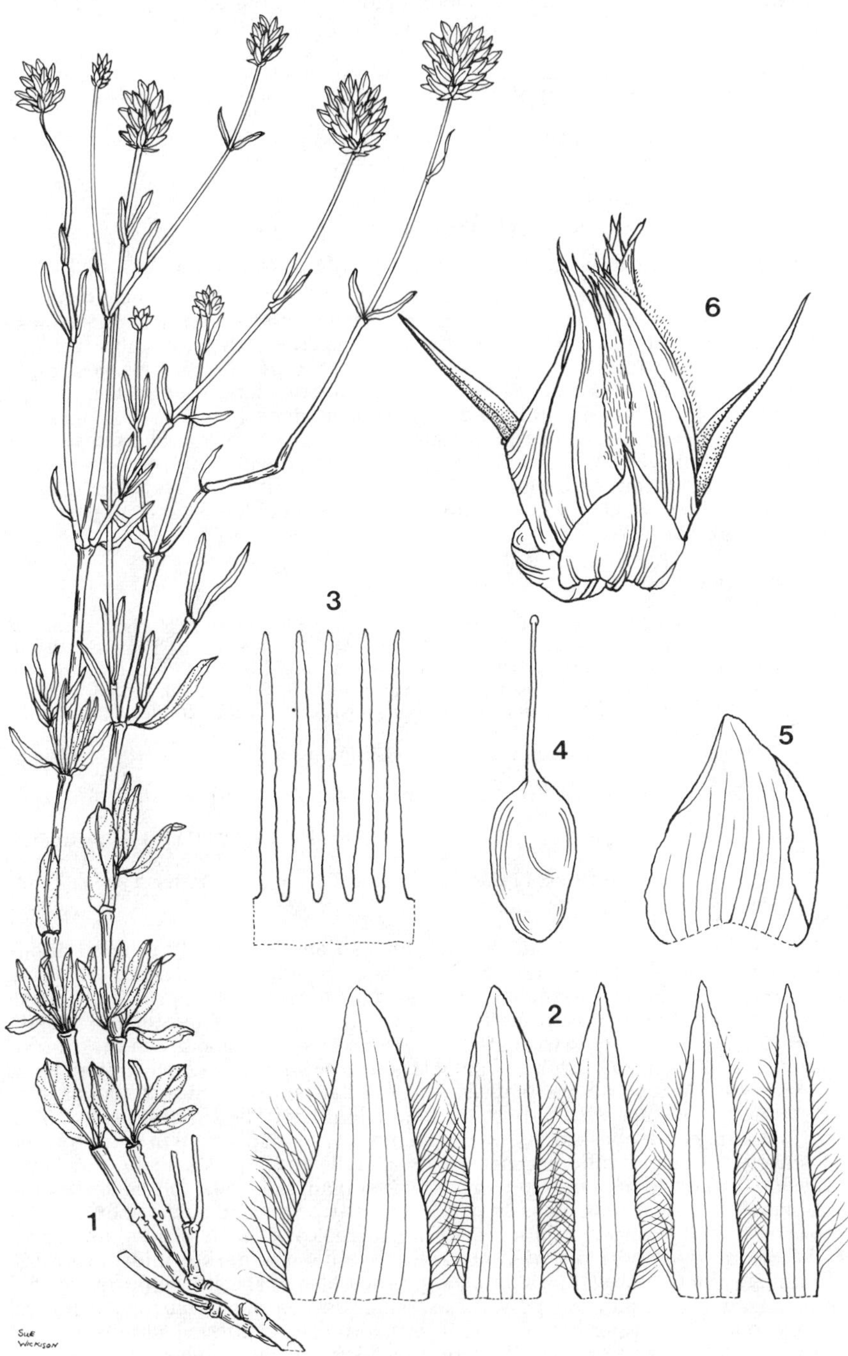

Tab. 16. CENTEMA SUBFUSCA. 1, habit ($\times\frac{2}{3}$); 2, tepals, inner surface ($\times6$); 3, androecium ($\times6$); 4, gynoecium ($\times6$); 5, outer bract ($\times6$); 6, mature flower ($\times6$), all from *Leach* 11760.

anthers narrowly oblong, c. 2 mm. long Fruit c. 3–3.5 mm. long; seed 2.75–3 mm. long, brown, shining, faintly reticulate.

Mozambique. GI: near. Moatize c. 42 km. NW. of Magul, 90 m., 30.ix.1963, *Leach & Bayliss* 11760 (K; LISC; SRGH); M: Marracuene, prox. de Umbeluzi, 16.xi.1944, *Mendonça* 2881 (K; LISC).

Also in S. Africa (Natal). On sand, loam and black cotton soil in grassland, at edge of thorn bush and in abandoned cultivated ground.

9. LEUCOSPHAERA Gilg

Leucosphaera Gilg in Engl. & Prantl Pflanzenfam. Nachtr. 3, **1a**: 152 (1897).

Dwarf shrub with entire opposite leaves and branches. Inflorescences capitate, almost sessile above the uppermost pair of leaves, white from the pilose hairs of the bracteoles, tepals and sterile flowers; partial inflorescences bibracteolate, of 1–2 sterile and 1–2 fertile flowers. Perianth segments 5, 3-nerved, more or less long-aristate. Stamens 5, without intermediate pseudostaminodes; anthers bilocular. Ovary uniovulate, pyriform, densely pilose above; style slender, stigma single, capitate. Capsule not compressed, the firm pilose apex flat, the hyaline wall glabrous and ruptured by the developing seed. It has not been determined from herbarium material whether fruit fall is by dropping seed, or whether the partial or whole inflorescences are deciduous, and no perfectly mature seed has been seen.

A monotypic genus.

Leucosphaera bainesii (Hook. f.) Gilg in Engl. & Prantl Pflanzenfam. Nachtr. 3, **1a**: 153 (1897).—Schinz in Viert. Nat. Ges. Zürich **56**: 250 (1911); in Engl. & Prantl Pflanzenfam. ed. 2, **16 C**: 45 (1934).—Cavaco in Mém. Mus. Hist. Nat. Paris Sér. B, **13**: 83 (1962).—Podlech & Meeuse in Merxm. Prodr. Fl. SW. Afr. **33**: 18 (1966). TAB. **17**. Type: Botswana, Ngamiland, Kobi Pan to N. Shaw Valley, *Baines* s.n. 1863 (K, holotype).

Sericocoma bainesii Hook.f. in Benth. & Hook. Gen. Pl. **3**, 1: 31 (1880). Type as above.

Sericocomopsis bainesii (Hook.f.) Schinz in Engl. Bot. Jahrb. **21**: 185 (1895). Type as above.

Leucosphaera pfeilii Gilg in Notizbl. Bot. Gart. Berl. **1**: 328 (1897). Type from S. Africa.

Marcellia bainesii (Hook.f.) C.B. Clarke in F.T.A. **6**, 1: 51 (1909). Type as for *L. bainesii*.

Dwarf shrub, 8–45 cm. tall, much-branched and frequently gnarled about the base; branches terete, striate, the older parts glabrescent with a greyish to brownish cortex, the young twigs densely canescent with appressed, whitish hairs. Leaves narrowly or broadly elliptic to lanceolate or ovate, 10–30 (40) × 3–12(17) cm., subacute to acute (obtuse to retuse occasionally in the inferior leaves), equally densely furnished on both surfaces with appressed whitish hairs or somewhat less so above (glabrescent and darkening with age), cuneate or attenuate at the base with a short (2–5 mm. long) petiole. Inflorescence 1.5–2 cm. in diam., hemispherical in flower and more or less spherical in fruit, formed of densely set partial inflorescences of 1–2 fertile and 1–2 sterile flowers (very occasionally a triad may be found in the largest plants). Bracts deltoid lanceolate, hyaline, (3)4–5 mm. long, densely pilose centrally below and at the apex, shortly aristate with the excurrent darker midrib. Bracteoles of the partial inflorescences (5)7.7.5 mm. long, narrowly deltoid-ovate, hyaline, the darker midrib forming a plumose-hairy arista (1)2–3 mm. long; bracteoles of the sterile flowers c. (5)7 mm. long, densely hairy except at the extreme base, basal half narrowly elliptical or lanceolate, the upper half an arista formed by the excurrent plumose-pilose midrib. Sterile flowers of a few (2–4) narrow, plumose-hairy, linear, bracteoliform processes. Outer 2 tepals of fertile flowers c. (5.5)7–9 mm. long, lanceolate, obscurely 3-nerved, the excurrent midrib forming an arista (2)3–4 mm. long, margins hyaline below; inner 3 tepals similar, scarcely shorter but more expanded below with broader hyaline margins; aristae of all tepals plumose-hairy. Ovary pyriform, c. 1 mm. long, furnished with dense, erect, white hairs

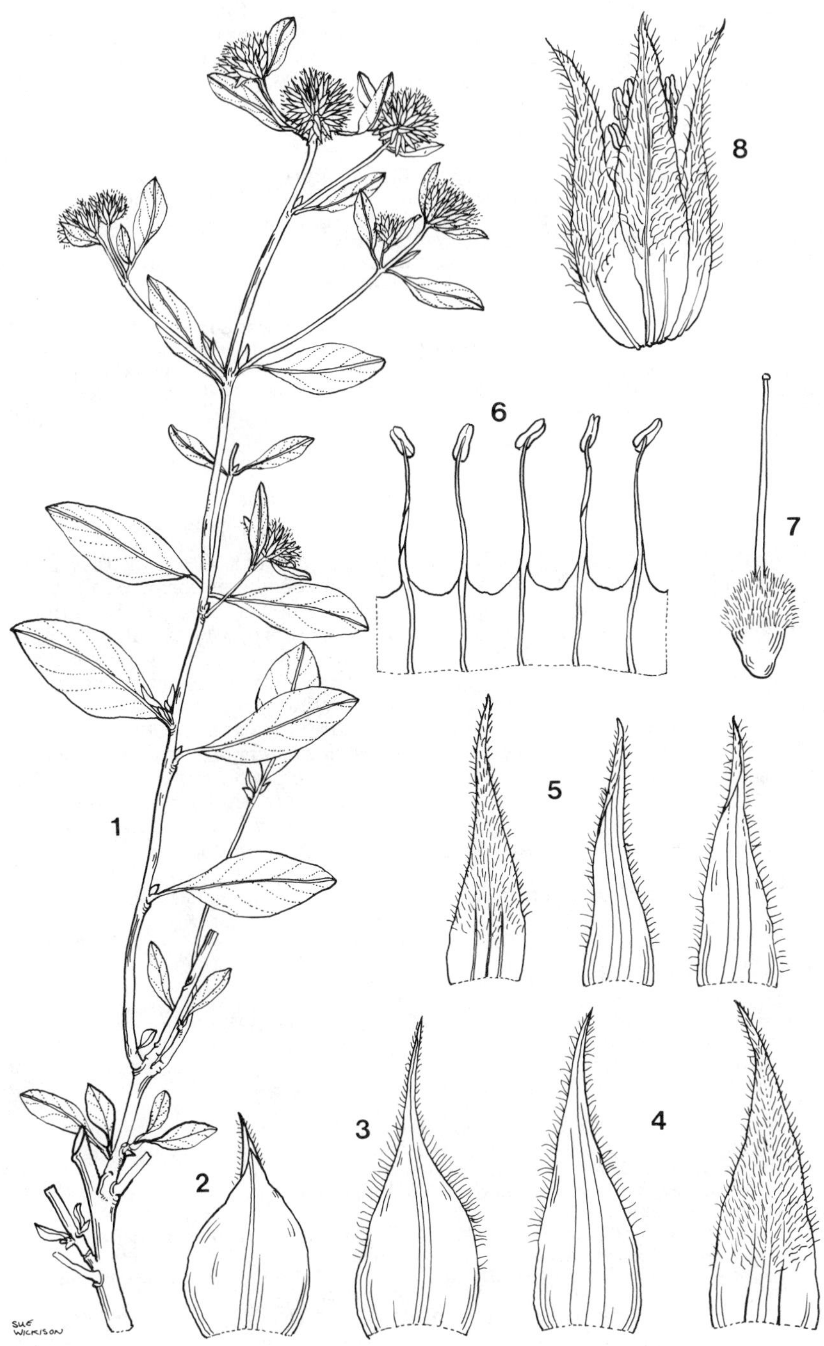

Tab. 17. LEUCOSPHAERA BAINESII. 1, flowering stem (×⅓); 2, outer bracteole (×6); 3, inner bracteole (×6); 4, outer tepals, one dorsal and one ventral surface (×6); 5, inner tepals, one dorsal and two ventral surfaces (×6); 6, androecium (×8); 7, gynoecium (×8); 8, single fertile flower (×6), all from *Lugard* 158.

except at the extreme base; style slender, c. 3–4 mm. long. Stamens (2.5)3.5–4 mm. long; filaments slender, abruptly expanded in about the basal one-third and fused into a cup, with acute sinuses between; anthers narrowly oblong, c. 1 mm. long. Fruit shortly cylindrical, (1.5)2–2.25 mm. long, densely erect-pilose on the firm, flat apex, the delicate walls glabrous; seed brown, c. 2 mm. long, shining, feebly reticulate (not perfectly mature).

Botswana. N: 113 km. W. of Nokaneng, 12.iii.1965, *Wild & Drummond* 6920 (K; LISC; SRGH). SW: Ghanzi, 20.ii.1960, *Wild* 5095 (K; SRGH). SE: Chukudu Pass, c. 370 km. NW. of Molepolole, 22.vi.1955, *Story* 4955 (K; SRGH). **Zimbabwe**. S: Beitbridge Distr., Shashe-Limpopo confluence, 22.iii.1959, *Drummond* 5951 (K; LISC; SRGH).
Also in Angola, Namibia and S. Africa (Cape Prov.). Chiefly on sandy or loamy soil in or near pans in open places in Acacia bush, in overgrazed areas, on alluvial flood plains.

The two specimens from Zimbabwe have among the smallest flowers seen, with short, squarrose awns; one has the basal cup of the androecium slightly separated, showing small lobules on each side of the filaments, but the other is quite normal in this respect. In size, specimens from Botswana grade into them, and there seems no need to regard them as a separate taxon.

10. CENTEMOPSIS Schinz

Centemopsis Schinz in Viert. Nat. Ges. Zürich **6**: 242 (1911).

Annual or (? short-lived) perennial herbs with entire, opposite leaves. Inflorescence terminal on the stem and branches, spiciform, capitate or fastigiate, bracteate, flowers solitary or paired in the axils of each bract. Modified sterile flowers absent, all flowers hermaphrodite and bibracteolate. Bracts persistent, finally weakly deflexed or deflexed-ascending, bracteoles and perianth falling with the fruit. Perianth segments 5, very shortly mucronate with the excurrent midrib, usually considerably indurate at the base in fruit. Stamens 5, the filaments delicate, shortly monadelphous at the base, alternating with distinct quadrate or oblong, fimbriate pseudostaminodes; anthers bilocular. Style slender, stigma capitate. Ovary with a single pendulous ovule, glabrous below with a dense ring of hairs centrally and more thinly pilose above, or entirely glabrous. Fruit a thin-walled capsule, irregularly ruptured by the developing seed. Seed compressed-reniform; endosperm copious.

A genus of 12 species, all in tropical Africa.

1. Ovary glabrous . . . . 3. *conferta*
– Ovary pilose 2
2. Inflorescence condensed-corymbiform . . . . 4. *fastigiata*
– Inflorescence spicate or capitate 3
3. Perianth 2.5–4 mm.; inflorescence finally long and slender, flowers solitary in the axils of the bracts . . . . 2. *gracilenta*
– Perianth (4)4.5–6 mm.; inflorescence shorter, stouter and finally more spiky, at least some bracts with pairs of flowers . . . . 1. *kirkii*

1. **Centemopsis kirkii** (Hook. f.) Schinz in Viert. Nat. Ges. Zürich **56**: 243 (1911); in Engl. & Prantl Pflanzenfam. ed. 2, **16 C**: 44 (1934). TAB. **18** fig. A. Type: Malawi, "W. shore of Lake Nyasa", *Kirk* s.n. (K, lectotype).
Centema kirkii Hook.f. in Benth. & Hook. Gen. Pl. **3**: 31 (1880).—Baker & Clarke in F.T.A. **6**, 1: 56 (1909). Type as above.
Achyranthes breviflora Baker in Kew Bull. **1897**: 280 (1897). Type: Malawi, between Khondowe and Karonga, *Whyte* s.n. (K, holotype).
Centema rubra Lopr. in Engl. Bot. Jahrb. **27**: 49 (1899); in Malpighia **14**: 442 (1901).—Baker & Clarke in F.T.A. **6**, 1: 56 (1909). Type from Kenya.
Centemopsis rubra (Lopr.) Schinz in Viert. Nat. Ges. Zürich **56**: 243 (1911).
Centemopsis clausii Schinz in Viert. Nat. Ges. Zürich **57**: 543 (1912). Type from Tanzania.

Annual herb (or probably short-lived perennial in more stable habitats in rocky areas or undisturbed ground), erect, 15–100 cm. Stem wiry, strongly striate with pale ridges, smooth and glabrous except for tufts of multicellular whitish hairs about the nodes to strongly scabrid and moderately pilose, much branched from near the base upwards (simple in poorly grown forms), the branches ascending and frequently with sterile axillary short shoots. Leaves linear-filiform to lanceolate-elliptic, 1.2–8.5 cm. × 0.75–16 mm., surfaces glabrous to more or less finely hairy, more or less scabrid along the incrassate margins and lower surface of the midrib, gradually narrowed above to a mucronate apex, attenuate and indistinctly petiolate below. Inflorescence mauve-pink to crimson or purplish-red (becoming stramineous in dried material), spiciform or reduced and capitate, 1.5–3.5 × 1–1.3 cm. in flower, rounded or conical at the apex, where more longly-aristate bracteoles frequently give it a bristly appearance; in fruit elongating to as much as 13 cm.; axis white-lanate, deeply sulcate, after the fall of the fruiting perianth often finally honeycombed with pits formed by the widening of the grooves above each flower scar, and densely clad with the persistent bracts. Bracts 2.5–5 mm. long, lanceolate to ovate-lanceolate, membranous with a dark, shortly excurrent midrib, glabrous to ciliate. Bracteoles deltoid-ovate, 2–4 mm. long, acute to truncate or in the uppermost flowers sometimes even incised above, membranous with a distinctly excurrent midrib, glabrous to ciliate. Flowers solitary or paired in the axils of the bracts. Tepals 4–6 mm. long, oblong to subpanduriform; outer 2 at anthesis with a deltoid-ovate, not or narrowly hyaline-margined, opaque, glabrous to sparingly floccose in the lower half, more or less strongly 3-nerved base more or less constricted above, the apical half being lanceolate, increasingly widely hyaline-margined upwards with the margin incurved and with the midrib shortly excurrent; inner 3 increasingly more delicate, narrower and more obviously constricted centrally, distinctly 3-nerved below with the hyaline margin descending much nearer the base; all tepals (but especially the outer) indurate and prominently 3-ribbed at the base in fruit (deeply sulcate between the ribs), giving the inflorescence a frequently spiky appearance. Filaments very slender, 2–3.5 mm.; pseudostaminodes 1–1.5 mm. long, oblong or spathulate, denticulate to fimbriate around the apex. Ovary ovoid, glabrous and hyaline below, firm above, pilose centrally and more thinly so above. Style 1.5–3.5 mm. long glabrous, slightly asymmetrically placed on the ripe capsule. Capsule 1.5–2 mm. long, formed as the ovary but somewhat compressed. Seed compressed-ovoid, brown, c. 1.25–1.75 mm. long, shining, minutely reticulate.

Malawi. N: Likoma Isl., Lake Malawi, 28.vi.1900, *Johnson* 181 (K). C: Salima Distr., Senga Bay, 470 m., 25.iv.1971, *Pawek* 4724 (K; MAL; SRGH) S: Elephant Marsh, xii.1887, *Scott* 5188 (K). **Mozambique**. T: Nhaluire, 12.iv.1972, *Macêdo* 5194 (LISC; SRGH).

Also in Kenya, Tanzania and Angola. Ecological data only available for two much-collected localities (Senga & Nkhata Bay), on beach sand and neighbouring dunes; c. 35–490 m., but obviously higher on the Nyika. In other regions along dry roadsides, in open woodlands, on rocky hills etc.

The inflorescence colour used in F.T.A., by Suessenguth and others to distinguish *C. kirkii* and *C. rubra* (straw-coloured in the former, red in the latter), was clearly taken from dried material without colour notes. Recent, well-annotated material of this affinity demonstrates that specimens appearing pale in the herbarium also have colour notes showing that they were red when gathered; conversely, no material bearing colour notes at all describes the inflorescence as other than same shade of red.

2. **Centemopsis gracilenta** (Hiern) Schinz in Viert. Nat. Ges. Zürich **57**: 547 (1913).—Hauman in F.C.B. 2: 42 (1951). Type from Angola.

Centema gracilenta Hiern, Cat. Afr. Pl. Welw. **1**: 890 (1900). Type as above.

Psilotrichum gracilentum (Hiern) C.B. Clarke in F.T.A. **6**, 1: 59 (1909). Type as above.

Centemopsis myurus Suesseng. in Bull. Jard. Bot. Brux. **15**: 61 (1938). Type: Zimbabwe, R. Kafue, *Lynes* 63b (BR, holotype).

Achyropsis oxyuris Suesseng. & Overk. in Bot. Archiv **41**: 74 (1940). Type: Zambia, near source of R. Isongailu, Mwinilunga, *Milne-Redhead* 3896 (K, holotype).

Annual herb (or probably short-lived perennial in undisturbed ground), erect, mostly 0.5–1.6 m. Stem wiry and tough, strongly striate with pale ridges, smooth and glabrous throughout or more commonly with whitish multicellular hairs about the somewhat swollen nodes, rarely slightly scabrid on the ridges below the nodes, much-branched and bushy from near the base upwards, the branches divaricate-ascending and frequently with sterile, axillary short shoots.

Leaves linear to linear-filiform, 1.5–9.5 cm. × 0.5–3 (4) mm., glabrous to sparingly pilose (especially when young), occasionally slightly scabrid on the pale and prominent lower surface of the midrib and on the revolute margins, sessile, sharply mucronate at the apex. Inflorescence carmine to pink or whitish, spiciform, 1.2–4 × 0.6–1 cm. in flower, rounded or conical at the apex, elongating in fruit to as much as 20 cm.; axis very densely white-lanate, strongly furrowed but not becoming honeycombed with pits after fruit-fall, densely clad with the persistent bracts. Bracts 1.25–1.5 mm. long, lanceolate to ovate-lanceolate, membranous with a dark brown, shortly excurrent midrib, glabrous or more commonly basally pilose and sparingly ciliate. Bracteoles deltoid-ovate, 1.25–1.75 mm. long, membranous with a distinctly excurrent brownish or red midrib, glabrous or sparingly pilose. Flowers solitary in the axils of the bracts. Tepals 2.5–3(4) mm. long, oblong; outer 2 at anthesis firm centrally with a slender, slightly excurrent midrib and 2 pairs of slender lateral nerves (these often branching above), glabrous, increasingly hyaline-margined upwards with the margins inflexed; inner 3 similar but slightly less firm, at least the innermost usually with only one pair of lateral nerves. Tepals slightly indurate and the outer especially somewhat more prominently 3-nerved in fruit, but never strongly ribbed as in that stage with *C. kirkii*, nor with the inflorescence appearing spiky. Filaments very slender, 1.5–3 mm. long, the pseudostaminodes finally c. three quarters the length of the filaments, oblong, fimbriate around the apex. Ovary ellipsoid with a dense beard of hairs around the middle, glabrous below and more thinly hairy above. Style slender, 1.25–2 mm. long, not asymmetrically placed on the ripe capsule, glabrous. Capsule 2.25–2.5 mm. long, pyriform, compressed. Seed compressed-ovoid, brown, 2–2.25 mm. long, shining, finely reticulate.

Zambia. N: Track off Rd. to Koa near Mbala, 1520 m., 6.iv.1955, *Richards* 5333 (K). W: Ndola, 16.v.1953, *Fanshawe* 11 (K; NDO). C: 8 km. E. of Chiwefwe, 1370 m. 15.vii.1930, *Hutchinson & Gillett* 15730 (K). E: Chipata, 15.iv.1951, *Gilges* 61 (SRGH). S: Zimba, 9.viii.1909, *Rogers* 8276 (K; SRGH). **Zimbabwe**. N: E. side of Umvukwe Mts. near Dawsons, 28/29.iv. 1948, *Rodin* 4465 (K; UC). W: Gwampa Forest Reserve, v.1956, *Goldsmith* 103/56 (K; LISC; SRGH). E: Mutare Gimbokki Farm, 27.iv.1976, *Meara* 324 (K; SRGH). S: Devuli R. on Gutu to Buhera Rd., 18.i.1973, *Cannell* 550 (K; SRGH). **Malawi**. C: Lilongwe Distr., Dzalanyama Forest Reserve, 1280 m., 26.iii.1977, *Brummitt, Seyani & Patel* 14936 (K; MAL; SRGH). S: Ntcheu Hills, 1060 m., 12.v.1937, *Lawrence* 414 (K). **Mozambique**. MS: c.20 km. W. of Dombe between Mucrera and Mevumosi Rivers, SE. foothills of Chimanimani Mts., c. 450 m., 23.iv.1974, *Pope & Müller* 1269 (K; SRGH). N: back of Mendone (?), iv.1921, *Johnson* 315 (K).

Also in Kenya, Tanzania, Angola and Zaire. On Kalahari Sand or other sand, or sandy loam, commonly along tracks in *Brachystegia/Uapaca/Isoberlinia* and *Cryptosepalum* woodland in grassland, also by roads, in bushland, on termitaria, or on dry rocky hills; 450–1680 m.

3. **Centemopsis conferta** (Schinz) Suesseng. in Mitt. Bot. Staatss. München **1**: 187 (1953).—Townsend in Publ. Cairo Univ. Herb. **7** & **8**: 69 (1977). Type from Tanzania.

Achyranthes conferta Schinz in Bull. Herb. Boiss. **4**: 420 (1896). Type as above.

Psilotrichum confertum (Schinz) C.B. Clarke in F.T.A. **6**, 1: 59 (1909). Type as above.

Centrostachys conferta (Schinz) Standley in Journ. Wash. Acad. Sci. **5**: 76 (1915). Type as above.

Achyropsis conferta (Schinz) Schinz in Engl. & Prantl Pflanzenfam. ed. 2, **16** C: 63 (1934).—Cavaco in Mém. Mus. Hist. Nat. Paris Sér. B, **13**: 153 (1962).

Annual herb 0.2–1 m. tall, simple below the inflorescence or with numerous divaricate-ascending branches in the lower part of the stem. Stem and branches wiry, terete below in large plants, more usually conspicuously striate with pale ridges and green furrows, glabrous or thinly pilose in the younger parts and about the nodes. Leaves narrowly linear to filiform, 2–7 × 0.1–0.4 cm., glabrous or thinly pilose, usually scabrid at least along the revolute margins, attenuate at each end, distinctly mucronate at the apex. Inflorescences solitary and terminal or 2–4 on lateral branches, stout and very dense, spicate, (1)1.7–8 × (0.6)0.8–1.2 cm., greenish-white to brownish-pink or red; axis densely lanate, almost concealed by the patent-ascending, dense, persistent bracts. Bracts concave, narrowly oblong and almost parallel-sided to spathulate, c.3–4 mm. long, membranous, abruptly narrowed to the short arista formed by the

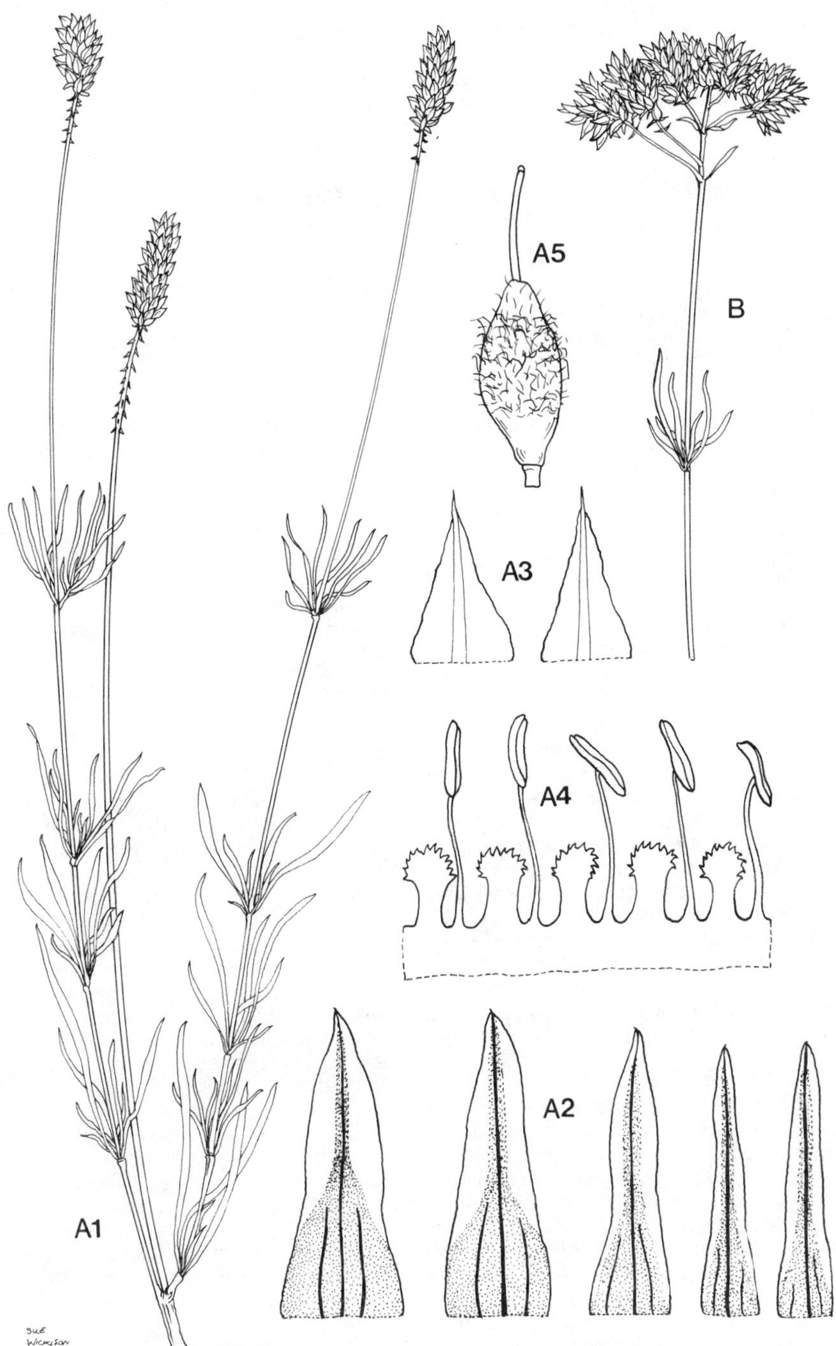

Tab. 18. A.—CENTEMOPSIS KIRKII. A1, flowering stem (×⅔); A2, tepals, inner surface (×8); A3, bracteoles (×8); A4, androecium (×10); A5, gynoecium (×10), A1–5 from *Pawek* 6539. B.—CENTEMOPSIS FASTIGIATA. B1, inflorescence (×⅔) *Sanane* 641.

excurrent nerve, glabrous or sparingly pilose. Bracteoles narrowly lanceolate-oblong, often notched above, sharply aristate, c., one half to three quarters as long as the perianth. Flowers solitary in the axils of the bracts, slightly compressed. Tepals 3—3.5 mm. long, oblong, gradually diminishing in width inwards, the 4 outer strongly 5-nerved, the innermost usually 3-nerved, all glabrous and broadly hyaline-margined above the middle, scarcely mucronate. Filaments delicate, c. 2 mm. long, the pseudostaminodes c. 1 mm. long, narrowly oblong, fimbriate at the apex. Ovary oblong-ellipsoid, glabrous. Style slender, c. 1.5 mm. long, symmetrically placed. Capsule oblong-ellipsoid, c. 1.5 mm. long. Seed oblong-ellipsoid, shining brown, feebly reticulate, c. 1.25 mm. long.

Zambia. N: Kasama, 4.vii.1964, *Fanshawe* 8801 (K; NDO; SRGH). **Mozambique**. N: Nampula, Serra de Chinga, 26.vii.1936, *Torre* 771 (COI; LISC). Z: Between Milange and Molungo, 3.viii.1943, *Torre* 5750B (LISC).

Also in Tanzania and Zaire. On sandy or light yellow soil along roadsides, in shaded bush and abandoned shambas or in *Brachystegia* woodland, in one case at the edge of a pan; 1520—1830 m.

To judge from herbarium material seen, this species has consistently longer, (to c. three quarters of the length of the perianth), more deeply notched bracteoles in Zambia than in Tanzania where they are often little more than half the length of the perianth and more or less tapering into the apical arista.

4. **Centemopsis fastigiata** (Suesseng.) Townsend in Publ. Cairo Univ. Herb. **7** & **8**: 70 (1977). TAB. **18** B. Type from Zaire.

Robynsiella fastigiata Suesseng. in Bull. Jard. Bot. Nat. Belg. **15**: 70 (1938).—Hauman in F.C.B. **2**: 52 (1951).—Cavaco in Mém. Mus. Hist. Nat. Paris Sér. B, **13**: 155 (1962). Type as above.

Centemopsis trichotoma Suesseng. in Mitt. Bot. Staatss. München **1**: 187 (1953). Type: Zambia, path to Koa village from Mbala, *Richards* 748 (K, holotype).

Perennial herb with numerous rigidly erect stems from the base, 0.2—1.3 (2) m. tall. Stems wiry and tough, strongly striate with pale ridges, smooth and glabrous throughout or scabrid below the nodes, simple or sparingly branched with long, ascending branches. Leaves narrowly linear to narrowly elliptic, 2.5—8.5 × 0.6—1.3 cm., glabrous or slightly puberulent when young, entirely smooth or scabrid along the prominent lower surface of the midrib and the revolute margins, attenuate both above and below, apex sharply mucronate. Inflorescence greenish-white or creamy to deep carmine-red, condensed, repeatedly branched and cymose-fastigiate, forming a more or less round-topped head up to 7 cm. across but often less; shorter upper branches more or less white-lanate, sulcate, after fruit-fall densely clad with the patent, persistent bracts. Bracts ovate-lanceolate, c.2 mm. long, membranous, mucronate with the shortly excurrent yellowish or reddish midrib, glabrous or very sparingly whitish-pilose. Bracteoles similar to the bracts. Flowers solitary in the axils of the bracts. Tepals 4—5.5 mm. long, oblong; outer 2 at anthesis firm centrally with a rather obscure slightly excurrent midrib and 2 faint lateral nerves, more or less white-pilose below, increasingly broadly hyaline-margined and delicate above with the margins inflexed; inner 3 increasingly shorter,narrower and more broadly hyaline-margined, the margins of the innermost wider than the narrow, firm centre. Tepals indurate at the base in fruit, slightly or markedly more prominently nerved. Filaments very slender, c. 5—6 mm. long, the pseudostaminodes c. 1—2.5 mm. long, oblong, fimbriate around the apex. Ovary ellipsoid, pilose in most of the upper part with the hairs sometimes ascending the base of the style at least on one side. Style slender, c. 3.5—4 mm. long, symmetrically placed. Capsule ovoid-ellipsoid, 2—2.5 mm. long. Seed ovoid, pyriform, shining brown, feebly reticulate, c. 2 mm. long.

Zambia. N: Kasama Distr., Chibutubutu, 1320 m., 26.ii.1960, *Richards* 12613 (K; SRGH).

Also in Tanzania and Zaire. On sandy or damp loamy soil by paths, in woodland or bush, in long grass in open woodland or in rough grassland; 1200—1520 m.

11. CYATHULA Blume

Cyathula Blume, Bijdr. Fl. Nederl. Ind.: 548 (1825) nomen cons. (non *Cyathula* Lour., Fl. Cochinch.: 93, 101 (1790)).

Annual or perennial herbs with entire, opposite leaves. Inflorescence terminal on the stem and branches, spiciform or capitate, bracteate, the ultimate division basically a triad of fertile flowers, the outer pair bracteolate and subtended laterally by 2 modified bracteolate flowers consisting of a number of sharply uncinately hooked (more rarely straight or glochidiate) spines or bracteoliform processes, but one or both of the outer pair sometimes absent, bracteoles also sometimes with a hooked arista; spines of the modified flowers at first small, rapidly accrescent, few to many, clustered with the clusters not or very shortly stalked. Bracts persistent, finally more or less deflexed; bracteoles and perianth falling with the fruit. Perianth segments 5, very shortly mucronate or some (especially the outer 2) hooked-aristate, serving with the bracteoles and modified flowers to distribute the fruit. Stamens 5, the filaments delicate, shortly monadelphous at the base, alternating with distinct, commonly toothed or lacerate pseudostaminodes; anthers bilocular. Style slender, stigma capitate. Ovary with a single pendulous ovule. Fruit a thin-walled utricle, irregularly ruptured by the developing seed. Seed ovoid, slightly compressed; endosperm copious.

About 25 species in the tropics of both Old and New Worlds.

1. Spines of the sterile modified flowers and the awns of their bracteoles straight, not uncinate-hooked at the apex - - - - - - - - - - - - 2
– Spines of the sterile modified flowers, and usually the awns of their bracteoles also, uncinate-hooked at the apex - - - - - - - - - - - - 3
2. Flowers in solitary or paired, terminal, more or less globose capitula - - - - - - - - - - - - - - - - - 6. *lanceolata*
– Capitula not solitary or paired, arranged in distinct terminal racemes - - - - - - - - - - - - - - - - - 5. *orthacantha*
3. Tepals 2.5–3 mm., inflorescence a narrow, elongating "spike" - - 1. *prostrata*
– Tepals at least 4.5 mm. long, inflorescence of globose heads or a stout spike at least 1.5 cm. wide - - - - - - - - - - - - - - - - 4
4. Leaves spathulate with a broad claw abruptly contracted to a short petiole, and the lamina abruptly contracted to a short acumen above; fruiting heads globose, woolly - - - - - - - - - - - - - - - - -7. *natalensis*
– Leaves not spathulate with the above peculiar form; fruiting heads,if globose, not woolly - - - - - - - - - - - - - - - - - 5
5. Outer 2 tepals of mature fertile flowers mucronate only, if occasionally an uncinate tepal present then the hook not nearly as pronounced as those of the bracteoles and the terminal part of the inflorescence dense and cylindrical - -4. *cylindrica*
– Outer 2 tepals of mature fertile flowers with the arista tipped with an uncinate hook (occasionally straight in some flowers), the hook as sharp as the hook of the bracteoles; inflorescence spherical, or a raceme of spherical or hemispherical partial inflorescences - - - - - - - - - - - - - - - - - - 6
6. Tepals glabrous - - - - - - - - - - - - 2. *uncinulata*
– Tepals more or less pilose with multicellular, minutely barbellate hairs - - - - - - - - - - - - - - - - - 3. *divulsa*

1. **Cyathula prostrata** (L.) Blume, Bijdr. Fl. Nederl. Ind.: 549 (1825).—Baker & Clarke in F.T.A. **6**, 1: 43 (1909).—Schinz in Engl. & Prantl Pflanzenfam. ed. 2, **16 C**: 47 (1934).—Hauman in F.C.B. 2: 62 (1951).—Cavaco in Mém. Mus. Hist. Nat. Paris Sér. **B**, 13: 87 (1962).—J.H. Ross Fl. Natal: 158 (1973). Type, Linnean specimen 287/13 (LINN, lectotype).

Achyranthes prostrata L., Sp. Pl. ed. 2: 296 (1762). Type as above.

Pupalia prostrata (L.) Mart. in Nov. Acta Acad. Caesar. Leop. Carol. **13**: 321 (1826). Type as above.

Desmochaeta prostrata (L.) DC., Cat. Hort. Monsp.: 102 (1813). Type as above.

Annual herb (? sometimes short-lived perennial), stems prostrate and rooting at the lower nodes to erect, 0.2–1.2 m., simple or usually branched up to about the middle, more or less swollen at the nodes, lower branches divaricate, the upper more erect; stem and branches bluntly 4-angled to subterete, striate or sulcate, subglabrous to more or less densely pilose (especially the lower internodes). Leaves 1.5–8 × 1–5 cm., mostly rhomboid to rhomboid-ovate, sometimes rhomboid-elliptic to shortly oval or subcircular, occasionally with the margin outline distinctly excavate below and/or above the middle, shortly acuminate at the apex, acute to rather blunt (more rarely rounded), shortly cuneate to cuneate-attenuate at the base, subglabrous to moderately pilose with strigose hairs on both surfaces, subsessile or distinctly (up to 13 mm. long) petiolate. Spikes terminal on the stem and branches, at first dense, soon considerably elongating to as much as 25 cm. but mostly c. 6–15 cm. with maturing lower flowers increasingly distant, 5–7 mm. wide, peduncle up to c. 10 cm. long, axis and peduncle thinly to more or less densely pilose;bracts and bracteoles membranous, lanceolate-ovate, c. 1.5–2 mm. long, mucronate with the shortly excurrent midrib, glabrous or ciliate; flowers in sessile or shortly pedunculate clusters (peduncles to c. 2 mm. long), cymose, of 2–3 hermaphrodite flowers, the two laterals subtended by two modified flowers, or the uppermost hermaphrodite flowers of the spike solitary, similarly subtended by modified flowers, bibracteolate. Tepals 2.25–3 mm. long, elliptic-oblong, 3–nerved, subglabrous to more or less densely white-pilose; the outer firmer with the lateral nerves more distinct and joining the shortly excurrent midrib just below the apex, usually more densely white-pilose than the inner; the inner sometimes more or less falcate, slightly shorter. Spines of modified flowers sharply uncinate, numerous, glabrous, reddish, c. 2 mm., fasciculate, in fruit scarcely exceeding the tepals of the fertile flower; 2- or 3-flowered clusters, more or less globose finally deflexed, falling as a unit to form a "burr" c. 5 mm in diam. Filaments very slender, c. 1.5 mm. long, the pseudostaminodes rectangular-cuneate with a truncate, dentate or excavate apex. Ovary with a pileiform cap. Style slender, c. 0.6 mm. long, often slightly swollen towards the base. Capsule ovoid, membranous save for the flat, firm apex, c. 1.5 mm. long. Seed ovoid, c. 1.5 mm. long, shining, brown, smooth.

Var. **prostrata**

Partial inflorescences sessile or with a short, thick peduncle not exceeding 1 mm. in length, shorter than the subtending bract. Outer tepals more or less densely white-pilose.

Zambia. B: Masese, 28.vi.1963, *Fanshawe* 7890 (K; LISC; NDO). **Zimbabwe**. E: Haroni/Makurupini Forest, 392 m., 3.xii.1964, *Wild* 6608 (K; LISC; SRGH). **Malawi**. N: Chombe Estate, Nkhata Bay, 9.ix.1955, *Jackson* 1753 (K).

The typical variety is practically pantropical in the Old World, south to the Queensland rain forests; it also occurs in C. & S. America, being apparently frequent in Brazil. Forest floors and in rubber plantations; 270–395 m.

Var. **pedicellata** (C.B. Clarke) Cavaco in Fl. Cameroun **17**: 46 (1974). Type from Uganda.
Cyathula pedicellata C.B. Clarke in F.T.A. **6**, 1: 46 (1909).—Cavaco in Mém. Mus. Hist. Nat. Paris Sér. B, **13**: 86 (1962). Type as above.
Cyathula prostrata f. *pedicellata* (C.B. Clarke) Hauman in F.C.B. **2**: 64 (1951). Type as above.

Partial inflorescences with a distinct, slender peduncle up to 2 mm. long, equaling or exceeding the subtending bract. Outer tepals usually thinly white-pilose to subglabrous.

Zimbabwe. E: Inyanga Distr., Pungwe Bridge, 15.iv.1970, *Chase* 8602 (K; SRGH). **Malawi**. S: Mulanje Mt., foot of Gt. Ruo Gorge, 870–1060 m., 18.iii.1970, *Brummitt & Banda* 9207 (K; LISC; MAL; PRE; SRGH; UPS). **Mozambique**. MS: Garuso forest, 1935, *Gilliland* 1819 (BM; K).

Also widespread in tropical Africa from Sierra Leone to Ethiopia and Zaire; forms somewhat divergent to this form occur in Indonesia, and especially in New Guinea. Forest floors; 870–1200 m.

A weak variety, which future workers might be content to drop, and less well defined in the Flora Zambesiaca area even than in E. Africa. But these slender-pedunculate forms are absent over the American and much of the Asiatic range of the species, and I have not been able to study it in the field in Africa to any useful degree.

2. **Cyathula uncinulata** (Schrad.) Schinz in De Wild., Pl. Bequaert. 5: 386 (1932); in Engl. & Prantl Pflanzenfam. ed. 2, **16 C**: 47 (1934).—Hauman in F.C.B. 2: 67 (1951).—White, F.F.N.R.: 45 (1962).—J.H. Ross Fl. Natal: 158 (1973). TAB. **19** fig. B. Type, Cultivated material from Goettingen Botanic Garden (LE, holotype, photo.!).

Achyranthes uncinulata Schrad. in Ind. sem. hort. Goett. **1833**: 1 (1833). Type as above.

Cyathula globulifera Moq. in DC., Prodr. **13**, 2: 329 (1849).—Baker & Clarke in F.T.A. **6**, 1: 44 (1909). Type from Madagascar.

Desmochaeta uncinulata (Schrad.) Hiern in Cat. Afr. Pl. Welw. **1**: 890 (1900). Type as for *C. uncinulata*.

Cyathula polycephala sensu Hauman in Bull. Jard. Bot. Brux. **18**: 109 (1946); in F.C.B. 2: 68 (1951) non Baker.

Erect and bushy or more commonly straggling to scandent perennial (?) herb, (0.3) 0.75—3 (6) m.; stem and branches terete and striate in the older parts, becoming bluntly tetragonous and finally sharply tetragonous-sulcate above, the older parts thinly to moderately pilose with patent or deflexed yellowish multicellular hairs, the younger parts densely pilose to thickly yellowish-tomentose or pannose; nodes distinctly swollen, when dry the stem and branches often shrunken above the nodes. Leaves broadly ovate to broadly elliptic-oblong, (2.5) 3.5—12 × (2) 2.5—8 cm., shortly cuneate to subcordate at the base, shortly acuminate at the apex, moderately furnished on the darker green superior surface with appressed, multicellular, barbellate hairs, on the inferior surface densely pilose to closely velutinous, especially along the nerves; petiole c. 1—3 cm. long. Inflorescences terminal on the stem and branches, each a dense globose head (occasionally slightly laxer, oblong and lobed) of agglomerated lateral compound cymes on a short tomentose axis, mostly 1.75—2.5 cm. in diam. when flowering but enlarging in fruit to form a burr up to 4 cm. across; the short inflorescence branches with white, lanate, matted hairs; peduncle (0.6) 3—4 (10) cm. long; bracts deltoid-ovate, glabrous, c. 2.5 mm. long, stramineous, distinctly mucronate with the brown, excurrent midrib; bracteoles lanceolate, c. 5—6 mm. long, the excurrent nerve forming a long, uncinate-tipped arista; ultimate divisions of lateral cymes formed mostly of a central fertile flower subtended on each side by a triad of one fertile and two lateral modified flowers - but variable and sometimes the central fertile or one or more of lateral fertile or modified flowers absent, and occasionally the "fertile" flowers with empty anther-sacs, abortive ovary, or both. Tepals glabrous, acute, narrowly lanceolate; 2 outer tepals frequently more or less transversely undulate, 5—6.5 mm. long, 1-nerved, the distinct arista usually uncinately hooked but sometimes not; 3 inner tepals shorter, 2.5—5 mm., 2—3-nerved with hyaline margins and a greenish centre, shortly mucronate. Modified flowers of 2 narrow, uncinate-tipped bracteoliform processes and 2 shorter uncinate spines. Filaments delicate and slender, c. 2—2.5 mm. long; pseudostaminodes subulate or linear-oblong to narrowly obcuneate, simple or denticulate to fimbriate, c. one third the length of the filaments. Style slender, 1.75—2 mm. long; ovary obovoid-turbinate, c. 1 mm. long. Capsule ovoid, c. 2—2.25 mm. long, membranous with a firm, flattish top. Seed c. 1.75—2 mm. long, ovoid, brown, almost smooth.

Zambia. N: Mbala Distr., Mwenzo, 1670 m., 10.vi.1951, *Bullock* 3962 (K). E: Makutu Mts., 28.x.1972, *Fanshawe* 11600 (K; NDO). **Zimbabwe**. E: Chirinda Forest, 1090 m., vi.1961, *Goldsmith* 42/61 (K; LISC; SRGH). **Malawi**. N: Viphya Plateau, Chikangawa, 6.vii.1976, *Pawek* 11481 (K; MAL; MO; SRGH; UC). C: 1 km. W. of Dedza 1520 m., 22.iv.1970, *Brummitt* 10039 (K; MAL; SRGH; LISC; PRE; UPS; EA). S: Maone Estate, 2 km. N. of Limbe, 1190 m., 11.vi.1970, *Brummitt* 1138 (K; MAL; SRGH; LISC; PRE; UPS; EA). **Mozambique**. T: Planalto de Angónia, 25.viii.1941, *Torre* 3324 (C; COI; LISC; LMU; MO). MS: Mavita, border of Rutanda, 28.iv.1948, *Barbosa* 1654 (EA; LISC; PRE; WAG).

Widespread in Africa from the Cameroons and Sudan southward to Cape Province, also occurring in Madagascar. Usually scrambling over herbage at forest and plantation margins and in hedges (sometimes forming dense thickets), also on roadside and stream banks; 1090—1670 m.

3. **Cyathula divulsa** Suesseng. in Fedde, Repert. **51**: 196 (1952). Type from Tanzania.

Cyathula schimperiana var. *burttiana* Suesseng. in Mitt. Bot. Staatss. München **1**: 189 (1953). Type from Tanzania.

Herb (duration unknown), erect and apparently little branched, lax in habit, 0.3–1 m.; stem and branches slender, terete and striate below to bluntly or clearly tetragonous above, greenish- to purplish-brown, more or less densely furnished with spreading, ascending or sometimes deflexed, brownish multicellular hairs, older parts more or less glabrescent; nodes distinctly swollen, stem and branches contracted above the nodes or sometimes not. Leaves large, broadly elliptical, 5.5–18 × 3.5–10 cm., acute or acuminate at the apex, at the base abruptly narrowed or cuneate into the 1–2 cm. long petiole, moderately pilose to tomentose on both surfaces with the hairs longer on the venation of the lower surface. Inflorescences terminal on the stem and branches, each a lax, spiciform thyrse formed of very shortly stalked globose or cuneate-based condensed cymes 1.5–2 cm. in diam., the entire thyrse c. 3 cm. wide and up to 20 cm. long (but often less), the lower cymes increasingly distant in fruit; peduncle very short or up to 7 cm. long, both it and the inflorescence axis densely brownish-pilose; bracts deltoid or lanceolate-ovate, 4–6.5 mm. long, membranous, densely pilose and scarcely shining, shortly aristate with the excurrent midrib; bracteoles similar but longer from the long-excurrent, uncinate-tipped midrib, 7.5–8 mm. long; ultimate divisions of the lateral cymes formed of a central fertile flower subtended on each side by a triad of one fertile and two lateral modified flowers, or one of the latter sometimes missing. Outer 2 tepals narrowly lanceolate, thinly to densely (and matted) long-pilose, 6.5–8 mm. long, both with a long, uncinate awn formed by the excurrent midrib, narrowly hyaline-margined, with or without 1–2 fainter lateral nerves at the base which are evanescent below or a little above the middle of the tepal; inner 3 tepals shorter, 5.5–7 mm. long, narrowly lanceolate, more broadly hyaline-margined, thinly long-pilose to densely lanate, 3–5(6)-nerved with the inner 2 nerves meeting the midrib below the apex but the outer much shorter, the midrib excurrent in a short, rather fine and sharp mucro. Modified flowers of 2–4 narrowly lanceolate bracteoliform processes and 2–4 uncinate spines. Filaments delicate, 2.75–3 mm. long; pseudostaminodes 1–1.5 mm. long, broadly cuneate-obovate, thinly to densely lanate-pilose around the margins, dentate at the plane or incurved apex, with a subulate or finely fimbriate dorsal scale. Style slender,2–4 mm. long; ovary obovoid, thickened above, 1.75–2 mm. long. Capsule shortly cylindrical, 2.5 mm. long, with a hardened rim around the concave apex; seed 2.25 mm. long, ovoid, brown, almost smooth.

Zambia. 14 km. N. of Muzombwe, W. side of Mweru-wa-Ntipa, 1000 m., 16.iv.1961, *Phipps & Vesey-Fitzgerald* 3237 (K; LISC; SRGH). **Zimbabwe**. E: edge of Chirinda Forest, 1150 m., 23.iv.1947, *Wild* 1916 (K; SRGH). **Mozambique**. MS: Garuso Forest, iv.1935, *Gilliland* 1818 (BM; LISC).

Also in Tanzania. Always in shade, in or at the edge of forests or among shrubs; 1000–1150 m.

Apparently always a scarce and scattered species throughout its limited range.

4. **Cyathula cylindrica** Moq. in DC., Prodr. **13**, 2: 328 (1849).—Baker & Clarke in F.T.A. **6**, 1: 46 (1909).—Schinz in Engl. & Prantl Pflanzenfam. ed. 2, **16** C: 47 (1934).—Hauman in F.C.B. **2**: 66 (1951); in J.H. Ross Fl. Natal: 158 (1973).—Townsend in Publ. Cairo Univ. Herb. **7** & **8**: 74 (1977). Type from Madagascar.

Cyathula schimperiana Moq. in DC., Prodr. **13**, 2: 328 (1849).—Baker & Clarke in F.T.A. **6**, 1: 45 (1909). Type from Ethiopia.

Cyathula mannii Baker in Kew Bull. 1897: 278 (1897).—Baker & Clarke in F.T.A. **6**, 1: 46 (1909). Type: Fernando Po, *Mann* 296 (K, lectotype).

Cyathula albida Lopr. in Engl. Bot. Jahrb. **27**: 53 (1899). Type from Angola.

Pupal huillensis Hiern in Cat. Afr. Pl. Welw. **1**: 892 (1900). Type from Angola.

Desmochaeta distorta Hiern in Cat. Afr. Pl. Welw. **1**: 891 (1900). Type from Angola.

Cyathula distorta (Hiern) C.B. Clarke in F.T.A. **6**, 1: 46 (1909).

Cyathula cylindrica var. *mannii* (Baker) Suesseng. in Mitt. Bot. Staatss. München **1**: 188 (1953).

Cyathula cylindrica var. *orbicularis* Suesseng. in Mitt. Bot. Staatss. München **1**: 77 (1951). Type: Zimbabwe, Marondera *Wild* 326 (K, isotype; M; SRGH).

Perennial herb, very variable in habit from bushy and c. 0.6–1 m. high to sprawling or decumbent and rooting at the lower nodes, or subscandent, or

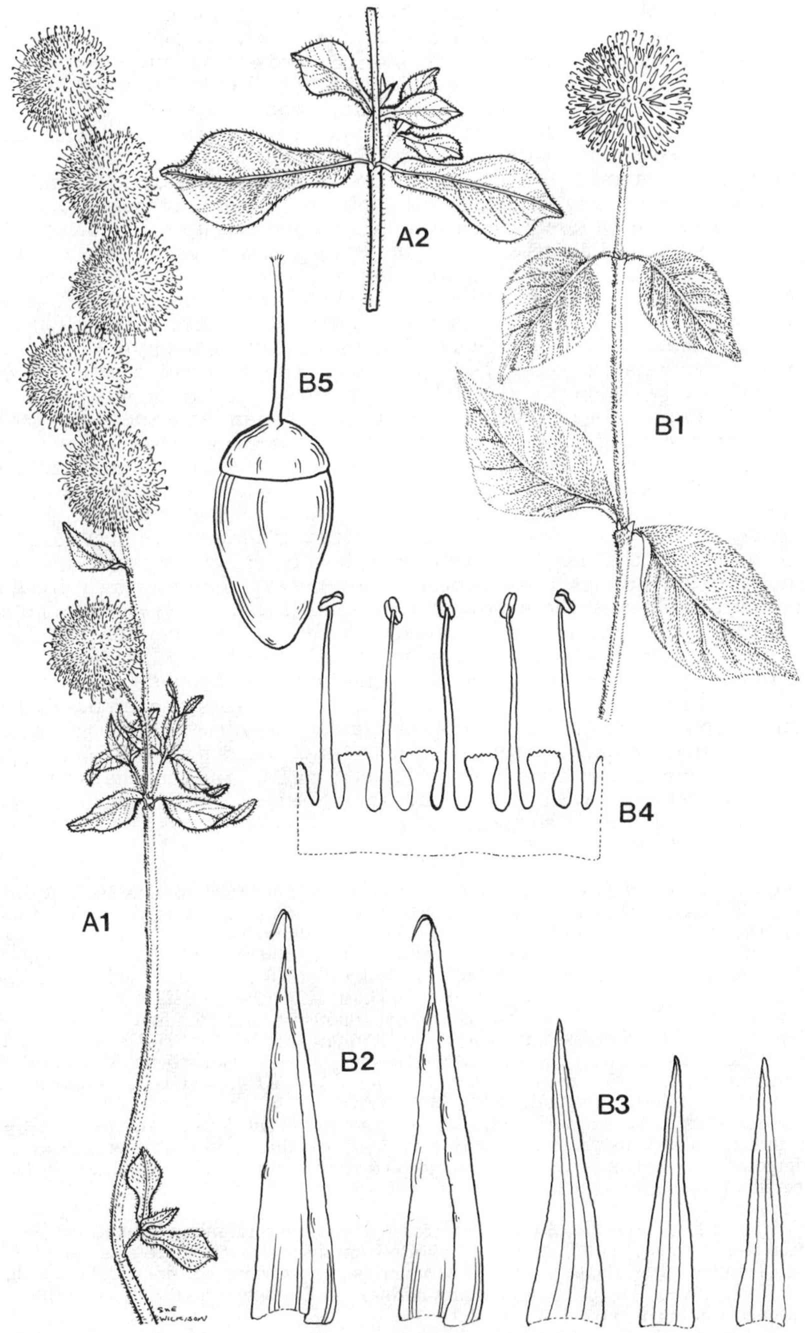

Tab. 19. A.—CYATHULA NATALENSIS. A1, flowering branch (×$\frac{2}{3}$); A2, pair of cauline leaves (×$\frac{2}{3}$), A1—2 from *Goldsmith* 110/56. B.—CYATHULA UNCINULATA. B1, flowering branch (×1$\frac{1}{2}$); B2, outer tepals (×8); B3, inner tepals (×8); B4, androecium (×8); B5, gynoecium (×16), B1—4 from *Brummitt* 10039.

rambling in forests to a height of 6 m. or more; stems and branches terete and striate in the older parts, becoming bluntly tetragonous and finally sharply tetragonous-sulcate above, glabrous or moderately to densely (especially upwards) furnished with long, spreading or more or less appressed fuscous, bristly, multicellular hairs; nodes distinctly swollen in life, in dried material the stem and branches commonly shrunken just above the nodes. Leaves very variable in size and shape, small and roundish (sometimes with undulate margins) to large and broadly oblong- or elliptic-ovate, 1–14 × 0.7–6 cm., subcordate to attenuate at the base, rounded to acuminate at the apex, glabrous to more or less densely furnished with long, appressed, multicellular hairs on both surfaces, more rarely tomentose; petiole distinct, up to c. 2.5 cm. long. Inflorescences terminal on the stem and branches, spiciform, 1.5–2 (2.5) cm. in diam., in robust plants elongate-cylindrical and up to c. 18 cm. long, or sometimes scarcely longer than broad; "spikes" formed of densely congested (more rarely a few of the lower distant) shortly pedunculate cymose clusters composed mostly of triads of fertile flowers each subtended by 1–2 modified flowers; bracts elliptic-oblong, 7–8.5 mm. long, stramineous or silvery, glabrous or furnished with long multicellular hairs about the tip, aristate with the excurrent midrib, the arista usually bent but not sharply uncinate; bracteoles broadly ovate, acuminate, 4.5–9 mm. long, glabrous or furnished with long multicellular hairs towards the apex, long-aristate with the excurrent midrib, the arista sharply uncinate at the tip or not. Tepals 4.5–7.5 mm. long, narrowly lanceolate-oblong, 3–(5) nerved; outer 2 tepals rather feebly nerved, gibbous dorsally at the base, broadly hyaline-margined, glabrous or almost so, acute to rather blunt, the midrib excurrent in a short mucro, rarely feebly uncinate; inner 3 tepals progressively more strongly nerved, more narrowly hyaline-margined and blunter, the innermost obtuse and often minutely lacerate-dentate at the apex with the midrib ceasing below the apex, all 3 moderately to densely furnished with long, white, multicellular, barbellate hairs. Modified flowers with a few, narrow, lanceolate, bracteoliform processes with uncinate apices, simple hooks, and shorter membranous scales within. Filaments slender, 3.5–5 mm. long, the pseudostaminodes cuneate-obovate, c. one quarter the length of the filaments, fringed above, frequently with a filiform dorsal scale. Style slender, long, 2–3 mm.; ovary obovoid, c. 1 mm. long, rather firm. Capsule ovoid, 2–3 mm. long, membranous save for the firm apex. Seed c. 1.5–2.75 mm. long, ovoid, brown, shining, almost smooth.

Zambia. N: Mbala, Distr., Nachilanga Hill above Kawimbe, 1800 m., 13.iv.1959, *Richards* 11213 (K; SRGH). **Zimbabwe**. W: Matobo Distr., Farm Quaringa, 1460 m., v.1959, *Miller* 5918 (K; SRGH). E: Inyanga Distr., Rukotso Mt., 2405 m. 9.iv.1977, *Grosvenor & Renz* 1317 (K; SRGH). S: Mt. Bukwa, Mberengwe Distr., 1500 m., 3.v.1973, *Biegel, Pope & Simon* 4282 (K; LISC; SRGH). **Malawi**. N: Nyika Plateau, 2180 m., 11.viii.1975, *Pawek* 9968 (K; MAL; MO; SRGH; UC). C: Dezda Distr., Chencherere Hill, Chongoni Forest Reserve, 1675–1800 m. 23.iv.1970, *Brummitt* 10059 (EA; K; LISC; MAL; PRE; SRGH; UPS). S: Blantyre Distr., Ndirande Mt., near. summit, 1580 m., 2.v.1970, *Brummitt* 10321 (K; LISC; MAL; PRE; SRGH). **Mozambique**. MS: Plateau above Mevumozi R., Chimanimani Mts., 1550 m., 17.iv.1960, *Goodier* 995 (K; LISC; SRGH). T: Monte Zóbuè, 3.x.1942, *Mendonca* 584 (COI; LISC; MO; WAG).

Widespread in Africa from the Cameroons and Sudan south to S. Africa (Cape Prov). In forests (evergreen rain forest to *Brachystegia* woodland), in open places or even in competition with rank vegetation, along shaded stream edges, on rocky hills and in rough grassland among rocks; 1150–2300 m.

Varies in form according to habitat. Forest plants are scrambling in habit with large, often thinly pilose, acuminate leaves; plants from drier situations are short and bushy with smaller, often densely pilose to tomentose, more rounded, not rarely undulate-margined leaves. The latter form is predominant in Zimbabwe, and Suessenguth's var. *orbiculata* is fairly typical of it.

5. **Cyathula orthacantha** (Hochst. ex Aschers.) Schinz in Engl. Pflanzenfam. **3**, 1a: 108 (1893).—Podlech & Meeuse in Merxm. Prodr. Fl. SW. Afr. **33**: 15 (1966). Type from Ethiopia.

Pupalia orthacantha Hochst. ex Aschers. in Schweinf., Beitr. Fl. Aeth.: 181 (1867). Type as above.

Kyphocarpa orthacantha (Hochst. ex Aschers.) C.B. Clarke in F.T.A. **6**, 1: 55 (1909).—Schinz in Engl. & Prantl Pflanzenfam. ed. 2, **16** C: 43 (1934).

Sericocomopsis orthacantha (Hochst. ex Aschers.) Peter in Fedde, Repert. Beih. **40**, 2: 230 (1932). Type as above.

Cyathula kilimandscharica Suesseng. & Beyerle in Fedde, Repert. **44**: 44 (1938). Type from Kenya.

Pupalia erecta Suesseng. in Fedde, Repert. **44**: 47 (1938). Type from Tanzania.

Cyathula orthacanthoides Suesseng. in Mitt. Bot. Staatss. München **1**: 4 (1950). Type from Tanzania.

Annual herb, usually much-branched, erect to prostrate, commonly straggling or sprawling, 0.3–1.5 m.; stem and branches coarse, terete and striate in the lower parts, becoming bluntly tetragonous or sharply angled and sulcate above, thinly to densely furnished with white, upwardly appressed or patent multicellular hairs, the older parts usually glabrescent; nodes distinctly swollen, usually densely pilose, in larger forms the stem and branches commonly considerably shrunken above the nodes when dry. Leaves variable in form and size, from broadly ovate to broadly or narrowly elliptic, lanceolate-oblong or narrowly lanceolate, 1–15 × 0.7–5.5 cm., acute or acuminate at the apex, at the base shortly cuneate to attenuate with a petiole 0.4–2 cm. long, surfaces thinly to densely pilose with appressed white hairs which in the more densely hairy forms are longer and more divergent especially along the venation of the lower surface. Inflorescences white to pale green, crimson or maroon, terminal on the stem and branches, each a spike-like or more rarely capitate thyrse of sessile condensed cymes c. 1.25–2 cm. in diam. at anthesis, the entire thyrse 1.25–8 cm. long with the lowest cymes somewhat distant or not; peduncle 0.6–6 (11.5) cm. long, both it and the inflorescence axis whitish-pilose; bracts 3–5 mm. long, lanceolate to deltoid-ovate with a long-excurrent midrib, sparingly to moderately pilose dorsally; bracteoles 3–5 mm., broadly deltoid-ovate, midrib excurrent in a long sharp, straight arista, sparingly pilose dorsally; ultimate divisions of lateral cymes formed of a central fertile flower subtended on each side by a triad of one fertile and two lateral modified flowers. Outer 2 tepals (3.5)4.5–7(9) mm. long, with 3 very strong and prominent nerves which meet just below the apex and are excurrent to form a short mucro, usually more or less densely furnished with matted, multicellular, barbellate, white hairs, but sometimes thinly hairy or almost glabrous, and then with a carmine colouration frequently developed; inner 3 tepals shorter, more faintly 3–5(6)-nerved with one margin usually wider below and the nerves on that side more widely separated, pilose chiefly about the apex or sometimes throughout; all tepals lanceolate-oblong, narrowly hyaline-margined. modified flowers of a few lanceolate-based, long-aristate bracteoliform processes and several simple yellowish or reddish spines. Filaments compressed, 2.5–5 mm. long; pseudostaminodes 1–2 mm. long, broadly cuneate-obovate, the dentate-fimbriate apex flat or incurved above, a dentate or furcate ("stags-horn") dorsal scale also present. Style slender, 2–5 mm. long; ovary obovoid-turbinate, c.1–1.5 mm. long. Capsule pyriform, membranous save for the strongly thickened rim around the apical depression, 2.5–3 mm. long. Seed 2–3 mm., ovoid, brown. In fruit the axis and branches of the lateral cymes become indurate-incrassate and concrescent, so that each cyme falls as a complete burr 1–1.5 cm. in diam. with 4–7 mm. spines; the hard base of the burr clad with the persistent bracteoles.

Botswana. N: Okavango R. below Mohembo, E. bank, 28.iv.1975, *Biegel, Müller & Gibbs-Russell* 5021 (K; SRGH). SE: Swaneng Hills School by Serowe, 1065 m., 1.iv.1967, *Mitchison* A26 (K). **Zambia**. C: Luangwa Game Reserve, 8.v.1965, *Mitchell* 2906. S: Lusitu, 19.v.1960, *Fanshawe* 5683 (K; NDO). **Zimbabwe**. N: Hunyani R., 10 km. S. of Danda Mission, 456 m., 15.v.1962, *Wild* 5752 (K; SRGH) W: Hwange, Matetsi, 23.v.1975, *Gonde* 25 (K; SRGH). E: Sabi Valley Chisumbanje, 456 m., 16.ii.1958, *Plowes* 2029 (K; SRGH). S: 12 km. SE of Tuli on Rd. to Shashi Irrigation Scheme, 22.iii.1959, *Drummond* 5915 (K; LISC; SRGH). **Malawi**. N: Mzimba Distr., S. side of Lake Kazuni, 1080 m., 20.v.1970, *Brummitt* 10945 (K; LISC; MAL; PRE; SRGH). **Mozambique**. T: Baroma Distr., Sisitso, 4.vii.1950, *Chase* 2767 (BM; SRGH).

Also in Sudan, Ethiopia, Uganda, Kenya, Tanzania, Angola and Namibia. Along roadsides and in disturbed or cultivated ground, in mopane woodland (probably always degraded), in secondary grassland, along river banks and at edges of pans; 460–1060 m.

Noted as a very bad weed by observers in various parts of its range.

6. **Cyathula lanceolata** Schinz in Engl. Bot. Jahrb. **21**: 188 (1895).—Podlech & Meeuse in Merxm. Prodr. Fl. SW. Afr. **33**: 15 (1966).—Townsend in Publ. Cairo Univ. Herb. **7** & **8**: 76 (1977). Type from "Ostafrika"

Cyathula crispa Schinz in Engl. Bot. Jahrb. **21**: 188 (1895). Type from Transvaal.

Cyathula merkeri Gilg in Engl. Bot. Jahrb. **36**: 207 (1905).—Baker & Clarke in F.T.A. **6**, 1: 47 (1909). Type from Tanzania.

Pandiaka deserti N.E. Br. in Kew Bull. **1909**: 134 (1909). Type: Botswana, near Chukutsa salt-pan, *Lugard* 221 (K; holotype).

Pandiaka lanceolata (Schinz) C.B. Clarke in F.T.A. **6**, 1: 68 (1909). Type as for *Cyathula lanceolata*.

Cyathula hereroensis Schinz in Viert. Nat. Ges. Zürich **66**: 222 (1921).—Podlech & Meeuse in Merxm. Prodr. Fl. SW. Afr. **33**: 15 (1966). Type from Namibia.

Cyathula deserti (N.E. Br.) Suesseng. in Fedde, Repert. **44**: 46 (1938). Type as for *Pandiaka deserti*.

Sericocomopsis lanceolata (Schinz) Peter in Fedde, Repert. Beih. **40**, 2: 229 (1932). Type as for *Cyathula lanceolata*.

Pandiaka wildii Suesseng. in Mitt. Bot. Staatss. München **1**: 63 (1950). Type: Zimbabwe, Lower Sabi, *Wild* 2313 (SRGH, holotype; K, isotype).

Perennial herb, much branched from the base upwards, erect and bushy to prostrate and sprawling, 15—90 cm.; stem and branches strongly striate-sulcate, terete or some of the upper internodes tetragonous, whitish to green or the striae brown, subglabrous to pilose with softer and more or less appressed to patent and substrigose multicellular hairs or sometimes white-woolly; nodes slightly swollen, the stem and branches sometimes slightly shrunken above them. Leaves firm in texture, oblong to narrowly lanceolate-elliptic, 1.2—5.5 × 0.3—1.5 cm., acute to obtuse at the apex with a frequently deciduous, firm, horn-like mucro up to 2 mm. long, at the base cuneate to abruptly rounded or subauriculate, sessile or with a petiole up to 5 mm. long, both surfaces more or less softly appressed-pilose or appressed - (more rarely more or less patent) - strigose when young, finally thinly hairy or glabrescent or with strigose hairs persisting particularly along the sometimes undulate-crispate margins and along the lower surface of the midrib and few primary nerves, sometimes lanuginose. Inflorescences terminal on the stem and branches, solitary, rounded-capitate (formed of condensed cymes), the heads 1.5—2 cm. in diam. in flower and scarcely larger in fruit, commonly sessile and subtended by a pair of leaves, or on a pilose peduncle up to 1(3) cm. long; bracts broadly ovate, broader than long, 3 mm. long, pale, glabrous, shortly aristate with the excurrent midrib; bracteoles broadly deltoid- ovate, 4—6 mm. long, similar but more longly aristate, glabrous or thinly pilose about the base; ultimate divisions of cymes of a central fertile flower subtended on each side by a triad of a central fertile and two lateral modified flowers, or sterile flowers solitary or absent. Outer 2 tepals narrowly oblong-lanceolate, 6.5—9 mm. long, firm, with broad, pale, opaque margins, with a narrow (one third of the width of the tepal or less) green vitta which ceases below the apex, 3—(5) nerved in the vitta with the lateral nerves and midrib sometimes forked, lateral nerves evanescent above, the midrib not or very slightly excurrent in an obscure mucro; inner 3 tepals similar but slightly shorter, progressively blunter and the innermost subcucullate at the apex, the midrib not excurrent in a mucro but occasionally protruding as a small dorsal cusp near the apex, green vitta in all 3 paler with the nerves more clearly defined; indument of all 5 tepals variable, commonly evanescent, from slightly pilose about the base or furnished with long, subappressed white hairs in the basal half (especially centrally along the pale margins) to rather densely floccose throughout. Modified flowers commonly lanate-hairy centrally, of a few narrowly bracteoliform processes and/or 2—8 very unequal stramineous to purplish spines which become accrescent and thickened at the base in fruit, the longest (3.5)7—10 mm. in length. Filaments firm, 3—4 mm. long, alternately longer and shorter; pseudostaminodes considerably fused to the filaments, 1.5—2 mm. long, oblong, lacerate-dentate at the plane apex with no dorsal scale. Stem slender, 3.5—6 mm. long, pale and firm; ovary squat, pyriform (onion-shaped), tapering into the c. 1 mm. long style. Capsule ovoid, c. 3 mm. long, firm in the upper half and rupturing when ripe in the hyaline lower half. Seed c. 2.75 mm. in diam., subglobose, brownish, smooth.

Botswana. N: Boteti delta area NE. of Mopipi, 850 m., 19.iv.1973, *Watts* 1 (K; SRGH). SE: Seleka Ranch, 22.ii.1977, *Hansen* 3046 (C; GAB; K; PRE; SRGH; UPS). **Zimbabwe**. E: Lower Sabi, E. bank, 28.i.1948, *Wild* 2313 (K; SRGH). S: Beitbridge, Shashe R. flood plain, 7.ii.1961, *Wild* 5310 (K; LISC; SRGH). W: Bulawayo, i.1898, *Rand* 203 (BM). **Mozambique**. GI: 5 km. W. of Zinave Camp at Banhine Vlei, vi.1973, *Tinley* 2946 (SRGH). M: Sábiè, near Mahel, 30.xi.1944, *Mendonça* 3155 (C; K; LISC; LMU).

Elsewhere in Tanzania, Namibia and S. Africa. In alluvial depressions in mopane woodland or on flood-plains, in wooded grassland or in one case recorded from calcium-rich clay along the margins of seasonal drainage line; 515–850 m.

Leaf form and indumentum is very variable in this species, as is the development of the spines of the sterile modified flowers.

7. **Cyathula natalensis** Sond. in Linnaea **23**: 97 (1850); in J.H. Ross Fl. Natal: 158 (1973). TAB. **19** fig. A. Type from Natal.

Cyathula spathulifolia Lopr. in Engl. Bot. Jahrb. **27**: 54 (1899); in Malpighia **14**: 444 (1900); in J.H. Ross Fl. Natal: 158 (1973). Type from Natal.

Perennial herb, much-branched with usually weak and slender branches, scrambling over other vegetation or prostrate and more or less matted, 0.6–1.4 m. (and probably more); stems thinly pilose and terete-striate, the upper part and branches more densely pilose or lanate, more or less tetragonous with pale angles above, purplish to green, nodes somewhat swollen. Leaves 1.7–7.5 × 1–4 cm; of highly characteristic shape, spathulate with the broadest part of the lamina abruptly narrowed above to a short, acute to rather blunt apex, and narrowed below to a broad claw which contracts abruptly into a short, 1–5 mm. long, usually densely pilose petiole; upper leaf surface darker with scattered long hairs and usually rather few short hairs, lower surface paler with long hairs along the primary venation and usually more plentiful shorter and finer hairs between. Inflorescences of dense heads (condensed cymes) terminal on the stem and branches, white-woolly or greenish, solitary or in a thyrse of up to 6 heads, each head c.1–1.5 cm. in diam. at anthesis, subsessile or on a peduncle up to c.5 mm. long; bracts broadly cordate-ovate, ciliate, whitish-membranous, c.3–4 mm. long, distinctly (usually brownish) aristate with the excurrent midrib; bracteoles lanceolate, c.4–5 mm. long, with long multicellular hairs at least along the margins and dorsal surface of midrib, which is excurrent in a brownish arista; ultimate divisions of the inflorescence of a central fertile flower subtended on each side by a triad of one fertile and two lateral modified flowers. Outer 2 tepals oblong-elliptic to lingulate, 5–6 mm., moderately to densely furnished with long, soft, white multicellular hairs, with broad hyaline margins and a green central band with a pair of veins on each side of the shortly excurrent midrib; inner 3 tepals scarcely shorter, glabrous or at least less densely pilose than the outer, wider, more narrowly hyaline-margined, the usually single pair of lateral nerves anastomosing into a reticulum above. Modified flowers scarcely visible as more than a minute bud at anthesis, rapidly accrescent, of mostly 4–6 bracteoliform and spiniform processes, all with uncinate apices, the outer with short, broad, hyaline bases; very reduced innermost processes also present. Stamens finally subequalling or c. three quarters the length of the tepals, filaments slender; pseudostaminodes c.1–1.5 mm., oblong to flabellate, finely lanate along the margins and across the back, with a short incurved ventral scale. Style slender, c.3–3.5 mm.; ovary obovoid-pyriform, c.0.75 mm., thickened above. Capsule c. 3 mm., oblong ellipsoid, firm; seed c. 2.75 mm., oblong-ellipsoid, brown, almost smooth. "burr" spherical, very prehensile, c. 1.5–3 cm. in diam., including the spines.

Zimbabwe. W: Gwampa Forest Reserve, ii.1956, *Goldsmith* 66/56 (K; LISC; SRGH). E: Gungunyana Forest Reserve., Chirinda Forest, 29.iv.1971, *Goldsmith* 14/71 (K; LISC; MO; PRE; SRGH). S: near Madzivire Dip, c. 6 km. N. of Lundi R. bridge, 3.v.1962, *Drummond* 7889 (K; LISC; SRGH). **Mozambique**. MS: Maringua Distr., Sabi R., 24.vi.1950, *Chase* 2556 (BM; SRGH). GI: Chibuto, estrada near Alto Changoene, 12.ii.1959, *Barbosa & Lemos* 8378 (COI; K; LMJ; SRGH). M: Maputo 46 m., 7.xii.1897, *Schlechter* 11640 (COI; K).

Also in Natal. In open forests and orchards, and on sandy beaches; sea level–910 m. Apparently a local species.

12. PUPALIA A. Juss.

Pupalia A. Juss. in Ann. Mus. Nation. Hist. Nat. Paris 2: 132 (1803) nomen cons.
Pupal Adans., Fam. Pl. 2: 268, 596 (1863).

Annual or perennial herbs or subshrubs with entire, opposite leaves. Inflorescence a spiciform bracteate thyrse terminal on the stem and branches, each bract subtending a single hermaphrodite flower on each side of which is set a bracteolate modified flower consisting of a number of sharply hooked spines, or, more commonly, each bract containing a hermaphrodite flower sub tended by two or more such triads each contained with a large bracteole. Spines of modified flowers at first very small, rapidly accrescent, finally disposed in 3 or occasionally more stalked clusters of 5—20 spines in 1—3 ranks, the clusters stellately or occasionally dendroidly set on a common stalk, subequalling to much exceeding the perianth and serving as a means of distributing the fruit. Bracts persistent, finally more or less deflexed, entire partial inflorescences falling intact in fruit. Perianth segments 5. Stamens 5, the filaments delicate to rather solid, fused at the extreme base on to a fleshy, lobed, disk-like cup into which the base of the ovary is narrowed; pseudostaminodes absent; anthers bilocular. Style slender, stigma capitate. Ovary with a single,pendulous ovule. Fruit a thin-walled capsule irregularly ruptured below the firm apex by the developing seed. Seed oblong-ovoid or ovoid, slightly compressed. Endosperm copious.

4 species in the tropics (extending to the subtropics) of the Old World from W. Africa to Malaysia and the Philippines.

1. Bracts each subtending a single fertile flower, i.e. one triad of one fertile and two modified sterile flowers; tepals 2.75—3.5 mm. - - - - - 1. *micrantha*
— Most or all of the bracts subtending more than one fertile flower; tepals only exceptionally as short as 3.5 mm. - - - - - - - - - - - - 2
2. Tepals narrowly oblong-lanceolate, (6)7—8 mm.; style long and slender, (2.75)3—3.5 mm. - - - - - - - - - - - - - - - 2. *grandiflora*
— Tepals oblong-ovate, (3)4—5 (6) mm.; style shorter, very rarely attaining 3 mm. and mostly less than 2 mm. - - - - - - - - - - - - 3. *lappacea*

1. **Pupalia micrantha** Hauman in Bull. Jard. Bot. Brux. **18**: 109 (1946); in F.C.B. 2: 61 (1951).—Townsend in Kew Bull. **34**: 133 (1979). TAB. **20** fig. C. Type from Zaire.
Pupalia psilotrichoides Suesseng. in Mitt. Bot. Staatss. München **1**: 64 (1950). Type: Mozambique, Namagoa, *Faulkner* 14 (K, holotype).
Cyathula prostrata var. *grandiflora* Suesseng., op. cit.: 77 (1951). Type from Tanzania.
Digera alternifolia sensu F.W.T.A. ed. 2, 1: 148 (1954) non. *Digera muricata* (L.) Mart. (= *alternifolia* (L.) Aschers.).

Annual herb c. 0.5—1.5 m. tall, bushy or straggling with many divaricate branches from the base upwards (small plants simple), stem and branches slender, terete, striate, thinly or in parts more densely furnished with whitish multicellular hairs, the older basal internodes finally glabrescent. Leaves lanceolate-ovate to elliptic, acuminate, those of the stem and main branches 5.5—15 × 2.9—5.4 cm. including the 0.5—1.8 cm. petiole, dark green and thinly rather long-pilose on the superior surface, paler and more densely and shortly pubescent beneath, subtruncate to cuneate at the base; leaves of upper part of stem and branches rapidly reducing in size. Inflorescences considerably elongating as the flowers open and finally up to c. 25 cm. long including the (to c.8 cm.) peduncle, solitary or paired in the leaf axils, up to 4—5 together at the ends of the stem and branches, simple or sometimes branched, axis thinly to densely pilose. Bracts lanceolate-ovate, 1—1.5 mm. long, membranous-margined, persistent, more or less thinly pilose, each subtending a single triad of one fertile and two modified bracteolate flowers. Bracteoles broadly deltoid-ovate, c. 2 mm. long, abruptly shortly acuminate with a short, sharp mucro formed by the excurrent midrib, broadly membranous-margined below, moderately pilose. Tepals 2.75—3.5 mm. long, oblong-ovate, broadly white-margined, 3-nerved in the green centre with the nerves confluent above to form a short, sharp mucro; outer 2 tepals slightly longer and broader, lanate chiefly about the base, the inner 3 lanate over most of the dorsal surface. Style short,

c. 0.5 mm. long. Sterile flowers of 3 branches each bearing c. 5—8 hooked setae to c. 2 mm. long, not forming a dense "burr" or concealing the fertile flower. Fruit an ovoid, somewhat compressed capsule c. 1.75—2 mm. long, rupturing irregularly at the thin-walled base; seed c.1.5—1.75 mm. in diam., circular, somewhat compressed, black, almost smooth, shining.

Botswana. N: Samocima, common, 24.iv.1975, *Biegel, Müller & Gibbs-Russell* 4979 (K; SRGH). **Zambia**. B/S: Machili, 10.iii.1961, *Fanshawe* 6408 (K; NDO). N: Samfya, 9.v.1958, *Fanshawe* 4416 (K; NDO). W: Ndola District, Sacred Lake, near St. Anthony's Mission c. 48 km. SW. of Luanshya, c.1200 m., 14.ii.1975, *Hooper & Townsend* 29 (K; LUS; PRE). C: Katondwe, 23.ii.1965, *Fanshawe* 9139 (K; NDO). S: near Zoo, in centre of Livingstone Game Park, c.910 m., 1975, *Hooper & Townsend* 1049 (K; LUS). **Malawi**. S: Ntcheu Distr., Dombole-Livulezi confluence, 31.i.1968, *Salubeni* 950 (K; SRGH). **Mozambique**. N: Cabo Delgado, c.3 km. from Montepuez near Nantulo, 7.iv.1964, *Torre & Paiva* 11705 (LISC). Z: Quelimane Distr., Namagoa, vii.1947, *Faulkner* Kl4 (K). MS: Gorongosa, Parque Nacional de Caca, 30.iv.1964, *Torre & Paiva* 12172 (LISC; SRGH).

Also in Nigeria, Ivory Coast, Zaire, Tanzania and Madagascar. Usually in lightly to densely shaded thickets or forests, often not far from water, also in tall *Rottboellia* grassland and as a ruderal; 40—1200 m.

2. **Pupalia grandiflora** Peter in Fedde, Repert. Beih. **40**, 2, Descriptiones: 22 (1932).— Hauman in F.C.B. **2**: 61 (1951).—Townsend in Kew Bull. **34**: 134 (1979). TAB. **20** figs. A5—6, B. Type from Tanzania.

Perennial (?) herb. often rather woody at the base, scandent or more rarely erect, 1—2(4) m.. much-branched, stem and branches weak, terete, striate, thinly to moderately furnished with whitish multicellular hairs, the older basal internodes finally glabrescent. Leaves lanceolate to broadly ovate, acuminate, those of the stem and branches 3.2—14 × 2.2—6 cm. including the 1—2.5 cm. petiole, dark green and thinly rather long-pilose on the upper surface, paler and more densely and shortly pubescent beneath (rarely tomentose on the midrib and principal veins), rounded to cuneate at the base; superior leaves of stem and branches rapidly reducing in size. Inflorescences thyrsoid, considerably elongating as the flowers open and finally up to 35(48) cm. long including the (up to 9 cm.) peduncle, solitary and terminal on the stem and branches, axis moderately spreading-pilose or densely tomentose. Bracts lanceolate, 3—4 mm., darkly membranous-margined, persistent, moderately pilose, each subtending a partial inflorescence of 3—7 fertile flowers, most of which are set between two modified sterile flowers, but the central solitary. Bracteoles of triads of 1 fertile and 2 sterile flowers broadly deltoid-ovate, c.4 mm. long, abruptly shortly acuminate with a sharp yellowish to dark mucro formed by the excurrent midrib, broadly membranous-margined below, moderately densely pilose dorsally. Bracteoles of sterile flowers ovate-lanceolate, c.4 mm. long, membranous with a green midrib which is excurrent in a distinct brownish arista, thinly to moderately pilose. Tepals (6)7—8 mm. long, 3-nerved in the green centre with the nerves confluent above to form a short, sharp mucro, narrowly oblong-lanceolate; outer 2 tepals slightly longer, more or less uniformly long-pilose, narrowly membranous-margined, the inner 3 more broadly pale-margined (not conspicuously so since the margins are incurved), more densely long-pilose. Style long and slender, (2.75)3—3.5 mm. long. Sterile flowers dendroidly branched with several divaricate branches each ending in (6)9—15(20) hooked setae up to c. 6 mm. long, usually brownish but occasionally yellow, forming a very dense globose "burr" c. 1.5—2.2 cm. in diam., concealing the fertile flowers. Fruit an oblong-ovoid capsule 2—2.25 mm. long, rupturing irregularly at the thin-walled base; seed c. 2 mm. long, ovoid, black, almost smooth, shining.

Mozambique. N: Malema, eastern slope of Serra Inago, 900 m., 20.iii.1964, *Torre & Paiva* 11291 (BM; LISC).

Also in Uganda, Kenya, Tanzania, Zaire, Rwanda, Sudan and Ethiopia. The Mozambique locality is in dry montane *Newtonia buchananii* forest on dark clay soil, elsewhere on forest edges, in rides and clearings, also in open woodland and bush and scrambling over rocks and shrubs along riversides.

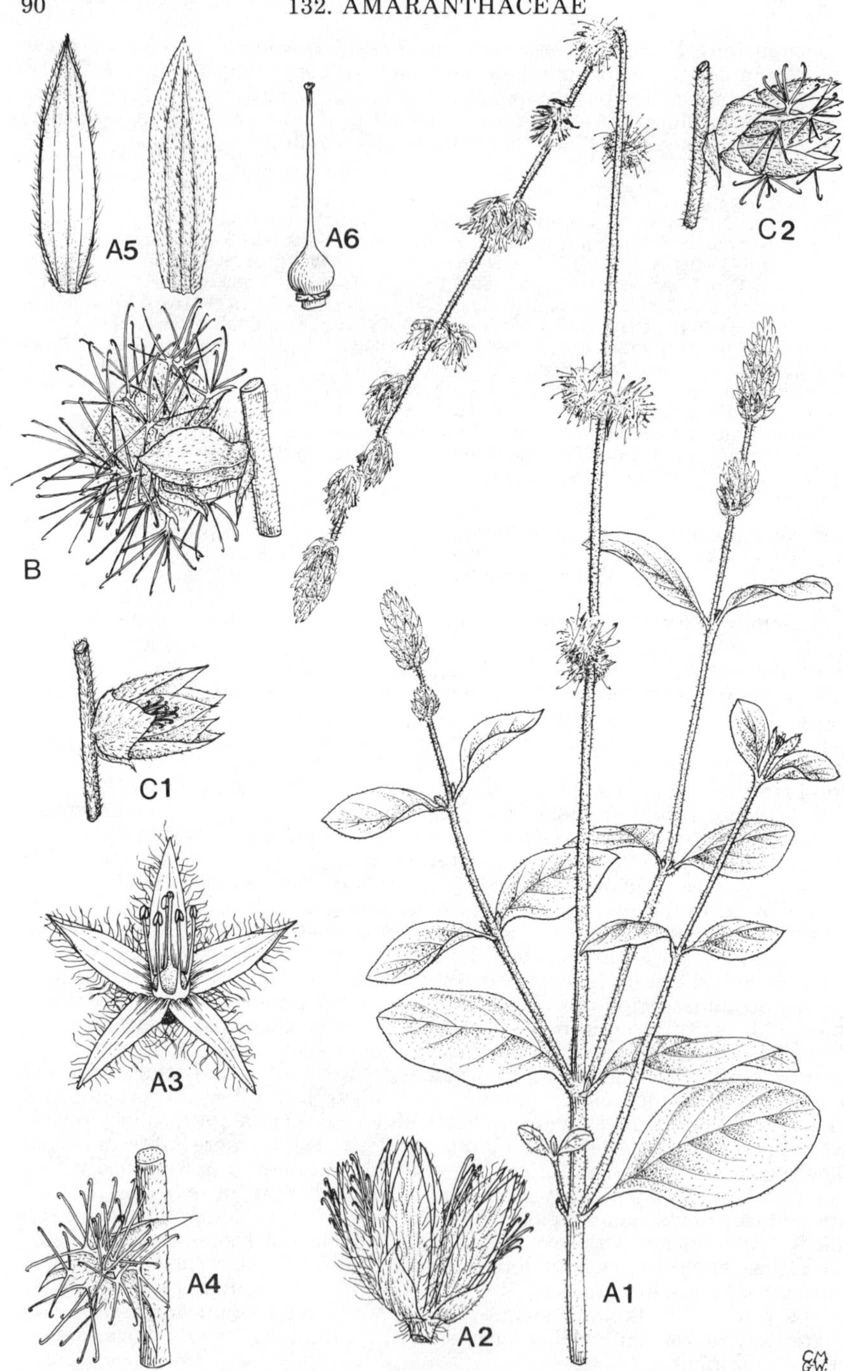

Tab. 20. A.—PUPALIA LAPPACEA VAR. VELUTINA. A1, flowering branch (×⅔) *Thomas* 3416; A2, triad of one fertile and two sterile flowers (×4); A3, flower with spread perianth (×4), A2—3 from *Gillett* 12888; A4, fruiting triad (×2) *Thomas* 3416. PUPALIA GRANDIFLORA. A5, outer tepals, ventral and dorsal surface (×4); A6, gynoecium (×4), A5—6 from *Mshana* 221. B1, fruiting partial inflorescence (×2) *Drummond & Hemsley* 1393. C.—PUPALIA MICRANTHA. C1, young triad of one fertile and two sterile flowers (×4); C2, fruiting triad (×4), C1—2 from *Fanshawe* 9139.

3. **Pupalia lappacea** (L.) A. Juss. in Ann. Mus. Hist. Nat. Paris **2**: 132 (803).—Baker & Clarke in F.T.A. **6**, 1: 47 (1909).—Schinz in Engl. & Prantl Pflanzenfam. ed. 2, **16** C: 48 (1934).—Hauman in F.C.B. **2**: 60 (1951).—White, F.F.N.R.: 45 (1962).—Cavaco in Mem. Mus. Hist. Nat. Paris Sér. B, **13**: 90 (1962).—Podlech & Meeuse in Merxm. Prodr. Fl. SW. Afr. **33**: 21 (1966); in J.H. Ross Fl. Natal: 158 (1973).—Townsend in Kew Bull. **34**: 135 (1979). TAB. **20**. Type from Sri Lanka.

Annual or perennial herb, more or less erect and c. 0.3—0.9 m. tall, or prostrate and sprawling, or subscandent and scrambling to as much as 2.5 m., stem generally much-branched and swollen at the nodes, branches opposite, divaricate or ascending, slender; stems and branches obtusely 4-angled to almost terete, thinly pilose to densely tomentose. Leaves variable to shape and size, from narrowly ovate-elliptic to oblong or circular, 2 10(14) × 1—5(7) cm., acuminate to obtuse-apiculate or retuse at the summit, shortly or more longly cuneate at the base, narrowed to a petiole 0.5—2.5(3.5) cm. long; indumentum of lamina varying from sericeous or tomentose to subglabrous with a few hairs running vertically along the lower surface of the primary venation, rarely quite glabrous, commonly moderately pilose with the hairs along the nerves divergent. Inflorescences at first more or less dense, elongating to as much as 0.5 m. in fruit with the lower flowers becoming increasingly remote, axis subglabrous to tomentose, peduncle c. 1—10 cm. long; bracts lanceolate, 1.5—2.5 mm. long, persistent, more or less deflexed after the fall of the fruit, subglabrous or pilose, sharply mucronate with the percurrent midrib; partial inflorescences mostly of one solitary hermaphrodite flower subtended on each side by a triad of one hermaphrodite and 2 modified flowers; bracteoles of each triad broadly subcordate-ovate, (2.75)3—5(6) mm., abruptly narrowed to the stramineous to dark arista forming by the excurrent midrib, membranous with a pale margin, thinly to very densely hairy; bracteoles of sterile flowers ovate-lanceolate, usually more shortly and less densely pilose. Tepals oblong-ovate to lanceolate-ovate, more or less quickly narrowed to a rather obtuse mucronate apex to gradually narrowed and acute-aristate, the outer two (3.5)4—5(6) mm. long, subglabrous to more or less tomentose dorsally, 3(5)-nerved, the midrib and 2 inner nerves confluent just below the apex and excurrent in the mucro or short arista, inner 3 slightly shorter and more densely pilose. Branches of sterile flowers 3, each terminating in (3)5—18(20) mm. long setae in 1—3 ranks; setae subglabrous to more or less villous in the lower half, yellowish to purple or red, (1.5)3—7 mm., the partial inflorescence falling intact to form a burr c. 8—18 cm. in diam. Style short to rather slender, (0.5)0.9—2(3) mm. long. Capsule ovoid. 2—2.5 mm. long. Seed oblong-ovoid with a prominent radicle, 2 mm. long, dark brown, shining, testa at first faintly reticulate but finally smooth or punctulate.

Widespread in the tropics of the Old World; throughout tropical Africa N. to Egypt, also in S. Africa and Madagascar, Arabia and Asia from India eastwards to Malaya, the Malayan Is. (Java, Celebes etc.), the Philippines and New Guinea. Introduced in Australia and elsewhere.

1. Tepals subglabrous in the upper half, with scattered hairs only except for lanate tufts at the basal angles, broad and rather blunt, not rarely with more than 3 nerves; vegetative parts glabrous or very sparingly hairy. Coastal - var. *glabrescens*
— Tepals more or less pilose throughout, moderately to very acute, 3-nerved; vegetative parts sparingly to very densely hairy - - - - - - - - - - 2
2. Plant constantly green, leaves thinly hairy above, paler and rather more densely so beneath; inflorescence slender, rather spiky from the more gradually narrowed, more or less acute tepals; setae of each branch of the sterile flowers numerous, rarely less than 8 at least on the middle branch, commonly 12 or more - var. *lappacea*
— Plant and indumentum variously coloured but the latter usually obvious; inflorescence generally stouter, less spiky, the tepals less sharply tapering; setae variable but mostly less than 8 even on the middle branch - - - - - - - var. *velutina*

Var. **lappacea**

Achyranthes atropurpurea Lam., Encycl., Méth. Bot.: 546 (1785). Type: specimen grown in Royal Gardens at Paris, Hb. Lamarck (P-LA, lectotype, IDC. microfiche of Lamarck Herbarium neg. 546 no. 4!).

Pupalia atropurpurea (Lam.) Moq. in DC., Prodr. **13**, 2: 331 (1849).—Baker & Clarke in F.T.A. **6**, 1: 48 (1909).—Schinz in Engl. & Prantl Pflanzenfam. ed. 2, **16** C: 48 (1934).—J.H. Ross Fl. Natal: 158 (1973).

Characters as in the key to varieties, leaves lanceolate or ovate-lanceolate; outer tepals usually 4–5 mm. long; setae of the sterile flowers reddish to yellow; inflorescence axis never densely hairy.

Mozambique. N: mouth of Rovuma R., 29.iii.1861, *Meller* s.n. on Livingstone's Zambezi Expedition (K). T: between Lupata and Tete, 7.ix.1859, *Kirk* s.n. on Livingtone's S. African Expedition (K). MS: Inhamitanga, 19.ii.1948, *Andrada* 1076 (LISC).GI: Gaza, Xai Xai, near the coast, Sepülveda, 14.viii.1957, *Barbosa & Lemos* 7840 (COI; K; LISC). M.: Inhaca Is., E. of Maputo, 11.ix.1964, *Mogg* 31747 (K).

Also occurs in S. Africa, S. India and Sri Lanka. In littoral woodland, sometimes ruderal there, or by riverside, on sandy or sandy-clay soil; sea level–200 m.

Intermediates between var. *lappacea* and var. *velutina* occur in Mozambique, and one has been seen from the Mutare region of Zimbabwe (*Eyles* 7078 E; K).

Var. **velutina** (Moq.) Hook. f. in Fl. Brit. India **4**: 724 (1885).—Townsend in Kew Bull. **34**: 138 (1979). TAB **20** fig. A1–4. Type from Burma.

Desmochaeta flavescens DC., Cat. Hort. Monsp: 102 (1813) nomen illegit.

Achyranthes mollis Thonn. in Schum., Beskr. Guin. Pl.: 137 (1827). Type from Ghana.

Pupalia mollis (Thonn.) Moq. in DC., Prodr. **13**, 2: 333 (1849). Type as above.

Pupalia velutina Moq. in DC., Prodr. **13**, 2: 332 (1849). Type from Burma.

Pupalia distantiflora A. Rich., Tent. Fl. Abyss. 2: 217 (1850). Type from Ethiopia.

Pupalia tomentosa Peter in Fedde, Repert. Beih. **40**, 2, Descriptiones 23 (1932). Type from Tanzania.

Pupalia lappacea var. *tomentosa* (Peter) Suesseng. in Mitt. Bot. Staatss. München 2: 31 (1954). Type as above.

Plant very variable in leaf shape and indumentum, from thinly pilose to densely villous or tomentose, but rarely silvery- sericeous. Inflorescence axis thinly hairy to densely furnished with patent or deflexed hairs. Outer tepals (3.5)4–5(6) mm. long, oblong-ovate, rather quickly narrowed to the mucronate apex, more or less densely pilose to lanate even in forms with thinly pilose foliage, 3-nerved. Branches of the sterile flowers terminated by (4)5–8(13) usually stramineous uncinate setae up to 7 mm. long.

Botswana. N: Maun, c.300 m. from river, ii.1967, *Lambrecht* 41 (K; PRE). SW: Kgalagadi, Ditatso Pan, c. 40 km. NW. of Tsabong, 940 m., 25.ii.1963, *Leistner* 3087 (K; SRGH). SE: Boteti delta area, NE. of Mopipi, 850 m., 18.iv.1973, *Standish-White* 22 (K: SRGH). **Zambia**. B: Sesheke Distr., near. Kazu Forest, Machili, 20.xii.1952, *Angus* 984 (FHO; K). N: Mbala Distr., Crocodile Isl., Lake Tanganyika, 780 m., 9.ii.1964, *Richards* 18976 (K). C: Luangwa Valley Game Reserve South, 5 km. S. of Lubi R. on Rd., 610 m., 6.ii.1967, *Prince* 130 (K). E: Beit Bridge, Luangwa R., 22.x.1967, *Mutimushi* 2171 (K; NDO). S: 1.6 km. S. of Mambo's Village, Gwembe Valley, 29.iii.1952, *White* 2359 (FHO; K). **Zimbabwe**. N: Gokwe Distr., Chief Nemangwe's area, near Sanyati Mission, 16.iii.1964, *Bingham* 1161 (K; SRGH). W: Lochview, Bulawayo, 1.xi.1975, *Cross* 268 (K; SRGH). C: Sebakwe R. near. Kwe Kwe 1380 m., 6.iii.1961, *Richards* 14541 (K). E: Inyanga, ad villam Cheshire, c.300 m., 15.i.1931, *Norlindh & Weimarck* 4372 (K; SRGH). E: Lower Sabi, Rupisi Hot Springs, 520 m., 28.i.1948, *Wild* 2305 (K; SRGH). **Malawi**. N: 32 km. S. of Karonga, 25.iv.1975, *Pawek* 9536 (K; MAL; MO; SRGH; UC). S: Tumbi Is. E., Monkey Bay, 470–500 m., 1.iii.1970, *Brummitt & Eccles* 8826 (K; LISC; MAL; PRE; SRGH). **Mozambique**. N: Angoche 22.i.1968, *Torre & Correia* 17328 (LISC). Z: Maganja da Costa, forest of Gobene, 12.ii.1966, *Torre & Correia* 14556 (LISC). MS: SW. of Beira road and Chicamba road junction, 700 m., 11.iii.1962, *Chase* 7655 (K; LISC; SRGH). GI: de Mabalane Mabote, Papai, 5.vi.1959, *Barbosa & Lemos* 8627 (COI; LISC).

The commonest variety, found practically throughout the range of the species. In a great variety of habitats, commonly in various types of woodland (light to very dense shade), grassland, thickets, roadsides, as weed of cultivation, on rock, hillsides, on sandy or pebbly lake margins etc., usually on light sandy or alluvial soils; 100–1370 m.

Var. **glabrescens** Townsend in Kew Bull. **34**: 139 (1979). Type from Zanzibar.

Pupalia atropurpurea (Lam.) Moq. in DC., Prodr. **13**, 2: 331 (1849) quoad pl. Pembae.

Vegatative parts of plant glabrous or very sparingly shortly hairy, dull green or blackish when dry. Stem usually red, at least about the nodes. Inflorescence axis glabrous or sparingly pilose. Outer 2 tepals 4.5–5 mm. long, rather broadly oblong-ovate, with tufts of lanate hairs at the basal angles. dorsal surface glabrous or sparingly pilose in the lower half, occasionally 4–5-nerved. Branches of the sterile flowers terminated with 5–11 uncinate setae up to 3 mm. long. Burrs 8–9 mm. in diam.

Mozambique. N: Goat Isl., 19.v.1961, *Leach & Rutherford-Smith* 10921 (K; SRGH). Elsewhere in Kenya, Tanzania, Zanzibar and Pemba. Habitat in Mozambique not recorded, elsewhere on sandy ground by the sea shore and a little inland, coral cliffs, arable land and waste places; sea level–6 m.

13. AERVA Forssk.

Aerva Forssk., Fl. Aegypt.-Arab.: 170 (1775) nomen cons.
Ouret Adans., Fam. Pl. 2: 268, 586 (1763).

Perennial herbs (sometimes flowering in the first year), prostrate to erect or scandent. Leaves and branches opposite or alternate, leaves entire. Flowers hermaphrodite or dioecious, sometimes probably polygamous, bibracteolate, in axillary and terminal sessile or pedunculate bracteate spikes, one flower in the axil of each bract. Perianth segments 5, oval, or lanceolate-oblong, membranous margined with a thin to wider green centre, the perianth deciduous with the fruit but bracts and bracteoles persistent. Stamens 5, shortly monadelphous at the base, alternating with subulate or rarely narrowly oblong and truncate or emarginate pseudostaminodes; anthers bilocular. Ovary with a single pendulous ovule; style very short to slender and distinct; stigmas 2, short to long and filiform (sometimes solitary and capitate, flowers then probably functionally male). Capsule thin-walled, bursting irregularly. Seed compressed-reniform, firm, black.

About 10 species in the tropics, chiefly centred on Africa.

1. Outer 2 tepals 2–3 mm. long, the midrib ceasing well below the apex; individual spikes up to c. 10 cm. in length - - - - - - - - - - 1. *javanica*
– Outer 2 tepals 0.75–1.5 (2) mm. long, the midrib excurrent in a short mucro; individual spikes mostly up to 4 (rarely to 6) cm. in length - - - - - - - - - 2
2. Spikes sessile, axillary, solitary or in clusters, forming an elongate inflorescence which is leafy to the apex of the stem and branches - - - - - - 2. *lanata*
– Spikes sessile or shortly pedunculate, towards the ends of the stem and branches forming leafless terminal panicles - - - - - - - - - - 3. *leucura*

1. **Aerva javanica** (Burm. f.) Juss. ex J.A. Schultes, Syst. Veg. ed. **16**, 5: 565 (1819).—Cavaco in Mém. Mus. Hist. Nat. Paris Sér. B, **13**: 100 (1962). TAB. **21** fig. B. Type, Hb. Burmann (G, holotype). Alleged to have originated in Java, but this is certainly an error, the species is unknown there.

Celosia lanata L., Sp. Pl. ed. 1: 205 (1753). Type from Sri Lanka.

Iresine javanica Burm. f., Fl. Ind.: 212 (sphalm. 312), t. 65 f. 2 (1768).

Iresine persica Burm. f., loc. cit. t. 65 f. 1 (1768).—Schinz in Engl. & Prantl Pflanzenfam. ed. 2, **16 C**: 51 (1934). Type, Hb. Burmann, Persia (G, holotype).

Aerva tomentosa Forssk., Fl. Aegypt.-Arab.: cxxii, 170 (1775).—Baker & Clarke in F.T.A. **6**, 1: 37 (1909). Type from Egypt.

Aerva persica (Burm. f.) Merrill in Philipp. J. Sci. **19**: 348 (1921). Type as for *Iresine persica*.

Perennial herb, frequently woody and suffruticose or growing in erect clumps, 0.3–1.5 m., branched from about the base with simple stems or the stems with long ascending branches. Stem and branches terete, striate, more or less densely whitish or yellowish-tomentose or pannose, when dense the indumentum often appearing tufted. Leaves alternate, very variable in size and form, from narrowly linear to suborbicular, more or less densely whitish or yellowish-tomentose but usually more thinly so and greener on the superior surface, margins plane or more or less involute (when strongly so the leaves frequently more or less falcate-recurved), sessile or with a short and indistinct petiole or the latter rarely to c. 2 cm. long in robust plants. Flowers dioecious. Spikes white to creamy-pink, sessile, cylindrical, dense and stout (up to c. 10 × 1 cm.), to slender and interrupted with lateral globose clusters of flowers and with some spikes apparently pedunculate by branch reduction; male plants always with more slender spikes (but plants with slender spikes may not always be male); upper part of stem and branches leafless, the upper spikes thus forming terminal panicles; bracts 0.75–2.25 mm. long, broadly deltoid-ovate, hyaline, acute or obtuse with the obscure midrib ceasing below the apex, densely lanate throughout or only about the base or apex, persistent; bracteoles similar, also persistent. Female flowers with outer 2 tepals 2–3 mm. long, oblong-obovate to obovate-spathulate, lanate, acute to obtuse or apiculate at the apex, the yellowish midrib ceasing well below the apex; inner 3 slightly shorter, elliptic-oblong, more or less densely lanate, acute, with a narrow green vitta along the midrib, which extends for about two-thirds the length of each tepal; style slender, distinct, with the two filiform, flexuose stigmas at least equalling it in length, style and stigmas together c. 1–1.5 mm. long; filaments very reduced, anthers absent. Male flowers smaller, the outer tepals 1.5–2.25 mm. long, ovate; filaments delicate, the anthers about equalling the perianth; ovary small, style very short, stigmas rudimentary. Capsule 1–1.5 mm. long, rotund, compressed. Seed 0.9–1.25 mm. in diam., circular, slightly compressed, brown or black, shining and smooth or very faintly reticulate.

Botswana. N: Toromoja, 21.iv.1971, 912 m., *Thornton* 2 (K; LISC; SRGH). SW: Kuke, 22.ii.1970, 912 m., *Brown* 8693 (K). SE: Boteti delta area, NE. of Mopipi, 18.iv.1973, 850 m., *Fry* 2 (K; SRGH). **Malawi**. S: Chiromo, 26.iv.1933, 61 m., *Lawrence* 90 (K). **Mozambique**. N: Pemba, 5 km. S. of the town, 14.iii.1960, *Gomes & Sousa* 4532 (COI; K). T: Baroma Distr., Msusa, 212 m., 25.vii.1950, *Chase* 2790 (BM; SRGH). MS: Chemba, Chiou, Estação Experimental do C.I.C.A., 12.iv.1960, *Lemos & Macuacua* 78 (COI; K; LMA; LMJ; SRGH).

Widespread in the drier parts of the tropics and subtropics of the Old World from Morocco south to Cameroon, across the drier regions of Africa to Egypt, Sudan and Somalia south to Madagascar, over Asia from Palestine and Arabia to Burma, India and Sri Lanka. Adventive in Australia and elsewhere.

2. **Aerva lanata** (L.) Juss. ex J.A. Schultes, Syst. Veg. ed. 16, 5: 564 (1819).—Baker & Clarke in F.T.A. **6**, 1: 39 (1909).—Schinz in Engl. & Prantl Pflanzenfam. ed. 2, **16** C: 52 (1934).—Hauman in F.C.B 2: 57 (1951).—Cavaco in Mém. Mus. Hist. Nat. Paris Sér. B, **13**: 103 (1962).—J.H. Ross Fl. Natal: 159 (1973).—Townsend in Kew Bull. **29**: 461 (1974). Type, Linnean specimen 290/6 (LINN, lectotype).

Achyranthes lanata L., Sp. Pl.: 204 (1753). Type as above.

Illecebrum lanatum (L.) L., Mant. Alt.: 344 (1771). Type as above.

Achyranthes villosa Forssk., Fl. Aegypt.-Arab.: 48 (1775). Type from Egypt.

Aerva mozambicensis Gandoger in Bull. Soc. Bot. Fr. **66**: 233 (1919). Type: Mozambique (no further details), *Carvalho* s.n. (LY, holotype; COI, isotype).

Aerva lanata var. *citrina* Suesseng. in Mitt. Bot. Staatss. München **1**: 66 (1950). Type: Zimbabwe, Mutare, *Chase* 176 (K; M; SRGH, isotype).

Aerva lanata var. *intermedia* Suesseng. in Bull. Jard. Bot. Brux. **15**: 57 (1938). Type from Zaire.

Aerva lanata var. *leucuroides* Suesseng., loc. cit. Type from Zaire.

Aerva sansibarica Suesseng. in Kew Bull. **1949**: 475 (1950). Type from Zanzibar.

Aerva lanata var. *leucuroides* Suesseng. in Mitt. Bot. Staatss. München **1**: 70 (1951) nom. illegit. Type: Zimbabwe, Marondera *Wild* 3304 (K, isotype).

Perennial herb, sometimes flowering in the first year, frequently more or less woody and suffrutescent below, erect to prostrate, decumbent or occasionally somewhat scandent, stiff or weak and straggling, (0.1) 0.3–2 m., with numerous stems from the base and these frequently also branched above. Stem and branches terete, striate, more or less densely lanate with whitish or yellowish, more or less shaggy hairs, more rarely tomentose or canescent. Leaves alternate, orbicular (not seen in the Flora Zambesiaca area) to lanceolate, spathulate or elliptic-ovate, shortly or more longly cuneate at the base with petioles from c. 2 cm. to obsolete, rounded and apiculate to acute at the apex, commonly densely lanate or canescent on the inferior surface and more thinly so above, rarely glabrous or (not seen in Flora Zambesiaca area) thickly lanuginose, those of the main stems c. 10–50 × 5–35 mm., those of the branches and inflorescence reducing and frequently becoming very small. Spikes sessile, solitary or usually in axillary clusters on the main stems or long to very short axillary branches, 0.4–1.5(2) × 0.3–0.4 cm., divergent, more or less cylindrical, silky, white to creamy, forming a long inflorescence which is leafy to the extreme apex and forms no terminal panicle even on the main stems; bracts (0.75) 1–1.25 mm., deltoid-ovate to oblong-ovate, hyaline with a short but distinct arista formed by the excurrent midrib, pilose, persistent; bracteoles similar or slightly smaller, also persistent. Flowers female, male or hermaphrodite. Tepals more or less

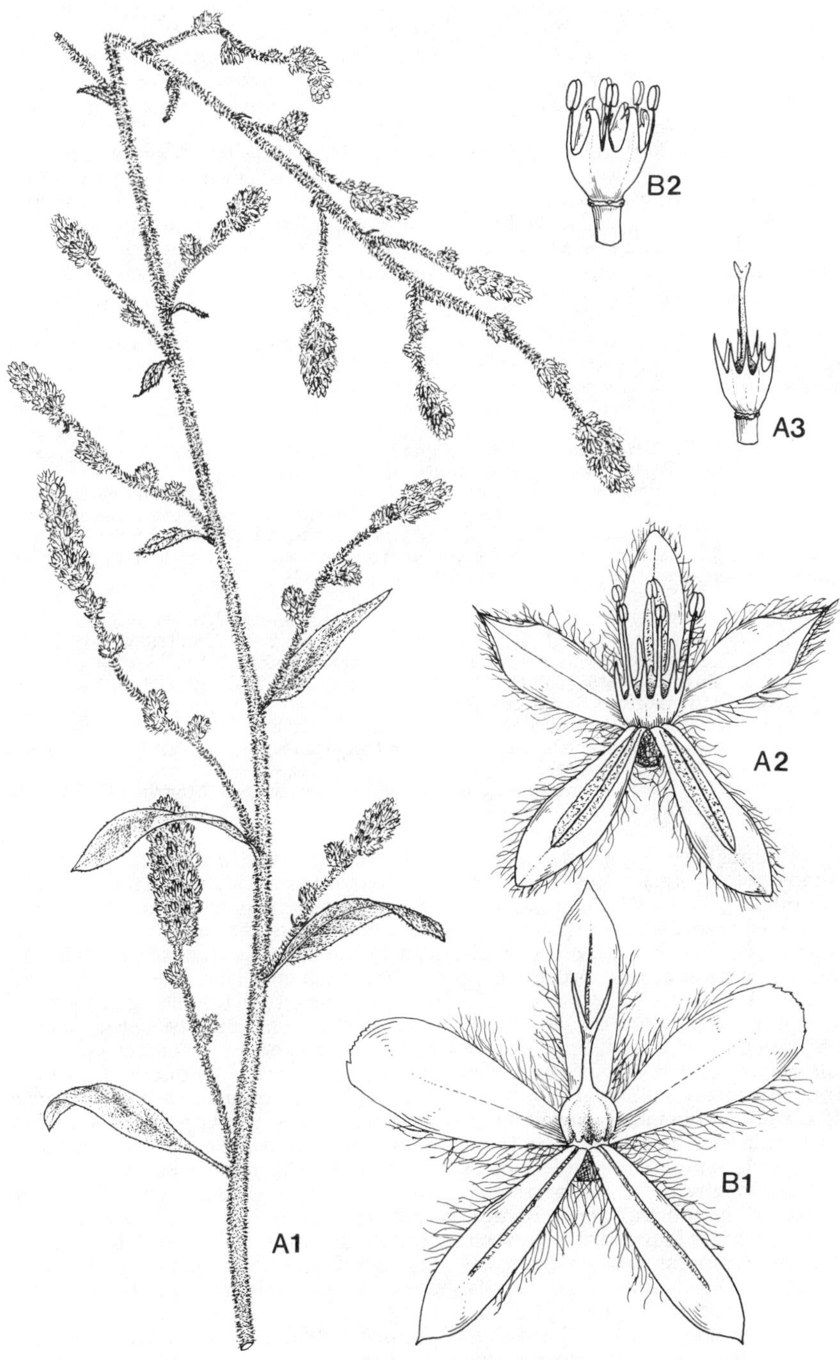

Tab. 21. A.—AERVA LEUCURA. A1, flowering branch (×⅔); A2, male flower (×12), A1—2 from *Richards* 11195; A3, female flower (tepals removed) (×12) *Anderson* 661. B.—AERVA JAVANICA. B1, female flower (×12) *Magogo* 1416; B2, male flower (tepals removed) (×12) *Bogdan* AB5636.

densely lanate dorsally. Hermaphrodite flowers: outer 2 tepals hyaline, elliptic-oblong, more or less abruptly contracted at the apex to a distinct mucro formed by the excurrent nerve, 1–1.75(2) mm. long; inner 3 slightly shorter and narrower, acute with a broad central green vitta which extends for about three quarters the length of each tepal and is usually furnished with a thickened border; style short; stigmas short, patent or divergent; anthers perfect. Male flowers similar but the stigmas reduced and capitate or obsolete, scarcely papillose. Female flowers also similar, or commonly with the tepals longer and narrower, tapering above, the outer to c. 2.25 mm. long; stigmas longer and frequently equalling the style, linear, divergent or suberect; filaments reduced, anthers absent. Capsule c. 1 mm. long, circular, compressed; seed c. 0.6–0.8 mm. long, reniform, black, shining, the testa almost smooth in the centre, faintly reticulate around the margin.

Zimbabwe. E: Chimanimani c. 8 km. from Haroni-Lusitu confluence upstream of Lusitu R., 22.xi.1967, *Ngoni* 30 (K; LISC; SRGH). **Malawi.** N: Mzimba Distr., Mzuzu, Marymount, 1365 m., 3.viii.1974, *Pawek* 8876 (UC; K; MO; SRGH; MA). S: Little Malosa R. near foot of Mulanje Mt., 15.viii.1971, *Leach, Rich & Whellan* 14812 (K; SRGH). **Mozambique.** N. Goa Isl., 5.v.1947, *Gomes e Sousa* 3510 (K). Z: Quelimane Distr., Namagoa, v–vii.? *Faulkner* K13 (K). MS: mountain slope beyond Penhalonga border, 1490 m., 8.ix.1957, *Chase* 6710 (K; SRGH). M: Bay of Maputo s.d., *Junod* 364 (BM; G).

Widespread in the tropics and subtropics of the Old World. In Africa from Sierra Leone across to Egypt, S. to S. Africa (rare) and Madagascar, also in Seychelles, Chagos Archipelago etc.; in Asia from Arabia E. to Malaysia, the Philippines and New Guinea. In the Flora Zambesiaca area in open forest on mountain slopes, on waste and disturbed ground, deserted cultivation and coastal scrub; sea level–1490 m.

3. **Aerva leucura** Moq. in DC., Prodr. **13**, 2: 302 (1849).—Baker & Clarke in F.T.A. **6**, 1: 39 (1909).—Schinz in Engl. & Prantl Pflanzenfam. ed. 2, **16 C**: 52 (1934).—Hauman in F.C.B. 2: 58 (1951).—Podlech & Meeuse in Merxm. Prodr. Fl. SW. Afr. **33**: 5 (1966); in J.H. Ross Fl. Natal: 159 (1973).—Townsend in Kew Bull. **29**: 461 (1974). TAB. 21 fig. A. Type from S. Africa.

Aerva ambigua Moq. in DC., Prodr. **13**, 2: 302 (1849). Type from S. Africa.

Aerva edulis Suesseng. in Mitt. Bot. Staatss. München **1**: 1 (1950). Type: Malawi, *Barker* 414 (EA, holotype).

Aerva leucura var. *lanatoides* Suesseng. in Mitt. Bot. Staatss. München **1**: 67 (1950). Type from Tanzania.

Perennial herb (apparently sometimes flowering in the first year), frequently woody below, erect or low-spreading to prostrate or occasionally more or less scrambling, 0.5–1.2 (1.5) m., with numerous stems arising from the base, stems simple or branched with short or long, ascending, slender branches. Stem and branches terete, striate, more or less densely tomentose or lanate with whitish hairs. Leaves alternate, broadly elliptic to linear-oblanceolate, those of the main stem c. 1.6–10 × 0.5–3 (3.6) cm., shortly cuneate to attenuate and more or less petiolate at the base with a petiole up to c. 1.5 cm. long, rather obtuse to very acute at the apex, moderately to thinly pilose on both surfaces or more sparsely so above, leaves of the branches and inflorescence reducing upwards. Inflorescence variable; always with a leafless lax or compact terminal panicle of simple or branched, sessile or pedunculate spikes, the upper leaf axils for a variable length of the stem with similar pedunculate panicles, or these reduced to a sessile, simple sometimes lobed spike or spikes. Ultimate spikes white, dense or occasionally laxer and more elongate, 0.6–5 (8) × 0.4–1.2 cm., divergent or ascending; bracts 1.25–1.75 mm. long, deltoid-ovate to oblong-ovate, hyaline with a short or longer arista formed by the excurrent midrib, pilose, persistent; bracteoles similar or slightly smaller, also persistent. Flowers hermaphrodite, female or functionally male. Tepals more or less densely lanate dorsally, more or less similar in all types of flower; outer 2 tepals hyaline, elliptic-oblong, (1.5) 2–2.5 mm. long, with a sharp mucro formed by the excurrent nerve; inner 3 tepals slightly shorter and narrower, acute with a central green vitta which extends for about three-quarters the length of each tepal and is often more or less furnished with a thickened border. Female flowers with linear stigmas somewhat shorter than or subequalling the style, divergent or suberect; anthers absent, filaments reduced. Male flowers with the stigmas very reduced or absent and only the truncate apex of the style slightly papillose, occasionally with short and apparently receptive stigmas but setting no seed; anthers longer. Hermaphrodite flowers with stigmas usually (but not invariably) shorter than the females, and anthers often somewhat smaller than in the males. Capsule and seeds as in *A. lanata*.

Botswana. N: near Nata R. delta, 14.iv.1976, *Ngoni* 496 (K; SRGH). SW: 35 km. S. of Takatshwane Pan, 21.ii.1960, *Wild* 5105 (K; SRGH). SE: Boteti delta area NE. of Mopipi, 17.iv.1973, 850 m., *Fry* 4 (K; SRGH). **Zambia**. B: Siwelewele, 8.viii.1952, *Codd* 7445 (K; PRE). N: Mbala, Kali Dambo near Kawimbi Mission, 1520 m., 6.v.1952, *Richards* 1621 (K; SRGH). W: Ndola, 23.vii.1953, *Fanshawe* 167 (K; NDO). C: Kabwe Distr., Kamaila Forest Station, 36 km. N. of Lusaka, 9.ii.1975, *Brummitt, Hooper & Townsend* 14285 (K; LUS; NDO) E: Lundazi Distr., Lukusuzi Nat. Park, 800 m., 12.iv.1971, *Sayer* 1173 (K; PRE; SRGH). S: Livingstone Distr., Katambora, 7.vi.1954, *Gilges* 624 (K; SRGH). **Zimbabwe**. N: Gokwe Distr., near Gwaye R., 16.iv.1962, *Bingham* 231 (K; SRGH). W: Hwange, Denda Farm, 945 m., 20.iii.1974, *Gonde* 77/74 (K; SRGH). C: Marondera 10.v.1951, *Wild* 3930 (K; SRGH). E: Mutare, Main Street, 31.v.1967, *Chase* 8452 (K). S: Gwanda Distr., Doddieburn Ranch, Umzingwane R., c. 720 m., 5.v.1972, *Pope* 646 (K; SRGH). **Malawi**. N: Mzimba District, S. side of Lale Kazuni, 1080 m., 20.v.1970, *Brummitt* 10941 (K; LISC; MAL; PRE; SRGH; UPS) C: Lilongwe, by Kandolo store, 1050 m., 26.vi.1970, *Brummitt* 11695 (K; MAL; SRGH). S: 5 km. NW. of Nsanje, near Nyamadzere Rest House, 180 m., 28.v.1970, *Brummitt* 11144 (K; LISC; MAL; SRGH). **Mozambique**. N; E. coast of Lake Malawi, 22.ix.1900, *Johnson* 17 (K). Z: De Mupeia near Aguas Quentes, 4.viii.1942, *Torre* 4490 (BR; EA; J; LISC; LUAI; M). MS: Beira Distr., Gorongosa Nat. Park, Chilengo area, viii.1970, *Tinley* 1982 (K; SRGH). T: 17 km. from Mágoè near Mágoè Velho, 2.iii.1970, *Torre & Correia* 18152 (LISC). M: Maputo, Porto Henrique, 17.iv.1948, *Torre* 7642 (LISC; PRE).

Also in Uganda, Kenya, Tanzania, Zaire, Angola, Namibia. In a large variety of habitats, frequently in bush along rivers, at edges of pans, dry patches in swamps and dambos, on termite mounds, in mopane forest, along roadsides, in grassland, dry rocky hillsides, waste ground and abandoned cultivation; in Kalahari Sand and other sandy soil, laterite, cultivated black soil, black basalt and on serpentine; 100–1520 m.

One herbarium sheet (*Mundy* 3171) notes that the leaves of this species are used as a vegetable; another (*Brummitt* 11144) observes that a rest house cook gave the information that it is used, presumably the soft inflorescences, for stuffing pillows.

14. NOTHOSAERVA Wight

Nothosaerva Wight, Ic. Pl. Ind. Or. **6**: 1 (1853).
Pseudanthus Wight, Ic. Pl. Ind. Or. **5**, 2: 3 (1852) non Sieb. ex Spreng., Syst. Veg. ed. 16, **4**, 2: 25 (1827).

Annual herb with opposite or alternate branches and leaves, leaves entire. Flowers small, solitary in the axils of scarious bracts, in dense, sessile, solitary or clustered spikes. Perianth segments 3–4(5), hyaline, the perianth subtended by two very small bracteoles. Stamens 1 or 2; filaments filiform, intermediate pseudostaminodes absent; anthers bilocular. Style short, stigma capitate. Ovary with a single, pendulous ovule, radicle ascending. Capsule delicate, irregularly rupturing. Seeds rounded, compressed, endosperm copious.

A monotypic genus.

Nothosaerva brachiata (L.) Wight, Ic. Pl. Ind. Or. **6**: 1 (1853).—Schinz in Engl. & Prantl Pflanzenfam. ed. 2, **16 C**: 52 (1934).—Cavaco in Mém. Mus. Hist. Nat. Paris Sér. B, **13**: 104 (1962). TAB. **22**. Type, Linnean specimen 290/1 (LINN. lectotype).
Achyranthes brachiata L., Mant.: 50 (1767). Type as above.
Illecebrum brachiatum (L.) L., Mant. Alt.: 213 (1771). Type as above.
Aerva brachiata (L.) Mart. in Nova Acta Acad. Caesar. Leop. Carol. **13**, 1: 291 (1826).—Baker & Clarke in F.T.A. **6**, 1: 40 (1909). Type as above.

Annual herb, (4) 10–45 cm., with many spreading branches from about the base upwards; stem and branches subterete, striate, glabrous or thinly hairy. Leaves narrowly to broadly elliptic, elliptic-oblong or ovate, thinly hairy to glabrous or almost so, obtuse, subacute at the apex, at the base gradually or more abruptly narrowed to a petiole about half the length of the lamina, lamina of the lower main stem leaves c. 10–40 (50) × 6–20 mm., upper and branch leaves becoming shorter and narrower. Flowers in dense, 3–15 × 2–2.5 mm. spikes which are clustered in the leaf-axils of the stem and branches or on very

Tab. 22. NOTHOSAERVA BRACHIATA. 1, flowering plant (×$\frac{2}{3}$); 2, single flower (×30); 3, bract (×60); 4, flower with one tepal removed (×30); 5, longitudinal section of ovary (×30); 6, fruiting perianth (×30); 7, seed (×30), all from *Drummond & Hemsley* 4067.

short axillary shoots; spikes sessile or the terminal spike of axillary shoots shortly (to c. 3 mm.) pedunculate; inflorescence axis thinly to rather densely pilose; bracts hyaline, minutely erose-denticulate, concave, acute or shortly acuminate, c. 0.5 mm., glabrous or very thinly hairy; bracteoles minute, hyaline. Perianth segments broadly oval-elliptic, c. 1.25 mm. long, subacute to shortly acuminate, villous on the outer surface, in the basal two-thirds with a thick greenish vitta and a single midrib. Stamens longer than the ovary and style. Capsule included, falling with the persistent perianth. Seed c. 0.4 mm. chestnut-brown, smooth and shining.

Zambia. S: 6.5 km. N. of Sinazongwe on Choma Rd., 515 m., 10.vi.1963, *Bainbridge* 832 (K; SRGH). **Zimbabwe**. N: Kariba Distr., Sengwa, 10.viii.1965, *Jarman* 221 (K; SRGH). E: Chipinge Distr., SW. of Mutema Pan, Mutema Common Land, Sabi R., 455 m., 18.iv.1965, *Chase* 8290 (K; SRGH). S: Mwenezi Distr., Urumbo Pan, between Fishans and Kapatenis, 25.iv.1962, *Drummond* 7710 (K; LISC; SRGH). **Malawi**. S: Chikwawa Distr., Lengwe Game Reserve, 100 m., 6.iii.1970, *Brummitt* 8907 (K; LISC; MAL; SRGH). **Mozambique**. MS: Sabi R., Meringua Distr., 30.vi.1950, *Chase* 2468 (BM; SRGH).

Distributed in tropical Africa from Senegal eastwards to Ethiopia and N. Somalia southwards to Flora Zambesiaca area and Angola; Mauritius; Pakistan, India from Punjab to Madras, Sri Lanka and Burma. Generally in places where water lies seasonally and then dries out, on sand or clay; 100–515 m.

15. MECHOWIA Schinz

Mechowia Schinz in Pflanzenfam. ed. 1, 3, **1a**: 110 (1893).

Perennial herbs with a stout rootstock and slender, erect stems. Leaves opposite, subopposite or some alternate, entire. Flowers bibracteolate, hermaphrodite;inflorescences dense, capitate with one flower in the axil of each bract; bracts persistent after fruit fall, bracteoles deciduous with the fruiting perianth. Perianth segments 5, free. Stamens 5, filaments free, delicate, flattened; pseudostaminodes shortly and narrowly oblong with an incised apex, or more commonly small and tooth-like,to obscure or absent; anthers bilocular. Ovary uniovulate, densely lanate in at least the upper two-thirds; style elongate (often exserted), stigma solitary and small. Capsule delicate, irregularly splitting with the ripening of the seed. Seeds yellow or reddish, somewhat compressed, feebly reticulate.

Two species in Central Africa.

Both species of this genus appear to occur on ground which is regularly burnt over. The very stout root with slender new shoots sent up every year, and burnt tops of old shoots frequently to be found on the bases of specimens, strongly indicate this fact. Notes seen on one specimen of each species give confirmation.

Outer tepals narrowly oblong, the hyaline margin at the middle not as wide as half the width of the well-demarcated, firm centre; style lanate at the base; leaves numerous and distinct even when narrow - - - - - - - - - - 1. *grandiflora*
Outer tepals broadly oblong-ovate, the hyaline margin at the middle at least as wide as the rather ill-defined, firmer centre; style glabrous throughout; leaves few, very reduced - - - - - - - - - - - - - - 2. *redactifolia*

1. **Mechowia grandiflora** Schinz in Engl. Bot. Jahrb. **21**: 186 (1895).—Baker & Clarke in F.T.A. **6**, 1: 36 (1909).—Schinz in Engl. & Prantl Pflanzenfam. ed. 2, **16 C**: 55 (1934).—Cavaco in Mém. Mus. Hist. Nat. Paris Sér. B, **13**: 109 (1962). TAB. **23** fig. A. Type from Angola.

Perennial herb with numerous, erect, slender annual stems (6.5) 12–30 cm. tall from a stout, deep, woody rootstock, stems striate with pale raised lines, almost glabrous to rather densely furnished with rather short multicellular hairs. Leaves numerous, opposite, subopposite or some alternate, broadly elliptic to very narrow and subfiliform, (10) 15–45 (55) × (0.75) 4–12 (18) mm., with a pale cartilaginous border, mucronate, glabrous to densely furnished with multicellular hairs (the more narrow-leaved a form is, for the most part the less pilose it is). Inflorescence capitate, globose or hemispherical, 1.25–1.6 cm. in diam., the axis somewhat elongating in fruit (to c. 2.25 cm.); axis deeply sulcate,

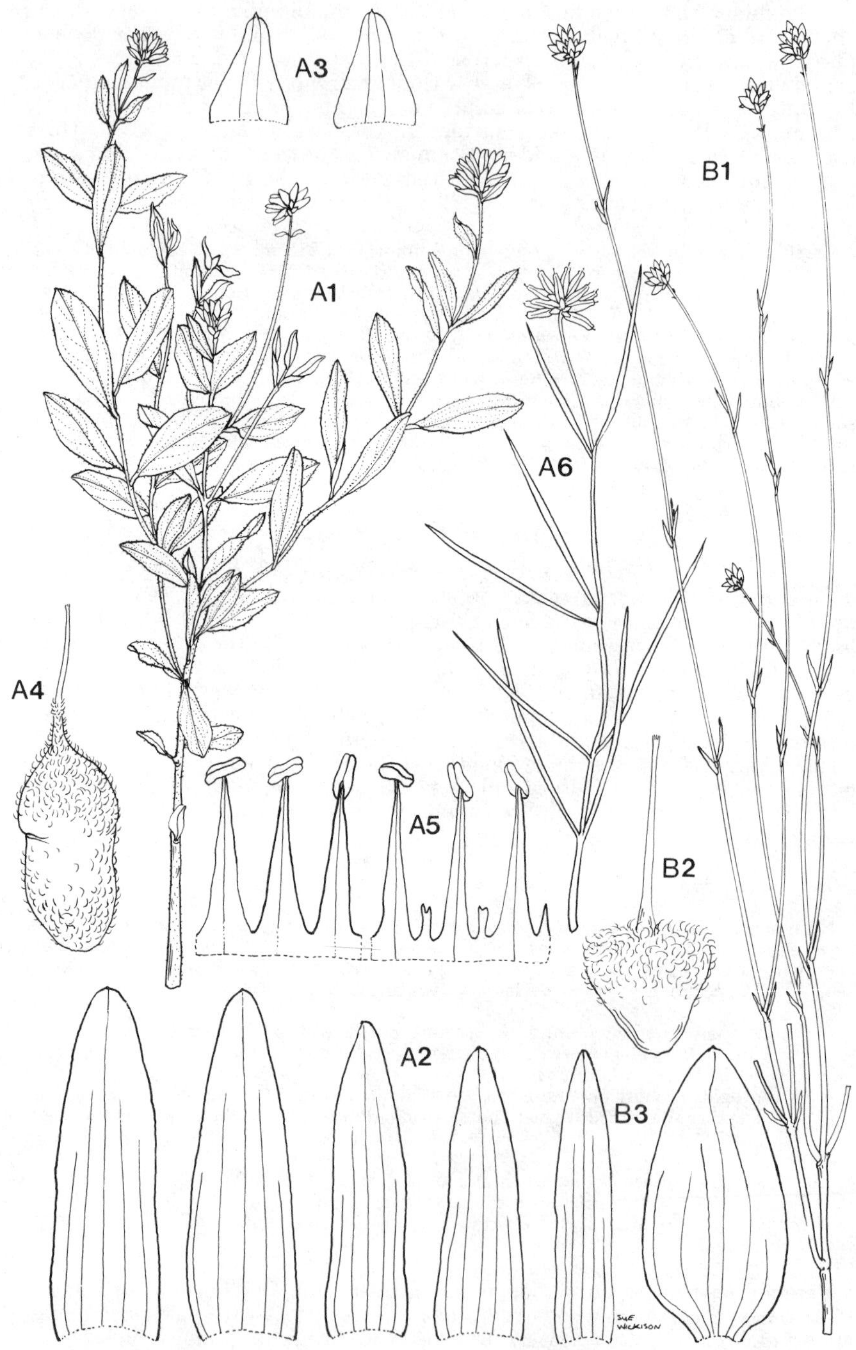

Tab. 23. A.—MECHOWIA GRANDIFLORA. A1, flowering stem (×$\frac{2}{3}$); A2, tepals (×8); A3, bracteoles (×8); A4, gynoecium (×8); A5, androecium (×8), A1—5 from *Fanshawe* 5975; A6, flowering branch of narrow-leaved form (×$\frac{2}{3}$) *Richards* 17183. B.—MECHOWIA REDACTIFOLIA. B1, flowering stem (×$\frac{2}{3}$); B2, gynoecium (×8); B3, outer tepal (×8), B1—3 from *Richards* 15388.

subglabrous to more or less densely furnished with multicellular hairs. Bracts ovate to lanceolate-ovate, 2–3 mm., glabrous or more commonly shortly pubescent at least along the margins, stramineous or sometimes reddish-tinged, the darker midrib excurrent in a very short mucro. Bracteoles oblong-ovate to ovate, usually ciliate at least along the basal margins, 2.5–3.5 mm. long, stramineous or tinged with red, sharply mucronate with the excurrent midrib. Perianth bright crimson. Outer 2 tepals narrowly oblong, 5–6 mm. long, the firm central part with 3–5 thick but obscure nerves (more distinct in fruit) and other finer veins, the midrib not excurrent; margins hyaline, inflexed above. Inner 3 tepals slightly shorter, gradually narrower and less firm, more widely hyaline-margined above. Filaments 6 mm. long; anthers narrowly oblong, 2–2.25 mm. long; pseudostaminodes absent, indistinct or small and tooth-like. Ovary obovoid, 2 mm. long, glabrous at the extreme base, but densely lanate above; style 2.5 mm., pilose at the base. Capsule oblong-ovoid, c. 3 mm. long, densely lanate except at the extreme base with whitish, multicellular hairs. Seeds subreniform, yellow to reddish, shining, c. 2.75 mm.

Zambia. B: Kataba, 12.xii.1960, *Fanshawe* 5975 (K; NDO). W: Mwinilunga Distr., S. of R. Kakema, 24.viii.1930, *Milne-Redhead* 953 (K).
Also in Angola, Zaire. On clay loam or sandy soil in *Brachystegia* woodland, by roadside and on rocky wooded hillside.

This species is extremely variable in leaf form, but even in the limited material available for study there is sufficient transition to indicate that the recognition of formal infraspecific taxa is of doubtful value.

2. **Mechowia redactifolia** Townsend in Publ. Cairo Univ. Herb. **7** & **8**: 73 (1977). TAB. **23** fig B. Type: Zambia, Kawambwa, Chimbwi, *Richards* 15388 (K, holotype; SRGH).

Perennial herb with numerous virgate annual stems c. 20–25 cm. tall from a stout, woody rootstock; stems striate with pale raised lines, all vegetative parts quite glabrous. Leaves very scattered and reduced, opposite or sometimes with the alternate branches subtended by a single leaf, narrowly linear, with a pale cartilaginous border, mucronate, the lowest up to 1 cm. long, the uppermost very reduced and bracteform. Inflorescence capitate, globose, up to 1.2 cm. in diam., the axis somewhat elongating in fruit (to c. 2.25 cm.); axis deeply sulcate, glabrous or thinly furnished with long, whitish multicellular hairs. Bracts lanceolate, glabrous, c. 3 mm. long, stramineous, the somewhat darker midrib excurrent in a very short mucro. Bracteoles ovate, glabrous, 3–3.5 mm. long, stramineous or tinged with red, sharply mucronate. Perianth pink or white flushed with pink. Outer 2 tepals broadly oblong-ovate, c. 4.5 mm. long, the firm central part with 3 thick but obscure nerves (more distinct in fruit) and several more slender veins, the midrib scarcely excurrent; margins hyaline, inflexed above. Inner 3 tepals slightly shorter, gradually less firm. Filaments 3.5 mm. long, delicate; anthers oblong, c. 1.25 mm.; pseudostaminodes linear and tooth-like to narrowly oblong with an incised apex, c. 1 mm. long. Ovary c. 2 mm. long, turbinate-ovoid, glabrous in the lower one-third but densely lanate above; style c. 3 mm. long, glabrous. Capsule oblong-ovoid, slightly compressed, c. 3 mm., densely lanate above with whitish, multicellular hairs. Seeds subreniform, yellowish, shining, c. 2.75 mm.

Zambia. N: Kawambwa-Mansa Rd, 1290 m., 27.xi.1961, *Richards* 15388 (K; SRGH).
Endemic. On sandy soil in grassland and *Syzygium* dambo.

16. PSILOTRICHUM Blume

Psilotrichum Blume, Bijdr. Fl. Nederl. Ind.: 544 (1825).
Psilostachys Hochst. in Flora 27, Beil.: 6, t. 4 (1844).

Perennial herbs or subshrubs, prostrate to erect or scandent, with entire, opposite or partly alternate leaves. Flowers hermaphrodite, in axillary and terminal bracteolate heads or spikes, solitary in the axil of each bract and bibracteolate; bracts persistent, finally spreading or deflexed, bracteoles falling with the fruit. Tepals 5, free, strongly to faintly nerved or ribbed (nerves 3 or more), the outer 2 tepals frequently finally more or less indurate at the base, usually differing in form and indumentum from the inner 2 with the middle tepal intermediate. Stamens 5, shortly monadelphous at the base, without or rarely with alternating pseudostaminodes; anthers bilocular. Ovary with a single pendulous ovule; style slender but rather short, stigma capitate. Utricle thin-walled, bursting irregularly. Seed ovoid, brownish.

About 16 species in tropical Asia and Africa, chiefly the latter.

1. Leaves linear - - - - - - - - - - - - - -3. *schimperi*
— Leaves not linear - - - - - - - - - - - - - - - - 2
2. Leaves silvery-sericeous beneath, at least when young; flowers in an open panicle, the axes of the ultimate spikes zigzag, very slender and capillary - - - - - - - - - - - - - - - - - 1. *sericeum*
— Leaves green on both surfaces; flowers in opposite axillary and terminal spikes, the axes of the spikes never zigzag or capillary - - - - - - 2. *scleranthum*

1. **Psilotrichum sericeum** (Koen. ex Roxb.) Dalz. in Dalz. & Gibson, Bombay Fl.: 216 (1861).—Verdcourt in Kew Bull. **17**: 491 (1964). TAB. **24** fig. B. Type, Smith's Herbarium, specimen 424.4 (LINN, lectotype).

Achyranthes sericea Koen. ex Roxb., Fl. Ind., ed. Carey 2: 502 (1824). Type as above.

Psilostachys sericea (Koen. ex Roxb.) Hook.f. in Benth. & Hook. Gen. Pl. **3**: 32 (1880). Type as above.

Psilotrichum boivinianum Baill. in Bull. Soc. Linn. Paris **1**: 622 (1889). Type from Zanzibar.

Psilotrichum nervulosa Baill., loc. cit. (1889); in F.T.A. **6**, 1: 61 (1909). Type, "E. African coast".

Psilotrichum filipes Baill., loc. cit. (1889); in F.T.A. **6**, 1: 61 (1909). Type from Zanzibar.

Psilotrichum axillare C.B. Clarke in F.T.A. **6**, 1: 60 (1909).—Schinz in Engl. & Prantl Pflanzenfam. ed. 2, **16 C**: 60 (1934). Type from Kenya.

Psilotrichum edule C.B. Clarke in F.T.A. **6**, 1: 61 (1909). Type from Tanzania.

Erect or occasionally prostrate or ascending annual herb, with numerous stems c. 0.3—1 m. tall, much-branched in the lower half and frequently also above, the lowest internodes of particularly the lower branches very divergent. Stem and branches striate, the lower internodes terete, the upper subquadrate and more or less sulcate, nodes somewhat swollen; indumentum very variable, from subglabrous or with sparse soft pubescence to densely pilose with a double indumentum of shorter soft hairs and long, fine, multicellular, spreading bristly hairs (even when absent elsewhere tufts of bristly hairs occur between the branches and leaves at least of the upper nodes, frequently forming an annulus there). Main stem and branch leaves cordate-ovate to broadly ovate or rarely cordate-lanceolate,obtuse to acuminate at the apex, 1.3—7 × 0.7—5.2 cm., the lowest on petioles up to c. 1 (1.8) cm. long, leaves towards the ends of the stem and branches reducing in size and almost sessile; indumentum of mature leaves very variable, from green on both surfaces and moderately softly appressed-pilose below and thinly so with long and shorter multicellular, strigose hairs above to moderately appressed-pilose above and densely silvery-sericeous on the lower surface (always so in young leaves). Inflorescence a large, open compound panicle, of axillary opposite panicles formed of simple or branched spikes, each with a flexuose, glabrous rhachis; peduncle and branches of the partial inflorescences capillary, glabrous or with long, strigose, multicellular hairs, the axes zigzag; bracts lanceolate, 0.75—1 mm. long, sparingly pilose, mucronate with the excurrent midrib; bracteoles similar but slightly shorter. Flowers sessile. Tepals, green, acute, prominently 3-nerved with the nerves confluent at the apex and excurrent in a minute mucro, glabrous or usually more or less furnished with spreading, white, multicellular, minutely denticulate hairs; outer 2 tepals 2—2.5 mm. long, narrowly hyaline-margined with all 3 nerves equally strong, if hairs present then all nerves white-pilose, hairs sometimes also present on the surfaces between; inner 2 widely hyaline-margined (the margin wider than the nerved central portion at halfway up the tepal), pilose only along the midrib, which is much stronger than the lateral nerves; middle tepal with one side as in the outer 2 and one as in the inner 2. Stamens delicate, c. 1.5—1.75 mm. long; pseudostaminodes none. Style short, c. 0.5 mm. long.

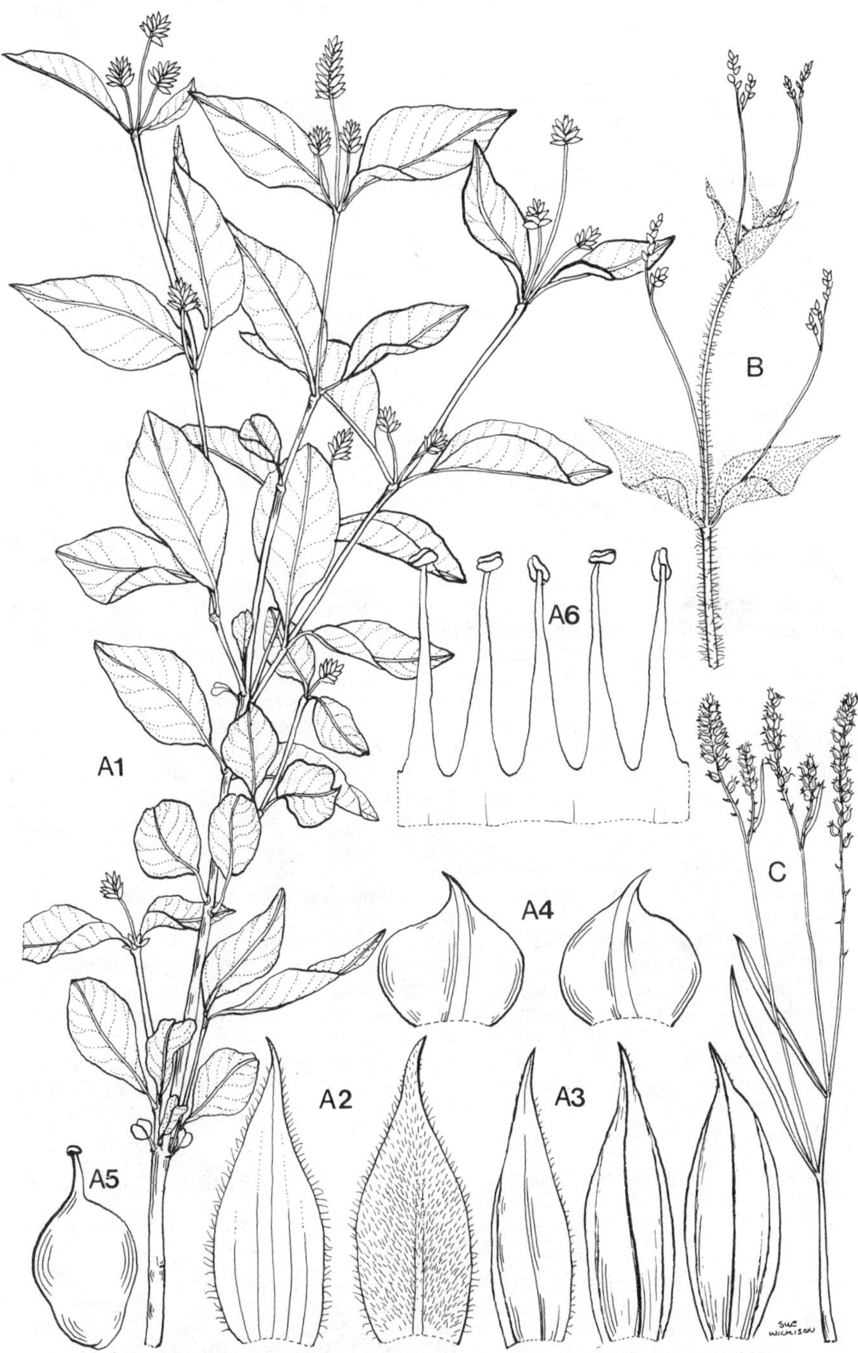

Tab. 24. A.—PSILOTRICHUM SCLERANTHUM. A1, flowering branch ($\times\frac{2}{3}$); A2, outer tepals, dorsal and ventral surface ($\times$10); A3, inner tepals, ventral surface ($\times$10); A4, bracteoles ($\times$10); A5, gynoecium ($\times$10); A6, androecium ($\times$16), A1–6 from *Eyles* 8282. B.—PSILOTRICHUM SERICEUM. B1, flowering branch ($\times\frac{2}{3}$) *Faulkner* 2799. C.—PSILOTRICHUM SCHIMPERI. C1, flowering branch ($\times\frac{2}{3}$) *Bogdan* 3389.

Capsule ovoid, c. 1.75–2 mm. long. Seed subglobose, black, shining, feebly reticulate, c. 1.5 mm. in diam.

Mozambique. N: Ilha de Mocambique, 31.viii.1942, *Mendonça* 1164 (BM; LISC) M: Maputo between Mata de Marracuene and the settlement Jafar, 1.x.1957, *Barbosa & Lemos* in *Barbosa* 7890 (K; LMU).

Also from E. African coastal and subcoastal region from Socotra and Somalia south to Tanzania (incl. Zanzibar and Pemba), also in India. Habitat in Mozambique not recorded, elsewhere in sandy places near the shore, roadsides, abandoned cultivation, in grassland and *Acacia* scrub; sea level–320 m.

2. **Psilotrichum scleranthum** Thw., Enum. Fl. Zeyl.: 248 (1861).—Verdcourt in Kew Bull. **17**: 492 (1964).—Schinz in Engl. & Prantl Pflanzenfam. ed. 2, **16 C**: 60 (1934). TAB. **24** fig. A. Type from Sri Lanka.

Psilotrichum africanum Oliv. in Hook., Ic. Pl. **16** t. 1542 (1886).—Baker & Clarke in F.T.A. **6**, 1: 58 (1909).—Schinz in Engl. & Prantl Pflanzenfam. ed. 2, **16 C**: 60 (1934).—Hauman in F.C.B. **2**: 37 (1951).—Cavaco in Mém. Mus. Hist. Nat. Paris Sér. B, **13**: 111 (1962).—J.H. Ross Fl. Natal: 159 (1973). Type from Tanzania.

Psilotrichum africanum var. *debile* Schinz in Engl. Bot. Jahrb. **21**: 185 (1895). Type: Malawi, Blantyre, *Last* s.n. 1887 (K, isotype).

Psilotrichum trichophyllum Baker in Kew Bull. **1897**: 279 (1897); in Baker & Clarke in F.T.A. **6**, 1: 58 (1909). Type: Mozambique, Shamo, *Kirk* s.n. 1859 (K, holotype).

Psilotrichum concinnum Baker, loc. cit. (1897); in Baker & Clarke in F.T.A. **6**, 1: 58 (1909). Type: Malawi, Blantyre, *Last* s.n. 1887 (K, holotype).

Woody perennial herb or small shrub, erect or rooting at the basal nodes, sometimes more or less scandent, 0.6–2 m., much-branched with the branches spreading at 45–90 degrees. Stem and branches in the older parts terete, striate and glabrescent, when young quadrangular and sulcate with pale, thick corners, more or less densely pilose with yellowish, subappressed hairs, slightly swollen at the nodes. Leaves ovate to elliptic or elliptic-oblong, 2–10 × 1–4.6 cm., acute to acuminate at the mucronate apex, shortly to longly cuneate at the base with a 2–5 mm. petiole, moderately but finely pubescent on both surfaces (generally more conspicuously so on the lower surface of the primary venation). Inflorescences rather short spikes, 7–8 mm. wide and finally elongating to 2–5 cm., terminal and generally 2 at each node in the axils of the opposite leaves of stem or branches, sessile or on peduncles up to c. 4 cm. long; bracts ovate-lanceolate, 2–2.5 mm. long, more or less densely appressed-pubescent, sharply mucronate with the excurrent midrib, finally spreading or deflexed; bracteoles whitish, broadly deltoid-ovate with the margins slightly overlapping at the base, c. 2 mm. long with a distinct 0.5 mm. long mucro formed by the excurrent midrib, glabrous or slightly pilose along the midrib and/or margins. Flowers sessile. Tepals white or greenish, very firm, faintly c. 5-nerved with still more obscure finer nerves; the 2 outer lanceolate-oblong, acute, sharply mucronate, 3.5–4.5 mm. long, finally indurate at the base, narrowly hyaline-bordered, shortly pilose over the entire dorsal surface; the 3 inner lanceolate-ovate, acute to acuminate,pilose mainly centrally, the hyaline border widened below. Stamens c. 2.5 mm. long; pseudostaminodes absent. Style c. 0.75 mm. long. Capsule oblong-ovoid, c. 2.5–3 mm. long. Seed ovoid, c. 1.75 mm. long, brown, shining, faintly reticulate.

Zambia. C: Katondwe, 24.ii.1965, *Fanshawe* 9140 (K; NDO). S: Kariba Hills near. Kariba access Rd., 2.ii.1958, *Drummond* 5446 (K; LISC; SRGH). **Zimbabwe**. N: Hurungwe, c. 21 km. ESE. of Chirundu Bridge, 31.i.1958, *Drummond* 5386 (K; LISC; SRGH). W: Hwange Nat. Park, Mandavu Dam, c. 16 km. SW. of Sinamatella Camp, c. 1030 m., 24.ii.1967, *Rushworth* 200 (K; LISC; SRGH). E: Chipinge Distr., below Birchenough Bridge, c. 530 m., i.1957, *Davies* 2394 (K; LISC; SRGH). S: Chiredzi, Chionjas, c. 490–550 m., 29.i.1957, *Phipps* 211 (K; SRGH). **Malawi**. N: Rumphi Gorge, 1060 m., 9.iii.1975, *Pawek* 9142 (K; MAL; MO; SRGH; UC). S: Blantyre Distr., Mpatamanga Gorge, E. bank of Shire R., 230 m., 24.ii.1970, *Brummitt* 8731 (K; LISC; MAL; PRE; SRGH; UPS). **Mozambique**. N: c. 53 km. from Pemba near Ancuabe, 21.xii.1963, *Torre & Paiva* 9642 (C; COI; LISC; MO; WAG) Z: Namagoa, 60–120 m., xi.1944, *Faulkner* 334 (K; PRE). T: between Lupata and Tete, ii.1859, *Kirk* s.n. on Livingstone's S. African Exp. (K). MS: 30 km. NE. of Inhamitanga, 5 km. S. of railway, 200 m., 5.xii.1971, *Müller & Pope* 1898 (K; LISC; SRGH). GI: near Chibuto, 11.ii.1942, *Torre* 3941 (BR; LISC; LUAI; P; PRE). M: Inhaca Isl., 31.xii.1956, *Mogg* 27096 (LISC).

Also in Kenya and Tanzania, eastern Zaire, Angola, Madagascar, S. India and Sri Lanka. An understorey species of mopane or evergreen forest, frequently near streams, also in thickets and on river banks, on sandy soil or granite; 90–1060 m.

3. **Psilotrichum schimperi** Engl. in Abh. Königl. Preuss. Akad. Wiss. Berlin 1891 (Hochgebirgsflora des trop. Afr.): 207 (1892).—Baker & Clarke in F.T.A. **6**, 1: 59 (1909).—Schinz in Engl. & Prantl Pflanzenfam. ed. 2, **16** C: 60 (1934). TAB. **24** fig. C. Type from Ethiopia.

Psilotrichum angustifolium Gilg in Notizbl. Bot. Gart. Berl. **1**: 328 (1897). Type from Tanzania.

Psilotrichum camporum Lebrun & Toussaint ex Hauman in Bull. Jard. Bot. Brux. **18**: 110 (1946).—Hauman in F.C.B. **2**: 39 (1951). Type from Rwanda.

Psilotrichum gramineum Suesseng. in Mitt. Bot. Staatss. München **1**: 111 (1952). Type from Kenya.

Psilotrichum schimperi var. *gramineum* (Suesseng.) Suesseng., tom. cit. **1**: 194 (1953).

Erect or decumbent annual, (15) 30–75 cm. tall, with numerous branches from the base upwards (or small plants unbranched below the inflorescence), vegetative parts glabrous or with scattered strigose hairs (especially along the margins and midrib of the upper leaves and on the upper internodes). Stem and branches strongly striate, subsulcate, or the basal internodes terete, nodes slightly swollen. Leaves linear, 3.5–12 × 0.2–1 cm., rather blunt at the apex with a distinct mucro formed by the confluence of the thickened margins and strong, excurrent midrib, attenuate and subsessile at the base. Inflorescences of axillary spikes c. 0.7 cm. wide and finally elongating to as much as c. 14 cm., on peduncles 0.2–6.5 cm. long, the upper often alternate; bracts deltoid-lanceolate, 1.5–2 mm. long, membranous except for the greenish, excurrent midrib, finally spreading; bracteoles similar but slightly smaller. Flowers sessile. Tepals green or sometimes suffused with red or brown, prominently 3-nerved and the nerves confluent with the midrib above but not excurrent in a mucro; outer 2 lanceolate-oblong, 3.5–4 mm. long, obtuse, with a very narrow hyaline border, shortly scabrid-pilose along the margins, nerves and often also between the nerves; inner 2 more broadly hyaline-bordered, the margins fringed with long, white hairs; middle tepal with one margin broadly hyaline and long-pilose, the other narrowly hyaline and scabrid. Stamens c. 2.5 mm. long, very delicate, almost free; pseudostaminodes none. Style c. 0.5 mm. long. Capsule ovoid, 2.75–3 mm. long. Seed ovoid, 1.5–2 mm., black, shining, faintly reticulate.

Zambia. S: Kafue Pilot Polder, Kafue Flats W. of Mazabuka, iii.1962, *Brochington* 10 (K; SRGH).

Also in Uganda, Kenya, Tanzania, Ethiopia and Rwanda, where it also occurs in other damp areas in seasonally wet grassland, in *Acacia* woodland, even in deep mud in a swamp, but apparently always on black clay; 970–1670 m.

All the material seen of this species from Zambia (4 gatherings) is from the same area, where one collector (*van Rensburg* 2677) states that it is a very abundant weed on heavy black clay soil with *Paspalum*.

A further species of this genus certainly occurs in the Flora Zambesiaca area. It is represented by herbarium material: Zambia, Mbala Distr., in shade in ground layer of "Itigi" type thicket of *Burttia*, *Pseudoprosopis* etc. near Mpulungu, Lake Tanganyika, 17.xi.1952, *Angus* 778 (FHO; K). This has been determined as *Psilotrichum trichophyllum* Bak. (i.e., *P. elliotii* Bak.), but is apparently not that species. The material is very poor, and the plant needs recollecting.

17. ACHYRANTHES L.

Achyranthes L., Sp. Pl. **1**: 204 (1753).

Herbs with opposite, petiolate, entire leaves. Inflorescence a more or less slender spike, terminal on the stem and branches, the flowers at first congested and patent, finally usually laxer and deflexed; bracts deltoid or ovate, the midrib excurrent in a spine. Flowers solitary in the bracts, hermaphrodite, bibracteolate. Perianth segments 4–5, 1–3(5) nerved, narrowly lanceolate, acuminate, mucronate with the excurrent midrib, indurate in fruit especially at the base. Bracteoles spinous-aristate with the excurrent midrib, the lamina forming short and free to longer and adnate membranous wings. Stamens 2–5, filaments filiform, monadelphous, alternating with quadrate to broadly quadrate-spathulate pseudostaminodes, these simple and dentate or fimbriate, or commonly furnished with a variably developed dorsal scale; anthers bilocular. Style slender, stigma small, truncate-capitate. Ovary with a solitary pendulous ovule, the ovary wall very thin in fruit. Entire flower with bracteoles falling with the ripening of the cylindrical seed, the deflexed bracts persistent. Endosperm copious.

A genus of about 6 species in the warm temperate and tropical regions of the world. Very close to *Pandiaka*, but with a very characteristic appearance owing to the generally elongate and rather lax inflorescence; the narrow, glabrous perianth with obscure nerves; the bracteoles with a relatively short lamina and a dorsally prominent midrib excurrent in a long, terete, spinous arista; and the perianth and bract commonly deflexed and closely appressed to the inflorescence axis in ripe fruit.

1. **Achyranthes aspera** L., Sp. Pl. **1**: 204 (1753).—Baker & Clarke in F.T.A. **6**, 1: 63 (1909).—Schinz in Engl. & Prantl Pflanzenfam. ed. 2, **16** C: 61 (1934).—Hauman in F.C.B. 2: 53 (1951).—Cavaco in Mém. Mus. Hist. Nat. Paris Sér. B, **13**: 114 (1962).—Podlech & Meeuse in Merxm. Prodr. Fl. SW. Afr. **33**: 4 (1966); in J.H. Ross Fl. Natal: 159 (1973).—Townsend in Kew Bull. **29**: 473 (1974). Type from Sri Lanka.

Perennial herb (sometimes woody and somewhat suffrutescent), occasionally flowering in the first year, 0.2–2 m., stiffly erect to subscandent or straggling and more or less prostrate, simple to much-branched, stems stout to very weak, distinctly to obscurely 4-angled, striate or sulcate, subglabrous to densely tomentose, the nodes more or less shrunken when dry. Leaves elliptic, oblong or ovate, acute or acuminate to almost circular and very obtuse, gradually or abruptly narrowed below, (2)5–22(28) × 1.3–8(10) cm., indumentum varying from subglabrous on both surfaces through subglabrous above and densely appressed-canescent below to more or less densely tomentose on both surfaces; petioles of main stem leaves 3–25(30) mm., shortening above and below. Inflorescences at first dense, finally elongating to (5)8–34(40) cm.; peduncles (0.6–1)1–6 (7.5) cm. long. Bracts lanceolate or narrowly deltoid-lanceolate, pale or brownish-membranous, 1.75–5(6) mm. long, glabrous. Bracteoles 1.5–4.5(6) mm., the basal wings one third to one quarter the length of the spine and more or less adnate to it (sometimes free above or tearing free), typically tapering off above but not rarely rounded or truncate. Perianth whitish or pale green to red or purple, segments 5, 3–7(10) mm. long, the outer pair longest, narrowly lanceolate to lanceolate, very acute, with a distinct midrib and 2 obscure to distinct lateral nerves, narrowly or moderately pale-margined. Stamens 5, the filaments 1.5–4.5(6) mm. long, alternating with subquadrate pseudostaminodes. Typically the apex of the latter curves slightly inwards as a narrow, crenate or entire. often very delicate flap, while from the dorsal surface arises a fimbriate-ciliate scale extending across the width of the pseudostaminode; not rarely, however, this is reduced to a "stag's horn" process at the centre of the dorsal surface or a shallow, dentate rim, or even becomes small and filiform, or else subapical or apical so that the pseudostaminode appears simple (this usually in small forms of var. *sicula*). Style slender, 1–4(6) mm. long. Capsule oblong-ovoid, 1–3(5) mm. long. Seed filling the capsule, oblong-ovoid, smooth.

1. Leaves broadly elliptic, broadly obovate to almost circular, blunt or abruptly apiculate; flowers short and rather plump, to c.4.5 mm. - - - var. *aspera*
– Leaves oblong to elliptic, bluntly to long-acuminate; flowers 3–7 mm. long, but less plump and if not exceeding 4.5 mm. the leaves frequently distinctly acuminate - - - - - - - - - - - - - - - - - - - 2
2. Leaves with variable indumentum but not greenish above and silvery canescent beneath; flowers mostly 5–7 mm. long - - - - - - var. *pubescens*
– Leaves distinctly acuminate, typically greenish above and, at least when young, silvery-canescent beneath; flowers mostly 3–4.5 mm. long - - var. *sicula*

Var. **aspera** TAB. 25 fig. A.

Achyranthes aspera var. *indica* L., Sp. Pl. **1**: 204 (1753). Type as for the species.
Achyranthes indica (L.) Mill., Gard. Dict. ed. 8 (1768).
Achyranthes obtusifolia Lam., Encycl. Méth. Bot. **1**: 545 (1785). Type source

unknown, Sonnerat s.n. (P-LA, holotype). (I.D.C. microfiche photo. 546, No. 11!).
Achyranthes obovata Peter in Fedde, Repert. Beih. **40**, 2, Descriptiones: 25 (1932). Type from Tanzania.
Achyranthes aspera var. *obtusifolia* (Lam.) Suesseng. in Mitt. Bot. Staatss. München: 152 (1952). Type source unknown.

Plant stout and robust, the leaves broadly elliptic to almost circular, blunt or abruptly apiculate, frequently tomentose at least on the lower surface, especially along the veins. Flowers 3—4.5 mm. long, usually shorter and plumper than in var. *pubescens*.

Mozambique. N: Nampula, 25.v.1935, *Torre* 794 (COI; LISC). T: Rd. Tete-Songo 16 km. to Marueira, 7.iv.1972, *Macêdo* 5157 (LISC; SRGH). MS: Nova Chupanga, 40 km. N. of Sena on the R. Zambezi. 30 m., 27 vii.1933, *Lawrence* 55 (K). GI: Guija. Missâo de S. Vicente de Paula, dry sandy soil, 23.vi.1947, *Pedro & Pedrógâo* 1186 (K; LMU). M: Maputo, 4.vi.1946, *Pimenta* 4320 (LISC). *Lawrence* 55 and *Pedro & Pedrógâo* 1186 cited above are rather poor; the upper leaves and short, plump fruiting perianths seem to be referable to var. *aspera*, one or two var. *aspera*/var. *pubescens* intermediates have been seen from Mozambique. In more northerly E. Africa the distribution of this variety favours its original presence along the coastal parts of Kenya and Tanzania (as well as Zanzibar and Pemba), or its introduction there from India, where this is the dominant variety. The inland localities are almost all around the eastern shore of Lake Victoria, to which the plant could have been taken in trading - the fruit clinging readily to clothing or to the hides of animals. The distribution in the Flora Zambesiaca area (near the coast or up the Zambesi) appears to show a similar pattern. The only habitat and altitude available for the variety in our area is given above.

Var. **pubescens** (Moq.) Townsend in Kew Bull. **29**: 473 (1974). TAB. **25** fig. B. Type from Mexico.
Achyranthes fruticosa var. *pubescens* Moq. in DC., Prodr. **13**, 2: 314 (1849).
Achyranthes aspera var. *nigro-olivacea* Suesseng. in Fedde, Repert. **51**: 194 (1942). Type from Tanzania.
Achyranthes aspera f. *robustiformis* Suesseng. in Mitt. Bot. Staatss. München **1**: 70 (1950). Type from Tanzania.
Achyranthes bidentata sensu Baker & Clarke in F.T.A. **6**, 1: 64 (1909).—Cavaco in Mém. Mus. Hist. Nat. Paris Sér. B, **13**: 125 (1962).—Townsend in Kew Bull. **28**: 145—6 (1973) quoad pl. afr. non Blume.
Achyranthes aspera f. *excelsa* Cavaco in Mém. Mus. Nat. Paris Sér. B, **13**: 118 (1962) nom. illegit. (sine descr. lat.).

Plant robust, the perianth-segments generally (4)5—7 mm. long. Leaves oblong to elliptic, not appressed-canescent on the inferior surface, subglabrous to pilose or tomentose, bluntly to distinctly acuminate.

Botswana. N: Kwara Bochai floodplain, 1.v.1973, *Smith* 561 (K; SRGH). SE: Boteti delta area, NE. of Mopipi, 850 m., 17.iv.1973, *Glanville* 3 (K; SRGH). **Zambia**. B: Sesheke, ii.1911, *Gairdner* 486 (K). N: Kasaba Sand Dune, Lake Tanganyika, 1050 m., 14.iv.1957, *Richards* 9219 (K). W: Kitwe, 4.ii.1964, *Mutimushi* 576 (K; NDO). C: 8 km. SE. of Lusaka, 22.ii.1957 *Noak* 122 (K; SRGH). S: Livingstone, 27.iv.1959, *Noel* 1895 (K; SRGH). **Zimbabwe**. N: Gokwe Distr., Sengwa Res. St. 11.v.1969, *Jacobsen* 657 (K; SRGH). W: Gwampa Vlei, vi.1956, *Goldsmith* 118/56 (K; SRGH). C: Harare 1914, *Craster* 129 (K). E: Chirinda Forest Reserve, 1090 m., vi.1967, *Goldsmith* 77/67 (K; LISC; SRGH). **Malawi**. N: Mzimba/Rumphi Distr., S. Rukuru R. bridge, 1060 m., . 8.vii.1974, *Pawek* 8804 (K; MAL; MO; SRGH; UC). C: 35 km. SE. of Lilongwe on Rd. to Dedza, near Kamphata, 1240 m., 21.iii.1970, *Brummitt* 9250 (K; LISC; MAL; PRE; SRGH) S: Zomba Distr., 3 km. W. of Lake Chilwa, by University Res. St. at Katchoka, 640 m., 1.vi.1970, *Brummitt & Williams* 11206 (K; LISC; MAL; SRGH). **Mozambique**. N: E. coast of Lake Malawi c.1900, *Johnson* 144 (K). Z: near Sena, 1859, *Kirk* s.n. (K). MS: c. 8 km. W. of Tambara, 14.vii.1969, *Leach & Cannell* 14332 (K; LISC; SRGH). GI: Guijá, in the area of Posto de Culturas at the edge of the R. Limpopo, 60 m., 7.vii.1948, *Myre* 42 (A; LMU; SRGH). M: Maputo, between Matutuine and Catembe, c. 4 km. from Matutuine, 3.vi.1964, *Balsinhas* 746 (LISC).

Practically throughout the general range of the species (see notes at end). In the Flora Zambesiaca area in numerous habitats from evergreen forest understory to *Acacia* scrub and floodplain (frequently on termitaria), in grassland, as a ruderal, or a weed of cultivation; 60—1460 m.

The African forms included here within *A. aspera* var. *pubescens* need cultivation, as well as cytological and field study to elucidate them. Suessenguth and Cavaco named various specimens, including from the Flora Zambesiaca region, as var. *porphyrostachya*

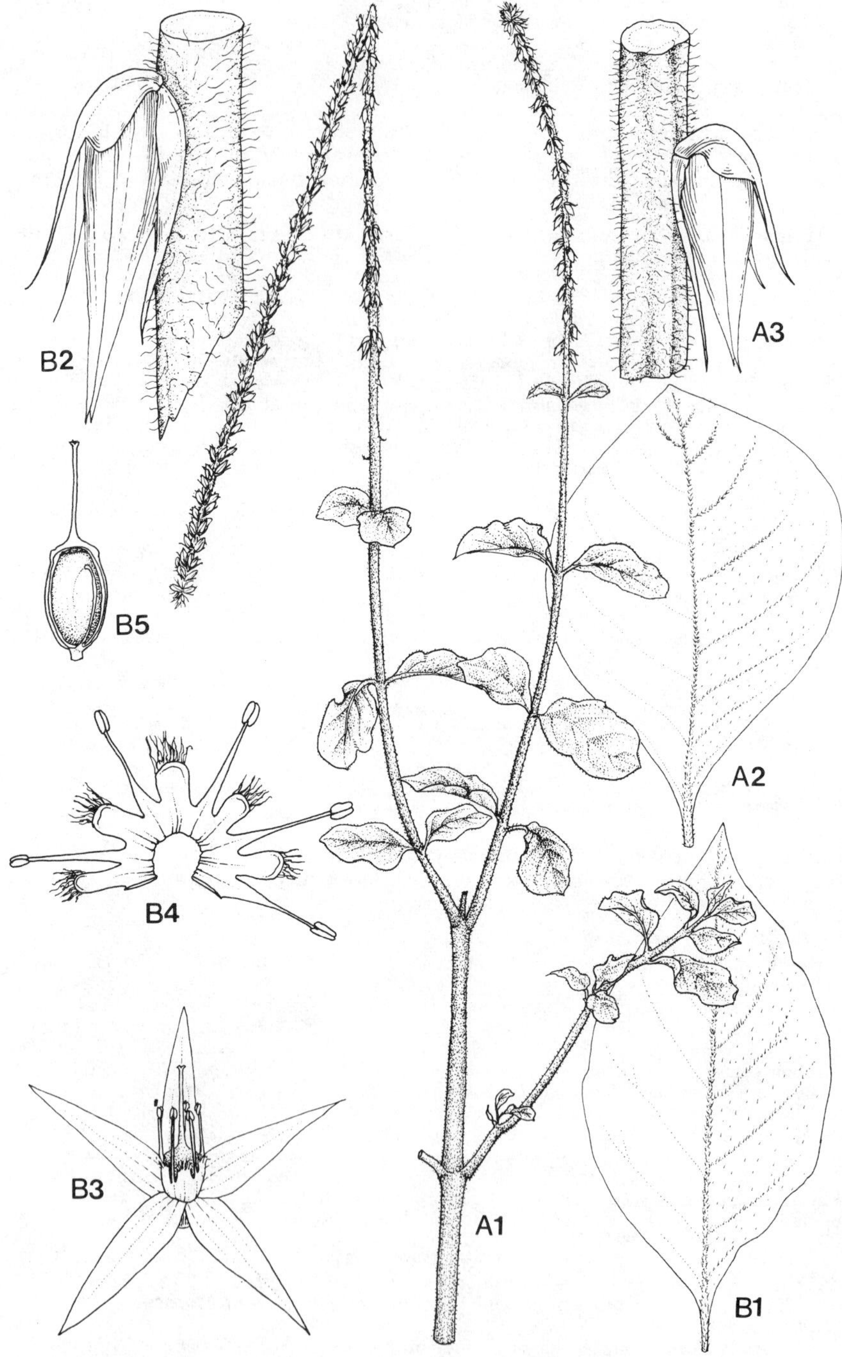

Tab. 25. A.—ACHYRANTHES ASPERA var. ASPERA. A1, flowering branch (×$\frac{2}{3}$) *Archbold* 14; A2, leaf (×$\frac{2}{3}$) *Faulkner* 3521; A3, single flower on stem (×6) *Archbold* 14. B.—ACHYRANTHES ASPERA var. PUBESCENS. B1, leaf (×$\frac{2}{3}$) *Mathenge* 100; B2, single flower on stem (×6); B3, flower with perianth segments spread (×4); B4, androecium (×6); B5, Gynoecium (×6), B2—5 from *Hancock* 121.

(Wall. ex Moq.) Hook.f., an almost glabrous plant described from India; this I believe to be a form of var. *pubescens*, of which it has quite the leaf shape but generally a laxer and more slender inflorescence. The plants needing special attention are those treated by Baker and Clarke in F.T.A., and subsequently by Suessenguth, Cavaco and myself, as belonging to *A. bidentata* Blume. I am now convinced that Hauman (F.C.B. 2: 54 (1951)) was correct in including these within *A. aspera*. They are subglabrous, robust plants with very large leaves drying blackish green and with often very reduced dorsal scales to the pseudostaminodes. They apparently occur constantly in relict-forests, and may possibly thus represent a truly indigenous African taxon worthy of formal infraspecific rank, which, however, I cannot feel able to give them from a study of herbarium material coupled with a field experience of such forms limited to western Kenya.

Var. **sicula** L., Sp. Pl. **1**: 204 (1753).—Cavaco in Mém. Mus. Hist. Nat. Paris Sér. B, **13**: 119 (1962). Type *"Amaranthus radice perpetua* tab. 9", Hb. *Boccone* (P, lectotype, photo.!).

Achyranthes sicula (L.) All. in Misc. Taur. Phil. Math. Soc. R. Turin. **5**: 93 (1774).—Podlech & Meeuse in Merxm. Prodr. Fl. SW. Afr. **33**: 4 (1966).—J.H. Ross Fl. Natal: 159 (1973). Type as above.

Achyranthes argentea Lam., Encycl. **1**: 545 (1785). Type: cultivated material from Paris.

Achyranthes aspera var. *argentea* (Lam.) C.B. Clarke in F.T.A. **6**, 1: 63 (190); in Hauman in F.C.B. 2: 55 (1951).

Achyranthes annua Dinter in Fedde, Repert. **15**: 82 (1951). Type from Namibia.

Achyranthes aspera f. *rubella* Suesseng. in Bull. Jard. Bot. Brux. **15**: 54 (1938). Type from Rwanda.

Plant more slender, the perianth small, generally 3–4.5 (5) mm. long. Leaves frequently silvery-sericeous below at least when young, green above (may be green on both surfaces in shade), rather long-acuminate.

Botswana. N: Okavango Swamps, Moremi Wildlife Reserve, Txatxanika camp site, 930 m., 3.iii.1972, *Biegel & Gibbs-Russell* 3841 (K; LISC; SRGH). SW: Ghanzi pan, Farm S9, 940 m., 11.iii.1970, *Brown* 8798 (K; SRGH). SE: Mochudi, Phutodikobo Hill, 910–1060 m., 10.iii.1967, *Mitchison* 40 (K). **Zambia**. N: Inono Valley, Pineapple Farm, 910 m., 13.iv.1955, *Richards* 5431 (K). W: Kitwe, 16.iv.1963, *Fanshawe* 7772 (K; NDO). C: Lusaka, 1210 m., 22.v.1955, *Best* 101 (K). S: Mumbwa, 1911, *Macaulay* 687 (K). **Zimbabwe**. W: Shangani/Bubi Distr., Gwampa Forest Reserve, 910 m., v.1956, *Goldsmith* 107/56 (K; LISC; SRGH) C: Makani Distr., Rusape Rd., 13.ii.1961, *Rutherford-Smith* 528 (K; SRGH). E: Inyanga, Rupango Mts., 1660 m., 8.ii.1952, *Chase* 4347 (K; LISC; SRGH). S: Mwenezi Experiment Station, 460 m., 20.i.1972, *Barnes* 146 (K; LISC; SRGH). **Malawi**. N: Mzimba Distr., 16 km. SW. of Mzuzu, Mbowe Dam, 1360 m., 30.vi.1974, *Pawek* 8767 (K; MAL; MO; SRGH; UC). C: foot of Dedza Mt. 2 km. NW. of town, c.1760 m., 1.iv.1970, *Brummitt* 9571 (K, LISC, MAL, PRE, SRGH, UPS). S: Zomba Plateau near Ku Chawe Inn, 1530 m., 19.iv.1970, *Brummitt* 9988 (K; LISC; MAL; SRGH). **Mozambique**. N: Lichinga, vi.1934, *Torre* 190 (COI). Z: Gurué, 19.ix.1944, *Mendonça* 2113 (C; LISC; LMU; WAG). M: Maputo, Marracuene, 23.iv.1947, *Barbosa* 168 (COI; LMA; SRGH).

This variety is more restricted in general distribution than var. *pubescens* but is equally widespread in tropical Africa. Its range of habitats in the Flora Zambesiaca region is similar to those of var. *pubescens*; 100–2150 m.

Intermediates between this variety and var. *pubescens* are relatively frequent though apparently not so in the Flora Zambesiaca region; the development of the dorsal scale of the pseudostaminode in var. *sicula* varies from normally developed as in var. *pubescens* to "stagshorn" or much more reduced, or sometimes apical or subapical so that the pseudostaminode appears simple.

Two Zambian specimens of this variety usefully and deliberately gathered in the same locality on the same day (Kitwe, *Fanshawe* 7771 & 7772 (K; NDO)) demonstrate well how the leaves of an *Achyranthes* become larger and greener with reduced indumentum under shaded conditions.

Achyranthes aspera is found almost throughout the world in tropical and warmer regions generally. Some forms may have originated in restricted regions (e.g. var. *sicula* in the Mediterranean region, where it is the only variety represented, and var. *aspera* in India) but if so then these have now become so widely dispersed as to make this purely conjectural.

18. CENTROSTACHYS Wall.

Centrostachys Wall. in Roxb., Fl. Ind. 2: 497 (1824).

Perennial herb with opposite leaves and branches. Leaves entire. Flowers bibracteolate in shortly pedunculate, elongate, bracteate spikes which are terminal on the stem and branches, each bract subtending a single flower; bracts persistent, hyaline; bracteoles more or less circular, hyaline. Perianth segments 5, somewhat spreading at anthesis, later closing together and considerably indurate at the base; upper tepal narrowest and longest, 1–3-nerved. Stamens 5, shorter than the perianth, shortly monadelphous at the base, alternating with spathulate pseudostaminodes furnished with fimbriate dorsal scales; anthers bilocular. Style filiform, stigma capitate. Ovary with a single ovule pendulous on a curved funicle, radicle ascending. Capsule thin-walled, tightly enclosing the seed, falling together with the persistent perianth and bracteoles. Endosperm copious.

A monotypic genus.

Centrostachys aquatica (R. Br.) Wall. ex Moq. in DC., Prodr. **13**, 2: 321 (1849).—Schinz in Engl. & Prantl Pflanzenfam. ed. 2, **16 C**: 62 (1934).—Hauman in F.C.B. **2**: 56 (1951).—Townsend in Kew Bull. **29**: 472 (1974). TAB. **26**. Type from Thailand.

Achyranthes aquatica R.Br., Prodr. Fl. Nov. Holl.: 417 (1810).—Baker & Clarke in F.T.A. **6**, 1: 64 (1909).—Cavaco in Mém. Mus. Hist. Nat. Paris Sér. B, **13**: 124 (1962). Type as above.

Aquatic or subaquatic perennial 0.3–1.5 m., prostrate to straggling or erect, usually much-branched, considerably rooting at the lower nodes with dense tufts of whitish rhizoids, stem near the base up to 2 cm. thick, spongy, hollow. Upper stem and branches sulcate-striate, glabrous for the most part but increasingly appressed-pilose towards the inflorescence. Leaves lanceolate to lanceolate-oblong or oblong-ovate, cuneate or usually attenuate to the base, acute to acuminate at the apex, moderately appressed-pilose on both surfaces (densely so when young), lamina (2.5) 7–15 × (0.8) 2–5 cm.; petiole 0.4–4 cm. long. Spike c. 4–12 cm. long in flower, elongating to as much as 25 cm. or occasionally even more in fruit; rhachis moderately to densely appressed-pilose; peduncle short, mostly c. 1.5 cm. long. Bracts deltoid-lanceolate, 3–4 mm. long, hyaline (drying pale brownish) with a single midrib, glabrous, finally deflexed below the hard callus left by the fallen perianth; bracteoles more or less circular hyaline, c. 1.5–2 mm. long, glabrous. Perianth 6–8 mm. long, the upper (outer) tepal, 1–3 nerved slightly longer than the remainder, with a sharper, often slightly recurved apex, and a somewhat narrower pale border; remaining tepals blunter, with up to 7 nerves. Filaments stout, 2–3 mm. long. Style 1.75–2.5 mm. long. Capsule c. 4 mm. long, slightly broader at the base but rounded above. Seed smooth, chestnut-brown.

Zambia. W: Kaputa Distr., Mweru-Wa-Ntipa, Mwawe R., 1050 m., 6.iv.1957 *Richards* 9054 (K). S: Chepezami Dam, 22 km. W. of Pemba, 1125 m., 21.iv.1954, *Robinson* 702 (K). **Zimbabwe**. N: Hurungwe Distr., Mana Pools, iii.1971, *Guy* 1620 (K; SRGH). S: Mweneze Distr., between Bubye drift and Dumela, 1.v.1961, *Drummond & Rutherford-Smith* 7648 (K; LISC; SRGH). **Mozambique**.

Also in tropical Africa from Nigeria to Sudan and Ethiopia south to Zaire, Tanzania and the Flora Zambesiaca area; in tropical Asia from India and Sri Lanka to Indonesia. In standing water and at the edge of a pan; the only recorded altitudes in the Flora Zambesiaca area are given above.

19. ACHYROPSIS Hook. f.

Achyropsis Hook.f. in Benth. & Hook. Gen. Pl. **3**: 36 (1880).

Annual or perennial herbs or low shrubs with entire, opposite leaves which may be solitary or fasciculate. Inflorescence subcapitate to spicate, bracteate, terminal on the stem and branches, flowers solitary in the axils of the bracts. Modified sterile flowers absent, all flowers hermaphrodite and bibracteolate, small. Bracts persistent, finally weakly deflexed or deflexed-ascending; bracteoles and perianth falling together with the fruit, bracteoles closely appressed to the tepals. Perianth segments 4–5, glabrous, firm, deeply concave and usually more or less cucullate at the apex, 1–3-nerved with the midrib ceasing below the muticous apex or excurrent in a minute mucro. Stamens 4–5, the filaments delicate, shortly monadelphous at the base, alternating with

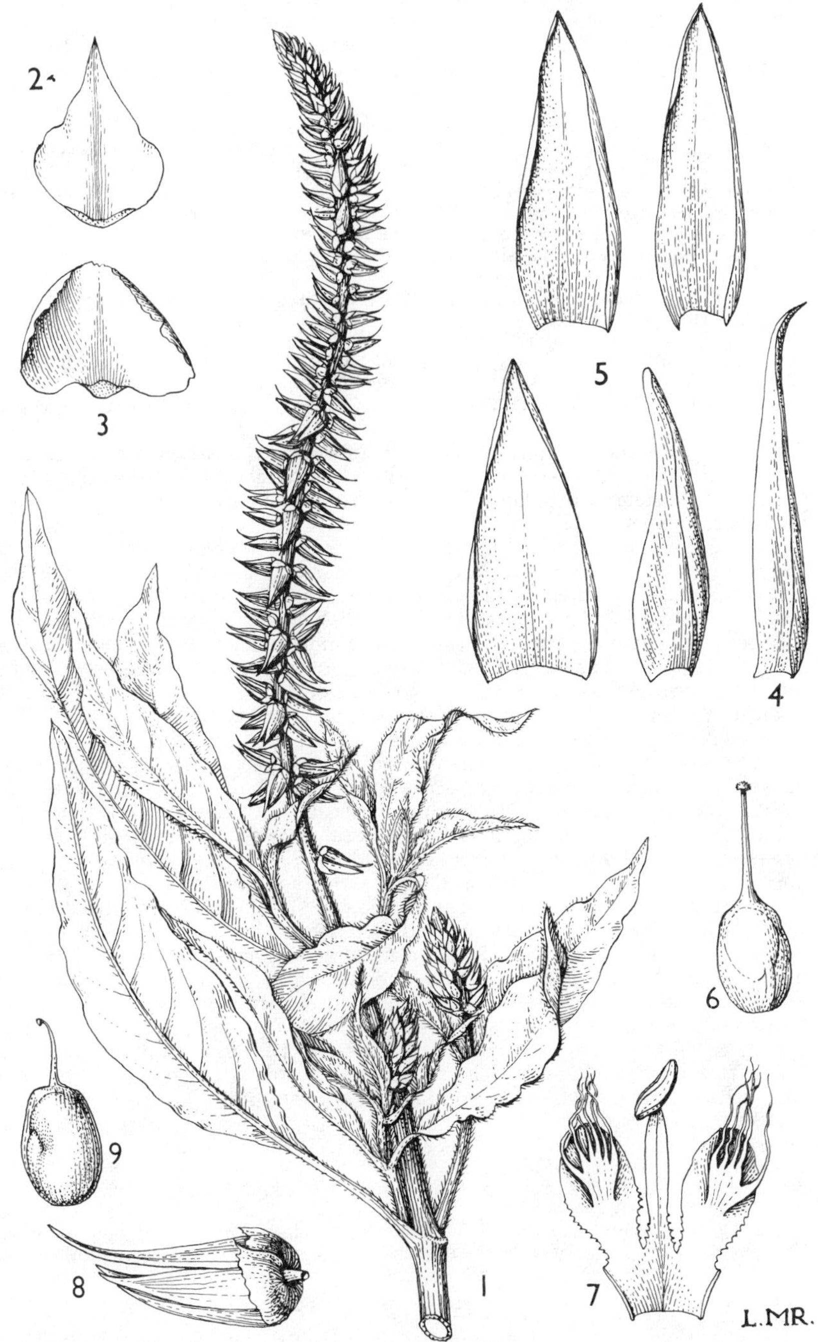

Tab. 26. CENTROSTACHYS AQUATICA. 1, flowering branch ($\times \frac{2}{3}$); 2, bract ($\times 6$); 3, bracteole ($\times 6$); 4, upper tepal ($\times 6$); 5, remaining tepals ($\times 6$); 6, gynoecium ($\times 6$); 7, part of androecium, dorsal surface ($\times 6$), 1–7 from *Greenway* 10146; 8, fruiting perianth ($\times 4$); 9, capsule ($\times 4$), 8–9 from *Greenway & Kanuri* 15446.

distinct pseudostaminodes with or without a dorsal scale; anthers bilocular. Style very short to slender, stigma capitate. Ovary with a single pendulous ovule, glabrous. Fruit a thin-walled capsule, irregularly ruptured by the developing seed. Seed globose or slightly compressed, endosperm copious.

A genus of 6 species in tropical and southern Africa.

1. Inflorescence nodding at maturity, the axis furnished with long, white hairs much exceeding the perianth in length - - - - - - - - - 1. *laniceps*
– Inflorescence erect, the hairs of the axis not as long as the perianth - - 2
2. Leaves elliptic-ovate, those of the main stem not more than 2.5 times as long as broad; tepals rather delicate, the midrib minutely but distinctly excurrent in a mucro - - - - - - - - - - - - - - - - - 3. *gracilis*
– Leaves linear to narrowly elliptic, at least three times as long as wide; tepals firm, the midrib not excurrent - - - - - - - - - 2. *leptostachya*

1. **Achyropsis laniceps** C.B. Clarke in F.T.A. **6**, 1: 66 (1909).—Schinz in Engl. & Prantl Pflanzenfam. ed. 2, **16** C: 63 (1934).—Cavaco in Mém. Mus. Hist. Nat. Paris Sér. B, **13**: 152 (1962). TAB. **27** fig. B. Type: Malawi, Chitipa, *Whyte* s.n., 1816 (K, holotype).

Annual with slender tap-root, 15—100 cm. tall, the smaller forms much reduced with an unbranched stem and a single terminal inflorescence, the larger with numerous slender branches diverging from the stem at about 45 degrees, each branch bearing 1(2) pairs of leaves below the inflorescence; stem and branches tetragonous, striate, furnished with upwardly-directed appressed white hairs. Leaves of main stem linear to linear-oblanceolate, (1)2.2—6(9) × (0.1)0.2—0.7(0.9) cm., narrowed to each end, not or indistinctly petiolate, mucronate at the apex, green and thinly furnished with appressed white hairs on the superior surface, generally whitish and more densely pilose below; upper stem and branch leaves much reduced. Peduncles slender, very variable in length (that of the terminal inflorescence up to c. 12 cm. long), indumentum similar to that of the stem. Inflorescence at first short and conical, becoming cylindrical, woolly, white or pink-tipped, 1—2.5 × 0.4—0.7 cm., nodding when mature, axis white-lanate with the hairs much exceeding the perianth and compressed together by the closely set flowers, so as to appear in tufts between bracts and perianth; bracts 2 mm. long, lanceolate, aristate with the excurrent midrib, hyaline, glabrous or more or less long-ciliate, persistent; bracteoles closely appressed to the perianth, subcircular, c. 0.6—0.75 mm., glabrous, nerveless, falling with the perianth. Tepals all very concave, similar, cucullate above, 1.75—2 mm. with a single very slender midrib which is not excurrent in a distinct mucro, glabrous, broadly white-margined with a green central vitta, occasionally flushed with pink. Filaments c. 1—1.5 mm. long, filiform, anthers subcircular; pseudostaminodes quadrate or rounded, with a narrow, truncate or rounded ventral flap and a narrowly oblong to linear dorsal scale which is fimbriate at the apex and equals or exceeds the filaments. Ovary obpyriform; style very short, to c. 0.4 mm. long. Capsule subglobose, c. 1.25 mm. in diam., delicate below with a small, firm apex; seed filling the capsule, globose, c. 1.25 mm. in diam., yellow-brown, somewhat shining, feebly reticulate.

Zambia. W: Kitwe, 20.iv.1957, *Fanshawe* 3212 (K; NDO). C: Serenje Distr., 1 km. from Kanona on Rd. to Kundalila Falls, c. 1650 m., 12.iii.1975, *Hooper & Townsend* 667 (K; NDO; EA). **Malawi.** N: Nkhata Bay Distr., Viphya, 56 km. SW. of Mzuzu, 1670 m., 15.v.1971, *Pawek* 4802 (K; MAL).

Also in Tanzania, Zaire and Burundi. On sandy or gravelly soil in *Brachystegia* woodland or grassland with sparse trees, also recorded from a rocky, thickety stream valley; 500—1520 m.

2. **Achyropsis leptostachya** (E. Mey. ex Meisn.) Baker & Clarke in F.T.A. **6**, 1: 66 (1909).—Schinz in Engl. & Prantl Pflanzenfam. ed. 2, **16** C: 63 (1934).—Cavaco in Mém. Mus. Hist. Nat. Paris Sér. B, **13**: 150 (1962); in J.H. Ross Fl. Natal: 159 (1973).—Townsend in Kew Bull. **34**: 432 (1980). TAB. **27** fig. A. Type from S. Africa.

Achyranthes leptostachya E. Mey. ex Meisn. in Hook., Lond. Journ. Bot. **2**: 548 (1843). Type as above.

Achyropsis alba Eckl. & Zeyh. ex Moq. in DC., Prodr. **13**, 2: 311 (1849) nom. illegit. superfl. Type as for *Achyropsis leptostachya* (e num., G, NY, isotype).

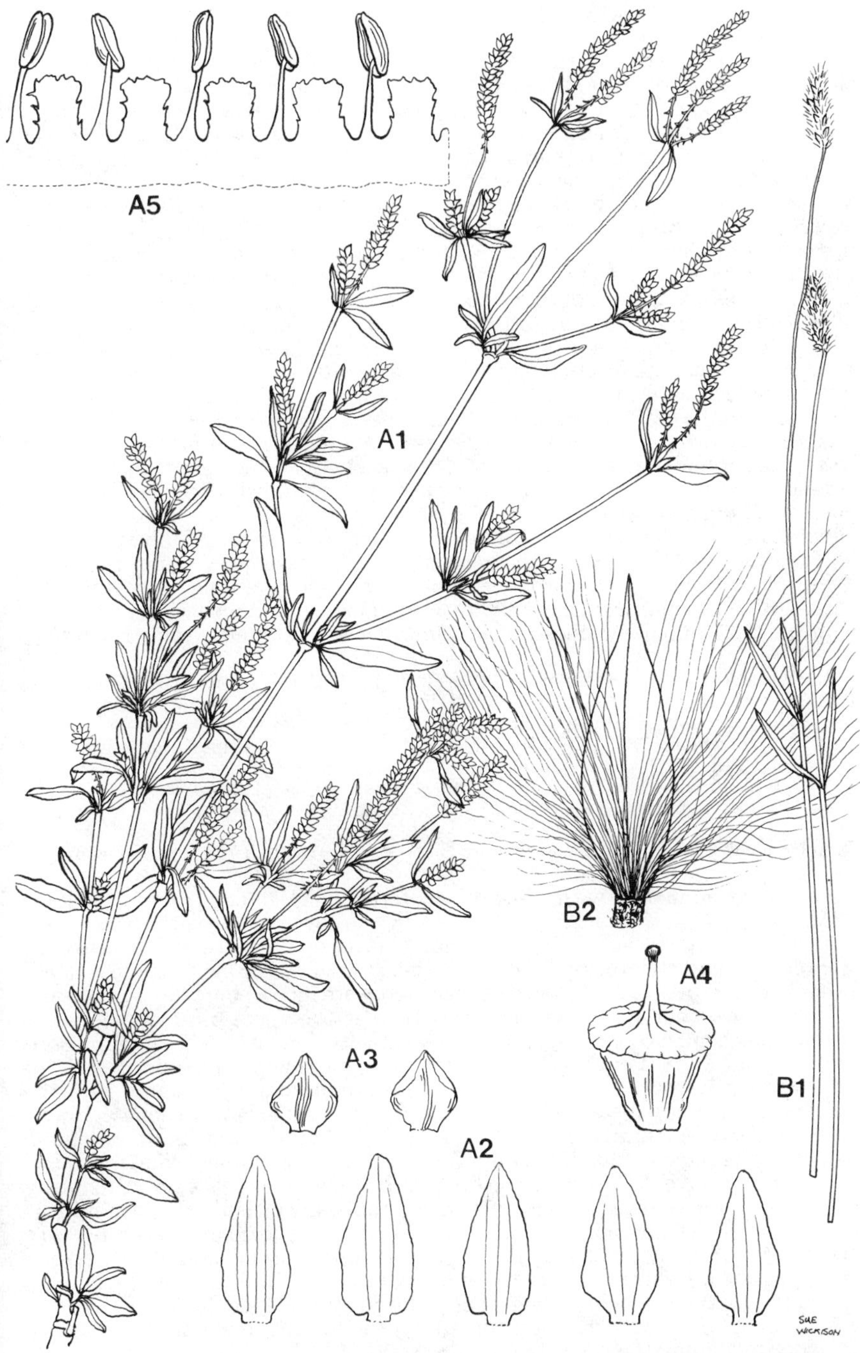

Tab. 27. A.—ACHYROPSIS LEPTOSTACHYA. A1, flowering branch (×⅔); A2, tepals, ventral surface (×10); A3, bracteoles (×10); A4, gynoecium (×24); A5, androecium (×24), A1—5 from *Senderayi* 40. B.—ACHYROPSIS LANICEPS. B1, flowering stem (×⅔); B2, bract (ventral surface) with hairs of inflorescence axis (×14), B1—2 from *Richards* 24488.

Psilotrichum densiflorum Lopr. in Engl. Bot. Jahrb. **30**: 110 (1901); in Malpighia **14**: 453 (1901). Type from S. Africa.

Bushy perennial herb with a tough rootstock, 15–60 cm. tall, with numerous slender branches diverging from the stem at c. 45 degrees; stem and branches tetragonous above with pale raised angles, the older parts terete and striate, furnished with more or less appressed upwardly-directed whitish hairs. Leaves of the main stem linear to narrowly elliptic, 10–55 × 3–6 (10) mm., densely furnished on the lower surface with appressed whitish hairs, more thinly so and dark (often drying blackish-green) above, obtuse to subacute at the apex, tapering to the base and in larger forms with a petiole up to c. 5 mm. long; upper and branch leaves similar but reducing in size; short sterile shoots or fascicles of leaves frequently present between the larger leaves and the stem and branches. Peduncles obsolete to as much as c. 4 cm. long (rarely to 7 cm.), indumentum similar to that of the stem. Inflorescence bluntly conical when young, soon cylindrical, c. 1–4.5 × 0.4 cm., the axis densely white-lanate; bracts 1–1.25 mm. long, glabrous, ovate-acuminate, whitish to stramineous, the stout midrib excurrent in a short but sharp arista, persistent; bracteoles closely appressed to the perianth, subrotund, c. 0.6–0.75 mm. long, pale-margined with a distinct midrib, falling with the perianth. Perianth c. 2 mm. long, more or less evenly rounded from base to apex; tepals all very concave, cucullate above, similar, glabrous, broadly white-margined with a midrib and pair of lateral nerves, all very slender. Filaments c. 1–1.5 mm. long, filiform, the anthers subcircular; pseudostaminodes more or less quadrate, shortly fimbriate, about half the length of the filaments. Ovary obpyriform; style very short, c. 0.5 mm. long. Capsule shortly and broadly ellipsoid, c. 1.25 mm. long, delicate below with a small, firm apex; seed filling the capsule, subglobose to very broadly ellipsoid, 1.25 mm. long, chestnut-brown, somewhat shining, feebly reticulate.

Botswana. SE: Gaberone campus, c. 985 m., 22.iii.1974, *Mott* 1806 (K; UBLS). **Zimbabwe**. W: Bulawayo, 24.x.1975, *Cross* 246 (K; SRGH). C: Shangani R. on Bulawayo-Gweru Rd., 4.v.1957, *Drummond* 5310 (K; LISC; SRGH). S: near. Lundi R., Masvingo-Beitbridge, 2.v.1962, *Drummond* 7860 (K; LISC; SRGH). **Mozambique**. M: Maputo, entre Umbeluzi e Porto Henrique, 16.iv.1948, *Torre* 7630 (COI; K; LISC; MO).

Also in S. Africa. Most frequently found in sandy grassy places, frequently in light shade of scattered small trees or bushes, also along watercourses.

3. **Achyropsis gracilis** Townsend in Kew Bull. **34**: 432 (1980). Type from Tanzania.

Spreading annual (?) with a slender rootstock, 30–40 cm., rooting at some of the lower nodes, sparingly branched in the lower part of the stem with branches divergent at 45 degrees and then ascending, stem and branches tetragonous and striate, moderately furnished with rather long, flexuose hairs. Leaves elliptic-ovate, 2–4 × 0.7–1.3 cm., cuneate into a short (2–4 mm. long) petiole below, shortly pointed and subacute at the apex, paler on the lower surface, both surfaces moderately furnished with fine, whitish, subappressed multicellular hairs. Peduncles 1–4.5 cm. long, indumentum similar to that of the stem. Inflorescence at first conical-capitate, becoming elongate with the flowers increasingly distant below, 2–6 × 0.6 cm., the axis white-lanate; bracts lanceolate-ovate, 2 mm. long, aristate with the excurrent midrib, finely ciliate, persistent; bracteoles subcircular-ovate, c. 1.25 mm. long, finely ciliate, hyaline, the midrib excurrent in a short,fine but distinct mucro. Tepals elliptic, reducing in width from the outer to the inner, membranous, whitish save for a narrow central green vitta, glabrous, 3 mm. long, the slender midrib excurrent in a very short mucro. Stamens c. 2 mm., the filaments filiform; pseudostaminodes oblong with a truncate ventral scale, the dorsal scale long-fimbriate with the cilia subequalling the filaments. Ovary obpyriform; style slender, c. 1 mm. Capsule 1.5 mm., subglobose, seed chestnut, filling the utricle, apiculate beneath the style-base, shining, feebly reticulate.

Mozambique. N: Between Mogincual and Quixaxa, 27.vii.1948, *Pedro & Pedrogão* 4709 (EAH). Z: 25 Miles Station, 61 m., 11.iv.1898, *Schlechter* Plantae Austro-Africanae Iter Secundum 12246 (K, sub *Achyranthes aspera*). M: Marracuene, Ricatla, v.1918, *Junod* 193 (G; LISC; LMU).

Known only from these three gatherings (the second of which was detected while revising Kew *Achyranthes* from the Flora Zambesiaca area) and the Tanzanian type.

20. **PANDIAKA** (Moq.) Hook. f.

Pandiaka (Moq.) Hook.f. in Benth. & Hook. Gen. Pl. **3**: 35 (1880).
Achyranthes Sect. *Pandiaka* Moq. in DC., Prodr. **13**, 2: 310 (1849).
Argyrostachys Lopr. in Engl. Bot. Jahrb. **30**: 108 (1901); in Malpighia **14**: 435 (1901).

Annual or perennial herbs with opposite, petiolate or sessile, entire leaves, rootstock slender to tuberous. Inflorescence spicate to capitate, elongating later or not, flowers spreading or in one species becoming deflexed after flowering. Flowers solitary in the bracts, hermaphrodite. Perianth segments 5, feebly to strongly 1–3 (5)-nerved, lanceolate to narrowly oblong, usually hairy (rarely glabrous), more or less mucronate to aristate with the excurrent nerve. Bracteoles 2, c. half the length of the perianth or more, the midrib excurrent in a short, glabrous and pungent or longer, flexuose and pilose arista, lamina firm, chartaceous or horny. Stamens 5, filaments filiform, monadelphous below, alternating with mostly quadrate pseudostaminodes, these glabrous to pilose, simple or with a highly developed, fimbriate dorsal scale; anthers bilocular. Style slender, stigma small, truncate-capitate. Ovary with a solitary pendulous ovule, fruiting wall very thin; apex at maturity firm, with a transverse crest passing through the base of the style, often forming a hump at each end where it meets the periphery, rarely without such a crest but with a circumferential rim, or almost flat. Perianth falling with the ripe seed, accompanied by the bracteoles or not; bracts persistent, spreading or rarely deflexed.

About 12 species, confined to tropical Africa.

1. Perianth glabrous, or more rarely thinly hairy in the upper half - 7. *carsonii*
— Perianth more or less densely hairy with upwardly directed or patent hairs, or densely lanate - - - - - - - - - - - - - - - - - 2
2. Perianth densely furnished with a long, matted, whitish indumentum more or less concealing the surface; leaves linear - - - - - - - - -6. *richardsii*
— Perianth furnished with straight, ascending to almost patent hairs not concealing the surface; leaves not linear - - - - - - - - - - - - - - 3
3. Leaves expanded and more or less auriculate-amplexicaul at the base, widest above the middle; annual with sharply aristate tepals - - - - - 2. *rubro-lutea*
— Leaves not expanded and more or less auriculate-amplexicaul at the base, if broadest above the middle then the plant a perennial without pungently aristate tepals - - - - - - - - - - - - - - - - - - 4
4. Bracteoles with the midrib excurrent in a long, almost always pilose arista c. half the length of the lamina or more; rootstock tuberous - - - - - - - 5
— Bracteoles with the midrib only shortly excurrent; rootstock not tuberous - - - - - - - - - - - - - - - - - - 6
5. Margins of lower part of filaments and reflexed margins of pseudostaminodes densely long-pilose - - - - - - - - - - - - - - 4. *confusa*
— Margins of filaments and flat margins of pseudostaminodes quite glabrous - - - - - - - - - - - - - - - - -5. *ramulosa*
6. Inflorescences sessile on a pair of broad-based leaves, globose to ovoid or rarely shortly cylindrical, perianth subequalled by the bracteoles - - - 1. *involucrata*
— Inflorescences not sessile on a pair of broad-based leaves, finally elongate-cylindrical; perianth distinctly longer than the bracteoles - - - - - 3. *welwitschii*

1. **Pandiaka involucrata** (Moq.) Jackson in Index Kewensis: 409 (1894).—Baker & Clarke in F.T.A. **6**, 1: 67 (1909).—Schinz in Engl. & Prantl Pflanzenfam. ed. 2, **16 C**: 63 (1934).—Cavaco in Mém. Mus. Hist. Nat. Paris Sér. B, **13**: 133 (1962). Type from Nigeria.
Achyranthes involucrata Moq. in DC., Prodr. **13**, 2: 310 (1849). Type as above.
Pandiaka involucrata var. *megastachya* Suesseng. in Fedde, Repert. **44**: 46 (1938). Type from Nigeria.

Annual herb, simple in very reduced plants, occasionally branched from the base with decumbent branches, but usually erect with many long, slender, erect branches, (0.25) 0.45–1.2 (2) m. tall; stems frequently red, terete below, more or less quadrangular above,striate, clad (as are the branches) with an indumentum of more or less dense, upwardly directed to subpatent whitish hairs. Leaves lanceolate-oblong to oblong-elliptic, those of the main stems 2–10 × 0.9–3 cm., more or less abruptly contracted at the base (superior leaves sessile, the lower usually very shortly petiolate), obtuse-apiculate to subacute at the apex, thinly to moderately furnished with upwardly-directed or subpatent hairs, the upper surface usually darker. Inflorescences subglobose to cylindrical, (1) 1.5–3.5 (5) × (1.25) 1.5–1.75 cm., the flowers very dense and concealing the white-pilose axis, sessile above 2–4 lanceolate to ovate, short leaves; commonly in groups of 3 (a terminal with two laterals on widely spreading branches). Bracts ovate-acuminate, c. 5–8 mm. long, membranous with a narrow or wider greenish central band, very shortly pilose along the midrib only to more generally long-pilose, the midrib excurrent in a distinct, sharp arista. Bracteoles lanceolate-ovate, 6–9 mm. long, densely long-pilose at least along the midrib dorsally, the midrib excurrent in a sharp mucro, the apex of one bracteole (and not rarely the entire flower) more or less curved. Tepals narrowly lanceolate, the outer 2 densely appressed-pilose, narrowly pale- margined with a firm green centre, 6–8.25 mm.; inner 2 c. 0.5–1 mm. shorter, less pilose, 3 nerves sometimes visible in transmitted light only; middle tepal intermediate; all with a rigid, spine-like apex. Stamens 2.5–5 mm. long, the filaments floccose-hairy near the base; pseudostaminodes flabellate, 0.75–1 mm. long, the margins more or less floccose, dorsal scale absent in all material examined. Ovary quadrate-obpyriform, slightly compressed, with a faint apical keel; style slender, (1) 2.5–3.25 mm. long Capsule turbinate ("drum-shaped"), c. 2.5 mm. long, with a broad, flat, firm apex; seeds oblong-ovoid, 2–2.25 mm. long, brown, shining, faintly reticulate.

Zambia. B: Mongu, 30.iii.1966, *Robinson* 6896 (K; SRGH). S: 40 km. NE. of Livingstone, 1140 m., 10.vii.1930, *Hutchinson & Gillett* 3506 (BM; K). **Zimbabwe**. W: Victoria Falls, 914 m., 13.iii.1932, *Eyles* 7298 (K; SRGH). **Malawi**. N: Nkhata Bay Distr., c. 56 km. S. on Chinteche Rd. from main Mzuzu Nkhata Bay Rd., 545 m., 4.iv.1971, *Pawek* 4588 (K; MAL). C: Salima, Lake Shore, by Grand Beach Hotel, 490 m., 29.iv.1970, *Brummitt* 10272 (K; MAL).

A very widespread species in West Africa eastwards to Kordofan, apparently absent in Zaire, in East Africa with only a single locality in extreme SW. Tanzania, and reaching its southernmost limit in the Flora Zambesiaca region. A weedy species, occurring in the Flora Zambesiaca area mostly in sandy *Brachystegia* woodland, and in one case on a sandy lake beach; in West Africa it has a much wider range of habitats, especially in grass savanna, along roadsides, as a weed of cultivation.

There is little if any doubt from the description that *Achyranthes nodosa* Vahl ex Schum. is the same plant as *P. involucrata*; but the International Code of Botanical Nomenclature aims at stability, and the making of a new combination for Schumacher's species under *Pandiaka* would be neither wise nor desirable. The absence of a type of *A. nodosa* gives a good reason for keeping the name by which the present species has long been known.

2. **Pandiaka rubro-lutea** (Lopr.) Townsend in Kew Bull. **34**: 428 (1980). TAB. **28**. Type from Zaire.

Achyranthes rubro-lutea Lopr. in Engl. Bot. Jahrb. **27**: 47 (1899).—Baker & Clarke in F.T.A. **6**, 1: 65 (1909).

Pandiaka andongensis Hiern in Cat. Afr. Pl. Welw. **1**: 895 (1900).—Baker & Clarke in F.T.A. **6**, 1: 70 (1909).—Hauman in F.C.B. 2: 43 (1951).—Cavaco in Mém. Mus. Hist. Nat. Paris Sér. B, **13**: 144 (1962). Type from Angola.

Annual herb, usually considerably branched (the branches divaricate-ascending) but simple in poorly developed specimens, 0.2–0.9 m.; stem and branches ridged, often reddish, moderately to rather densely pilose with upwardly directed, more or less appressed hairs. Leaves broadly obovate to linear, 1.5–8.5 × 0.2–3.5 cm., thinly to moderately pilose, sessile and in the broader-leaved forms commonly constricted c. a third up and then expanded to an auriculate base, apex blunt to subacute, obscurely mucronate. Inflorescence green or the apices of bracts, bracteoles and tepals frequently pink to carmine, conical when young but finally subglobose to cylindrical, 1.2–6.5 × 1.2–1.4 cm., very dense, on a short to long (up to 11 cm.) peduncle or more rarely sessile through branch-condensation and then often with short lateral inflorescences at the base, axis villous. Bracts ovate-lanceolate,3–5 mm.

long, white-membranous, sparingly pilose along the central dorsal surface or glabrous, gradually tapering to the acute, 1–1.5 mm. long arista formed by the excurrent midrib. Bracteoles narrower, lanceolate, 2.75–5 mm. long, more densely pilose, the arista 1.75–2.5 mm. long. Flowers truncate at the base, where they are indurate, fused to the bracteoles, and attached to the inflorescence axis by a knob-like stalk. Tepals lanceolate-subulate, greenish centrally, firm and opaque except for the narrowly hyaline margin, the stout midrib excurrent in a very sharp, pale to reddish arista, 1–2 mm. long, erect or divergent at the apices; 2 outer 5.5–7 mm. long, more or less densely furnished with upwardly directed, more or less spreading white hairs, with 5 nerves which are often white and obvious near the base but obscured above; inner 2 c. 1 mm. shorter, narrower, 3-nerved, hairs mostly confined to the central and upper parts. Perianth and bracteoles falling in one unit in fruit. Stamens 2–3 mm. long; pseudostaminodes c. 0.75–1 mm. long, oblong, truncate with an incurved lobule at the apex, dorsal scale filiform to considerably fimbriate. Ovary subcircular, c. 1 mm. in diam, with a firm, transversely ridged apex; style slender, 1–1.5 mm. long. Capsule oblong-ovoid, 2–2.25 mm. long, the firm, flat apex with a transverse crest on each side of the style; seed c. 2 mm. long, oblong-ellipsoid, brown, shining.

Zambia. B: Zambezi 25.ii.1964, *Fanshawe* 8337 (K; NDO; SRGH). N: Uningi Pans, Mbala, 1500 m., 5.iii.1965, *Richards* 19735 (K). W: Mwinilunga Distr., Kalene Hill, 23.ii.1975, *Hooper & Townsend* 345 (K). C: Kabwe Distr., Kamaila Forest St., 36 km. N. of Lusaka, 9.ii.1975, *Brummitt, Hooper & Townsend* 14298 (K; LUS; NDO). **Zimbabwe**. W: Shangani Distr., Gwampa Forest Reserve, iii.1955, *Goldsmith* 102/56 (K; LISC; SRGH). C: Chegutu Distr., 1.ii.1952, *Hornby* 3245 (BM; SRGH). **Malawi**. N: Rumphi Distr., 21 km. NE. of Rukuru R. bridge, Livingstonia Rd., 1200 m., 12.iv.1976, *Pawek* 10956 (K; MAL; MO). **Mozambique**. N: Mocambique, Meconta, c. 16 km. towards Corrane near Liúpo, c. 150 m., 28.iii.1964, *Torre & Paiva* 11421 (C; COI; EA; K; LISC; LMU; MO; SRGH; WAG).

Also in Angola, Zaire (Shaba Prov.), Rwanda, Burundi and Tanzania. Most commonly on yellowish sandy soil along roadsides and woodland tracks, in open *Brachystegia* or *Julbernardia* forest, in grassland or waste ground, occasionally in thin soil over or near granite rocks, or banks by lakes or watercourses; 150–1520 m.

3. **Pandiaka welwitschii** (Schinz) Hiern in Cat. Afr. Pl. Welw. **1**: 894 (1900).—Baker & Clarke in F.T.A. **6**, 1: 69 (1909).—Schinz in Engl. & Prantl Pflanzenfam. ed. 2, **16 C**: 64 (1934).—Cavaco in Mém. Mus. Hist. Nat. Paris Sér. B, **13**: 140 (1962).—Townsend in Kew Bull. **34**: 426 (1979). Type from Angola.

Achyranthes welwitschii Schinz in Engl. Bot. Jahrb. **21**: 187 (1895). Type as above.

Achyranthes schweinfurthii Schinz in Bull. Herb. Boiss. **4**: 421 (1896). Type from Sudan.

Psilotrichum debile Baker in Kew Bull. **1897**: 279 (1897). Type from Angola.

Pandiaka debilis (Baker) Hiern in Cat. Afr. Pl. Welw. **1**: 984 (1900).—Baker & Clarke in F.T.A. **6**, 1: 1 (1909). Type as above.

Pandiaka schweinfurthii (Schinz) C.B. Clarke in F.T.A. **6**, 1: 69 (1909).—Hauman in F.C.B. **2**: 48 (1951).—Cavaco in Mém. Mus. Hist. Nat. Paris Sér. B, **13**: 141 (1962). Type as for *Achyranthes schweinfurthii*.

Pandiaka welwitschii var. *debilis* (Hiern) Suesseng. in Fedde, Repert. **44**: 47 (1938).

Pandiaka kassneri Suesseng. in Bull. Jard. Bot. Brux. **15**: 67 (1938).—Hauman in F.C.B. **2**: 46 (1951).—Cavaco in Mém. Mus. Hist. Nat. Paris Sér. B, **13**: 145 (1962). Type from Zaire.

Perennial herb, (15) 50–120 cm., taller forms much-branched with the lower branches increasingly widely divaricate; stem and branches more or less densely furnished with whitish to yellowish spreading to upwardly ascending hairs, quadrangular, the older parts more or less glabrescent and terete. Leaves usually obovate and rounded-apiculate, sometimes elliptic and subacute, broadly tapering below, those of the main stem and branches 1.2–8 × 0.6–3.5 cm., moderately appressed-pilose on both surfaces with hairs usually of varying lengths, darker green above and paler beneath. Inflorescences terminal on the stem and branches, sometimes more or less clustered at the apex of the stem by branch reduction, pinkish, conical when young but finally cylindrical, 3–10 × 1.25–1.5 cm. long, very dense, on a 0.5–4 cm. densely tomentose peduncle. Bracts lanceolate-ovate, 4–6 mm. long, membranous with an obscurely to distinctly green centre, more or less densely pilose centrally, diminishing to glabrous at the margins, tapering to the short acute arista formed by the

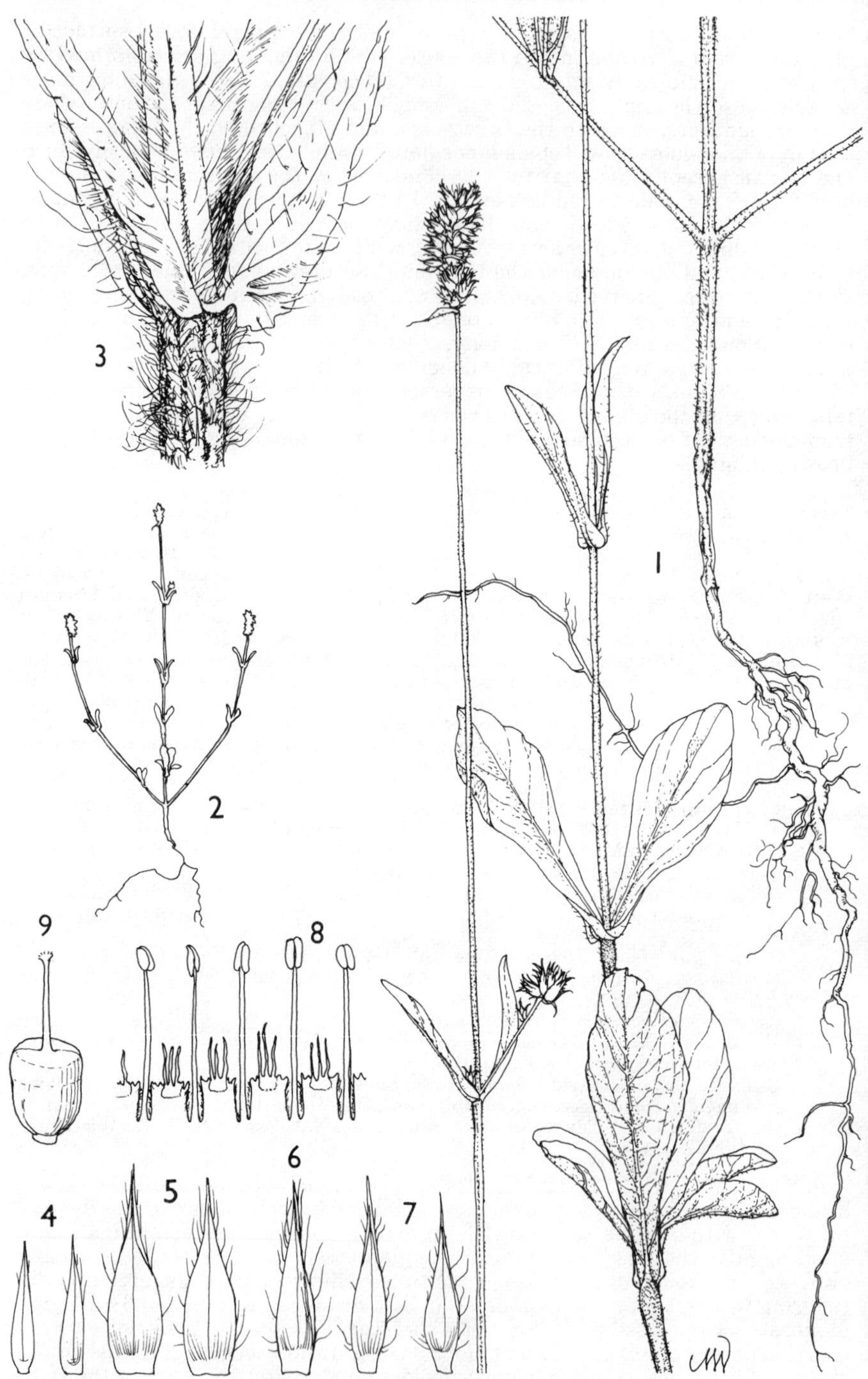

Tab. 28. PANDIAKA RUBRO-LUTEA. 1, flowering plant ($\times \frac{2}{3}$); 2, habit (much reduced); 3, leaf-bases at node (×6); 4, bracteoles (×5); 5, outer tepals (×5); 6, intermediate tepal (×5); 7, inner tepals (×5) (all tepals ventral surface); 8, androecium, ventral surface (×8); 9, gynoecium, (×10), all from *Taylor & Milne-Redhead* 9422A.

excurrent midrib. Bracteoles similar or slightly narrower, 3—5 mm. long, more densely pilose, more longly (c. 0.5—1 mm.) aristate. Flowers truncate at the base and indurate, attached to the inflorescence axis by a white, annular callus. Tepals narrowly lanceolate; outer 2 greenish and more or less densely pilose centrally, 5—8 mm. long, with 3 (5) white nerves, the 1 (2) outer pairs much shorter than the midrib, which is excurrent in a short arista; inner 3 similar but slightly shorter and progressively somewhat narrower and less pilose, 3-nerved or the inner most occasionally with only the midrib showing; all usually slightly upwardly curving near the tip. Perianth and bracteoles falling together in fruit. Stamens 2.5—5 mm.; pseudostaminodes 1—1.5 mm., oblong to flabellate, fimbriate (or dentate only at the apex with the margins fimbriate), dorsal scale broader and fimbriate to subulate-dentate. Ovary obpyriform, c. 1.5 mm.; style slender, 2.75—4.25 mm. Utricle oblong-ovoid, c. 2 mm., the firm, flat apex with a transverse crest on each side of the style; seed c. 1.75 mm., oblong-ellipsoid, brown, shining.

Zambia. B: 6 km. W. of Kabompo, 24.iii.1961, *Drummond & Rutherford-Smith* 7266 (K; LISC; SRGH).

Also in Sudan, Cameroon, Zaire, Angola, Tanzania. Mixed woodland on Kalahari sand; outside the Flora Zambesiaca region in grassland, along riversides, on rocky hillsides ., apparently always on sandy soil.

4. **Pandiaka confusa** Townsend in Kew Bull. **34**: 427 (1980). Type: Zambia, Mwinilunga Distr., just S. of Matonchi Farm, *Milne-Redhead* 2841 (K, holotype).

Pandiaka schweinfurthii var. *compacta* Suesseng. & Overkott in Bot. Archiv. **41**: 78 (1940). Type as above.

Pandiaka schweinfurthii var. *parvifolia* Suesseng. & Overkott, loc. cit. Type as above.

Pandiaka schweinfurthii var. *minor* Suesseng. & Overkott, loc. cit: 79. Type: Zambia, 0.8 km. N. of Mwinilunga, 23.xi.1937, *Milne-Redhead* 3358 (K, holotype).

Perennial herb with a tuberous rootstock, much-branched from the base with numerous, erect stems, these sometimes simple in shorter forms but usually also more or less branched, 15—50 cm. tall; stem and branches quadrangular above, terete below, striate, moderately to rather densely furnished with upwardly directed or more or less patent whitish hairs. Leaves opposite or the upper occasionally ternate, elliptic or narrowly elliptic, 2—7.2 × 0.4—2.5 cm., gradually or more abruptly narrowed at base and apex, more or less acute, moderately to densely furnished on both surfaces with appressed hairs, rarely sublanate. Inflorescence cylindrical, 2—12 × 1.5—2 cm., dense, the axis densely white-pilose, sessile or with a peduncle up to 4 cm. long. Bracts deltoid-ovate, 5—6 mm. long, subglabrous, ciliate or sparsely (the midrib sometimes more densely) pilose, membranous except for the greenish midrib, which is excurrent in a long more or less pilose arista. Bracteoles very similar, not falling with the perianth, the arista about half the length of the lamina. Tepals all rather narrowly lanceolate, narrowly hyaline-margined, obscurely 3—5-nerved in the central green portion, more or less densely pilose with fine, patent or subpatent greenish-white hairs; outer 2 tepals 7—10 mm. long, the inner 3 gradually less pilose and slightly more broadly margined. Stamens 3—4 mm. long, the filaments densely pilose below; pseudo-staminodes 1—2 mm. long, flabellate with reflexed, densely hairy margins, dorsal scale absent or very small and concealed in the marginal hairs. Ovary obpyriform, c. 1 mm. long, with a faint apical keel; style slender, 3—4 mm. long. Ripe capsule oblong-ovoid, c. 3 mm. long; seeds oblong-ovoid, c. 2.75 mm. long, brown, shining, faintly reticulate.

Zambia. W: Kitwe, 6.ii.1964, *Mutimushi* 596 (K; NDO; SRGH).

Also in Angola. Occurs on laterite in open ground or *Brachystegia* woodland.

5. **Pandiaka ramulosa** Hiern in Cat. Afr. Pl. Welw. **1**: 894 (1900).—Baker & Clarke in F.T.A. **6**, 1: 68 (1909).—Schinz in Engl. & Prantl Pflanzenfam. ed. 2, **16 C**: 63 (1934).—Cavaco in Mém. Mus. Hist. Nat. Paris Sér. B, **13**: 135 (1962).—Townsend in Kew Bull. **34**: 429 (1980). Type from Angola.

Pandiaka polystachya Suesseng. in Bull. Jard. Bot. Brux. **15**: 67 (1938). Type from Zaire.

Pandiaka incana Suesseng. & Overkott in Bot. Archiv. **41**: 75 (1940). Type: Zambia,

Mwinilunga Distr., S.W. of Dobeka Bridge, 13.x.1937, *Milne-Redhead* 2738 (K, holotype).

Pandiaka polystachya var. *incana* (Suesseng. & Overkott) Cavaco in Not. Syst. Paris **16**: 98 (1960).

Much-branched perennial herb 14—36 cm. tall, with a woody, nodose rootstock terminating in a large tuber; stem and branches slender, terete or bluntly quadrangular, more or less densely furnished with long, whitish hairs which may be patent, ascending, or a mixture of both. Leaves sessile or very shortly petiolate, 0.6—3 × 0.4—1.3 cm., lanceolate to lanceolate-ovate, shortly narrowed to subcordate at the base, acute to subacute at the apex, moderately to densely furnished on both surfaces with shorter and subappressed to longer and patent hairs. Inflorescence sessile or almost so on the uppermost pair of leaves, greenish to whitish, shortly oblong-ovoid, (sometimes subglobose on the lowest branches), 1—4.5 × 1—1.4 cm., axis white-pilose. Bracts lanceolate, 4—6.5 mm. long, more or less densely furnished with ascending whitish hairs at least centrally, whitish-membranous except for the green midrib, which is excurrent into an arista at least half the length of the lamina. Bracteoles 5—7.5 mm. long, ovate-lanceolate, indumentum similar, the terminal arista commonly longer. Flowers finally somewhat indurate at the base, not fused to the bracteoles. Tepals all lanceolate, broadly hyaline-margined, the green central portion with three distinct white nerves; outer 2 tepals 4—6 mm. long, the midrib excurrent in a short often glabrous mucro, more or less densely appressed-pilose along the green central portion; inner 3 slightly shorter, more pilose about the apex, the mucro absent or obscured by hairs. Perianth apparently falling without the bracteoles. Stamens 2—2.5 mm. long, filaments glabrous; pseudostaminodes small, oblong to narrowly spathulate, c. 1 mm. long, minutely denticulate with no dorsal scale, quite glabrous. Ovary obpyriform, c. 1 mm. long, with a faint apical keel; style slender, c. 2 mm. long. Ripe utricle oblong-ovoid, c. 3 mm. long, narrowly truncate at the apex; seed c. 2.75 mm. long, oblong-ovoid, yellowish brown, shining, obscurely reticulate or almost smooth.

Zambia. W: Mwinilunga Distr., SW. of Dobeka Bridge, 13.x.1937, *Milne-Redhead* 2738 (K, holotype of *P. incana*).

Also in Angola and Zaire. On termite mounds, at the dry fringe of a dambo, and in watershed grassland, perhaps confined to "black cotton" soil on the evidence available, and at least in the Flora Zambesiaca area.

6. **Pandiaka richardsiae** Suesseng. in Mitt. Bot. Staatss. München **1**: 192 (1953).—Cavaco in Mém. Mus. Hist. Nat. Paris Sér. B, **13**: 137 (1962). Type: Zambia, Mbala Distr., by Old Katwe Rd., above Inono Valley, Escarpment above Chilongowelo., c. 1520 m., *Richards* 65 (K, holotype).

Short perennial herb, 12—17 cm., considerably branched below with numerous stems arising from a tough rootstock, the stems simple or very sparingly branched, quadrangular upwards but terete below, more or less furnished with whitish, lanate hairs. Leaves thinly to moderately lanate, more so on the lower surface, linear, 12—40 × 1.5—3 mm., sessile, acute to subacute at the apex, dark green (at least in the dry state). Inflorescence subglobose to shortly cylindrical, 2—4 × 1.5 cm.,axis densely white-lanate, shortly but distinctly pedunculate with a 1—2.5 cm. long peduncle which is increasingly lanate upwards. Bracts ovate to ovate-lanceolate, 4.5—6 mm. long, subglabrous to more or less pilose over most of the surface, the green central portion almost as wide as each pale, membranous margin, midrib shortly excurrent in a rather blunt arista. Bracteoles ovate-lanceolate, 4—5 mm. long, almost entirely membranous, glabrous to thinly pilose, the midrib excurrent in a short (c. 1 mm.) sharp arista. Tepals lanceolate-oblong, all similar in form, densely furnished over the entire surface with long, appressed-matted whitish hairs, broadly white-margined, the green centre strongly 3-nerved, midrib not excurrent in an arista, the rather blunt apices very densely pilose; outer 2 tepals 5—7 mm. long, the inner 3 slightly shorter. Stamens 3.5—4 mm. long, filaments pilose near the base; pseudo-staminodes 1—1.25 mm. long, flabellate, shortly pilose all round, the well-developed dorsal scale divided into very long, filiform fimbriae. Ovary squatly turbinate, c. 1 mm. long, with a faint apical keel; style slender, 3.5–4 mm. long. Mature fruit unknown.

Zambia. N: Chimbwi, Kawambwa, 20.ix.1963, *Mutimushi* 427 (K; NDO; SRGH).
Not known from elsewhere; growing on damp sandy ground in or alongside dambos or amid short grass under *Uapaca* or *Protea* trees.

7. **Pandiaka carsonii** (Baker) Clarke in F.T.A. **6**, 1: 70 (1909).—Schinz in Engl. & Prantl Pflanzenfam ed. 2, **16** C: 64 (1934).—Hauman in F.C.B. **2**: 48 (1951).—Cavaco in Mém. Mus. Hist. Nat. Paris Sér. B, **13**: 138 (1962).—Podlech & Meeuse in Merxm. Prodr. Fl. SW. Afr. **33**: 20 (1966).—Townsend in Kew Bull. **34**: 429 (1980). Type: Zambia, Fwambo, Lake Tanganyika, *Carson* (1893) 8 (K, holotype).
Achyranthes carsonii Baker in Kew Bull. **1897**: 280 (1897). Type as above.
Argyrostachys splendens Lopr. in Engl. Bot. Jahrb. **30**: 109 (1901); in Malpighia **14**: 436 (1901). Type from Tanzania.

Perennial herb, simple in reduced forms but usually with numerous divaricately branched stems arising from the base; stem and branches slender, drying dark green to blackish, glabrous or thinly pilose in the upper part, striate; rootstock a tuber, spherical to large and parsnip-shaped, up to c. 15 cm. long. Leaves very variable, from very narrowly linear to broadly obovate, sessile, (6) 20–70 × 1–18 mm., thinly pilose to somewhat floccose-hairy or frequently glabrous or almost so,often darkening on drying, the margins sometimes more or less crispate, acute to obtuse or apiculate at the apex. Inflorescence cylindrical (or occasionally capitate in small forms), 1.5–8 (10) × 1.25–1.75 cm., not involucrate, on a slender peduncle 1–15 cm. long, axis brownish or usually white-pilose. Bracts very variable in size; from c. 2 mm. long, deltoid-ovate and scarcely differing from the bracteoles to broadly lanceolate, up to 7 mm. long and subequalling the perianth, white, firm, glabrous to minutely ciliate, the midrib shortly excurrent. Bracteoles deltoid-ovate, 1.5–3 mm. long, the nerve shortly to long-excurrent, glabrous or ciliate, falling with the perianth. Tepals more or less equal in length, 5–7 mm. long, with a narrow green central band, a pale midrib and two shorter white nerves, broadly white-bordered (the border of firmer texture in the two outer tepals), shortly and finely mucronate, glabrous or thinly pilose above especially along the firm margins of the central green portion and/or the midrib or extreme apex. Stamens 4–6 mm. long; pseudostaminodes flabellate, 1–1.5 mm. long, with a long-fimbriate dorsal scale. Ovary obpyriform, c. 1 mm. long, firm in the upper half; style slender, 3.5–4.5 mm. long. Capsule oblong-ovoid with a narrower firm apex, 2.5–3 mm. long. Seeds oblong-ovoid, 2.25–2.75 mm. long, brown, shining, faintly reticulate.

Var. **carsonii**.

Leaves broadly linear to broadly ovate, those of the central part of the main stem mostly more than 4 mm. wide.

Zambia. N: Kaputa Distr., Mweru-Wa-Ntipa, Kabwe Plain, c. 1000 m., 14.xii.1960, *Richards* 13699 (K; SRGH). W: Ndola, 22.xi.1959, *Wild* 4880 (K; LISC; SRGH). S: Choma Distr., Mochipapa Agric. St., 14.i.1960, *White* 6253a (FHO; K). **Zimbabwe**. N: Miami Distr., "Vlei settlement area", 22.xi.1944, *Hopkins* in SRGH 13056 (K; SRGH). **Malawi**. C: Kasungu Nat. Park, 1000 m., 23.xii.1970, *Hall-Martin* 1358 (K; SRGH).
Also in Zaire. Frequently in dambos or seasonally wet grassland, also in *Brachystegia/Julbernardia* or *Brachystegia/Isoberlinia* woodland, and one record from a roadside; 1000–1670 m.

Var. **linearifolia** Hauman in F.C.B. **2**: 49 (1951). Type from Zaire.
Pandiaka milnei Suesseng. & Overk. in Bot. Archiv. **41**: 76 (1941). Type: Zambia, Mwinilunga Distr., Kalenda Plain, *Milne-Redhead* 2835 (K, holotype).
Pandiaka carsonii var. *milnei* (Suesseng. & Overk.) Cavaco in Mém. Mus. Hist. Nat. Paris Sér. B, **13**: 139 (1962).

Leaves very narrowly linear to filiform, those of the central part of the main stem rarely attaining 3 mm. in width.

Zambia. B: c. 6 km. N. of Kalabo, 14.xi.1959, *Drummond & Cookson* 6456 (K; LISC; SRGH). N: Nsomba to Luwingu, 1200 m., 14.x.1947, *Greenway & Brenan* 8214 (EAH;

K). W: Mwinilunga District, c. 19 km. beyond Samuteba on Mwinilunga-Solwezi road, c. 1400 m., 26.ii.1975, *Hooper & Townsend* 407 (K; LUS,;NDO; SRGH). C: Chaongwe River, W. of Kasisi Mission, 30 km. NE. of Lusaka, 1150 m., 3.xii.1972, *Strid* 2635 (K).

Also in Zaire. Drier parts and edges of dambos, grassy plains on Kalahari Sand (Barotse sand), damp sand and black soil by rivers, damp shallow soil on sandstone, and in *Brachystegia* woodland.

A rather poor variety, connected with the type with certain intermediates; but the extreme forms are very striking. Of the above, the most extreme is *Hooper & Townsend* 407, with filiform leaves up to 7 cm. long, none of which exceed 1 mm. in width.

Certain species described from Zaire, *P. glabra* (Schinz ex Suesseng.) Hauman, *P. lanata* (Schinz) Hauman and *P. obovata* Suesseng. appear to be at most varieties of *P. carsonii*.

21. GUILLEMINEA Kunth

Guilleminea Kunth in H.B.K., Nov. Gen. Sp. **6**: 40, pl. 518 (1823), emend. Mears in Sida **3**: 137–8 (1967).

Gossypianthus Hook., Ic. Pl. 3, t. 251 (1840).

Brayulinea Small, Fl. S.-E. U.S.: 394 (1903).

Perennial herbs, rarely somewhat woody below, with entire, opposite leaves. Inflorescences sessile, axillary, densely spiciform, often densely fasciculate at the nodes. Flowers hermaphrodite, solitary within the axil of each bract, bibracteolate. Perianth segments elliptic or ovate, almost entirely delicate and hyaline or firmer with 3 green nerves, free or fused to about halfway; perianth and bracteoles falling with the fruit, bracts persistent. Stamens 5, the filaments fused into a tube, the tube free (not in Africa), or adnate to the perianth tube; anthers unilocular. Ovary with a single pendulous ovule; style usually short (longer in one West Indian species); stigma shortly bilobed. Utricle thin-walled, bursting irregularly. Seed compressed, firm.

A genus of 5 species, all natives of the Americas from southern U.S.A. to Argentina; one an increasingly widespread tropical weed.

Guilleminea densa (Willd.) Moq. in DC., Prodr. **13**, 2: 338 (1852).—Mears in Sida **3**: 140–144 (1967). TAB. **29**. Type, "Habitat in America meridionali".

Illecebrum densum Willd. ex Roem. et Schult., Syst. Veg. ed. **15**, 5: 517 (1819). Type as above.

Guilleminia illecebroides Kunth in H.B.K., Nov. Gen. Sp. **6**: 40, pl. 518 (1823). Type from Ecuador.

Brayulinea densa (Willd.) Small, Fl. S.-E. U.S.: 394 (1903).—Schinz in Engl. & Prantl Pflanzenfam. ed. 2, **16 C**: 64 (1934).—Cavaco in Mém. Mus. Hist. Nat. Paris Sér. B, **13**: 156 (1962).—Podlech & Meeuse in Merxm. Prodr. Fl. SW. Afr. **33**: 9 (1966).—Guillarmot, Fl. Lesotho: 167 (1971); in J.H. Ross Fl. Natal: 159 (1973).

Prostrate or sometimes decumbent, mat-forming perennial herb with a rootstock considerably thickened for up to c. 5 cm. below the ground and then abruptly more slender, mat from c. 7–70 cm. across. Stems numerous from the base, much-branched, branches opposite (or alternate by reduction of one of the pair), more or less densely white-lanate. Leaves variable in size and shape, the lamina narrowly elliptic to broadly ovate, mostly 5–22 × 1.5–14 mm., acute to subacute at the apex, rapidly narrowed below to a broad petiole up to c. 8 mm. long, glabrous or subglabrous on the upper surface, more or less densely lanuginose with long, matted, white hairs on the lower surface, especially when young. Inflorescences dense, ovoid, of up to c. 10 flowers, whitish, to c. 6 mm. long, the axis long-pilose; lower flowers frequently minutely pedicellate below the bract, the upper sessile. Bracts hyaline, very delicate and concave, c. 1.5–2 mm. long, frequently splitting with age, glabrous, persistent; bracteoles similar but slightly shorter. Tepals united for about half their length, densely sinuose-lanuginose, the lobes very delicate, hyaline except for a pale, firmer midrib, whole perianth c. 2–2.5 mm. long at maturity. Staminal tube completely adnate to the perianth tube, the filaments indicated only by short, triangular teeth; anthers very small, c. 0.25 mm., ovoid. Ovary ellipsoid, firm only at the extreme apex; style very short, c. 0.25 mm. long. Capsule c. 1–1.25 mm. long, ellipsoid. Seed c. 1 mm. long, compressed-ellipsoid, chestnut-brown, faintly reticulate.

Tab. 29. GUILLEMINIA DENSA. 1, flowering branch (×½); 2, leaf on branch (×8); 3, lateral view of flower (×16); 4, plan view of flower (×16); 5, flower opened up and flattened (×16); 6, oblique view of part of inside of flower (×18); 7, gynoecium (×18), all from *Drummond* 5135.

Botswana. N: Shakawe, 25.iv.1975, *Biegel, Müller & Gibbs-Russell* 5003 (K; SRGH). SW. Ghanzi, 30.i.1970, *Brown* 8274 (K; SRGH). SE: Mahalapye, 24.ii.1977, *Camerik* 77 (K; PRE). **Zambia**. S: Machili, 21.ix.1969, *Mutimushi* 3790 (K; NDO). **Zimbabwe**. W: Hwange Nat. Park, main camp near Guvalala Pan, 11.xi.1968, *Rushworth* 1256 (K; SRGH). C: Regina Mundi, 8 km. N. of Gweru, i.1967, *Biegel* 1806 (K; SRGH). E: 33 km. E. of Mutare on Harare Rd., 6.i.1969, *Biegel* 2738 (K; SRGH). S: Mwenezi Distr., Mapwe R., Masvingo-Beitbridge Rd., 2.v.1962, *Drummond* 7861 (K; LISC; SRGH). **Malawi**. S: Blantyre, city centre, 12.iii.1970, *Brummitt* 9033 (K; MAL; SRGH; LISC). **Mozambique**. MS: Chicamba Dam, 18.i.1968, *Wild* 7673 (K; LISC; SRGH). M: Maputo, Vasco da Gama's garden, 7.vii.1972, *Balsinhas* 2366 (K; SRGH).

A native of the warmer regions of the Americas from the southern U.S.A., to northern Argentina. Introduced into Australia (Queensland) and spreading in S. and tropical Africa. In sandy and gravelly soil and red clay along roadsides and rail tracks, as a lawn weed, and in an overgrazed trodden area in *Brachystegia/Upaca* woodland; 385–992 m. recorded.

The only member of the *Amaranthaceae* in the Flora Zambesiaca region in which the tepals are not free, this species has the mat-forming habit of an *Alternanthera* combined with the woolly inflorescences of an *Aerva*. The wavy, *Gomphrena*-like hairs of the perianth are, however, quite unlike those of *Aerva*.

The genus *Brayulinea* was intended as a replacement for *Guilleminea* Kunth (1823) non Necker (1790). Since Necker's names are not recognised as generic under Article 20 of the International Code of Botanical Nomenclature, this substitution is superfluous.

22. ALTERNANTHERA Forssk.

Alternanthera Forssk, Fl. Aegypt-Arab.: 28 (1775).

Annual or perennial herbs, prostrate or erect to floating or scrambling, with entire, opposite leaves. Inflorescences of sessile or pedunculate heads or short spikes, axillary, solitary or clustered, bracteate. Flowers hermaphrodite, solitary in the axil of each bract, bibracteolate, bracts persistent, the perianth falling with the fruit, bracteoles persistent or not. Perianth segments 5, free, equal or unequal, glabrous or furnished with smooth or denticulate hairs. Stamens 2–5, some occasionally without anthers, the filaments distinctly monadelphous at the base into a cup or tube, alternating with large and dentate or laciniate to very small pseudostaminodes (rarely these obsolete), anthers unilocular. Style short, stigma capitate. Ovary with a single pendulous ovule. Fruit an indehiscent capsule, thin-walled or sometimes corky, seeds more or less lenticular.

A large genus of c. 200 species, chiefly in the New World tropics.

1. Tepals very dissimilar in form - - - - - - - - - - - - - 2
– Tepals all similar in form, subequal or the inner 2 slightly shorter - - - 3
2. Abaxial tepals very long-aristate (the awn c. one third of the total tepal length), with tufts of barbellate hairs near the basal angles; adaxial tepal strongly denticulate - - - - - - - - - - - - - - - - 1. *pungens*
– Abaxial tepals only shortly aristate (the awn less than one quarter of the total tepal length), with barbellate hairs in the basal half or more; adaxial tepal almost entire - - - - - - - - - - - - - - - - 2. *caracasana*
3. Outer tepals prominently 3-nerved in the lower half, with numerous barbellate hairs - - - - - - - - - - - - - 3. *tenella* var. *bettzickiana*
– Outer tepals 1-nerved with the prominent midrib only - - - - - - 4
4. Ripe fruit yellowish with tumid margins on each side of the seed, only about half the length of the usually 3–4 cm. long perianth segments- - - 4. *nodiflora*
– Ripe fruit dark, thin-margined with only a narrow yellowish rim, almost as long to sometimes exceeding the 1.5–2.5 mm. long perianth segments - - 5. *sessilis*

1. **Alternanthera pungens** Kunth in H.B.K., Nov. Gen. Sp. **2**: 206 (1817).—Schinz in Engl. & Prantl Pflanzenfam. ed. 2, **16 C**: 74 (1934).—Podlech & Meeuse in Merxm. Prodr. Fl. SW. Afr. **33**: 5 (1966); in J.H. Ross Fl. Natal: 159 (1973). TAB. **30** fig. B. Type from Colombia.

Achyranthes repens L., Sp. Pl.: 205 (1753). Type, Eltham Gardens, ex Herb. Dillenius (OXF, lectotype★).

Illecebrum achyrantha L., Sp. Pl. ed. 2: 299 (1762). Type as above.

Alternanthera achyrantha (L.) Sweet, Hort. Suburb. London: 48 (1818). Type as above.

Alternanthera echinata Sm. in A. Rees, Cyclop.: 39, Add. & Corrig., sub.

Alternanthera No. 10 (1819).—Baker & Clarke in F.T.A. **6**, 1: 74 (1909). Type from Uruguay.

Alternanthera repens (L.) Link, Enum. Hort. Berol. Alt. **1**: 154 (1821).—Hauman in F.C.B. **2**: 76 (1951).—Cavaco in Mém. Mus. Hist. Nat. Paris Sér. B, **13**: 161 (1962). Type as for *Achyranthes repens*.

★ Fide Melville in Kew Bull. **13**: 172—3 (1958). The specimen has not been seen. No Linnean specimen exists of this plant; those under the name *Achyranthes repens* in Linnaeus' herbarium at Stockholm appear from IDC. microfiche Neg. 101 nos. 9 & 11 to be *A. repens* Gmel. - i.e. *A. sessilis*.

Prostrate, mat-forming perennial herb with a stout vertical rootstock, also rooting at the lower nodes, much-branched from the base outwards, mats up to c. 1 m. across. Stem and branches terete, striate, stout to more slender, more or less densely villous with long, white hairs but frequently glabrescent with age. Leaves broadly rhomboid-ovate to broadly elliptic or obovate, 1.5—4.5 × 0.3—2.7 cm., rounded to subacute at the apex with a mucro which in the young leaves is often fine and bristle-like, narrowed below to a petiole up to 1 cm. long, glabrous or thinly appressed-pilose on both surfaces, especially on the lower surface of the primary venation. Inflorescences sessile, axillary, solitary or more commonly 2-3 together, globose to shortly cylindrical, 0.5—1.5 cm. long and 0.5—1 cm. wide; bracts membranous, white or stramineous, 4—5 mm. long, deltoid-ovate, glabrous, marginally ciliate or dorsally pilose, distinctly aristate with the excurrent midrib, more or less denticulate around the upper margin; bracteoles similar but smaller, 3—4 mm. long, falling with the fruit. Tepals extremely dissimilar; the 2 outer (abaxial) deltoid-lanceolate, 5 mm. long, denticulate above, very rigid, 5-nerved below (the intermediate pair much shorter and finer), outer 2 nerves meeting above to join the pungently excurrent midrib, which forms a long arista c. one third of the length of the entire tepal; inner (adaxial) tepal oblong, flat, 3 mm., blunt and strongly dentate at the apex, 3-nerved below but the nerves meeting well below the apex and the apical mucro short and fine; lateral tepals c. 2 mm. long, sinuate in lateral view with the two sides of the lamina connivent and denticulate above, sharply mucronate; abaxial and adaxial petals with small tufts of glochidiate and barbed whitish bristles about the basal angles, the lateral tepals each with a large tuft about the centre of the midrib. Stamens 5, all with anthers, at anthesis subequalling or slightly exceeding the ovary and style, the alternating pseudostaminodes broad, subquadrate or shorter, entire to dentate. Ovary compressed, squat, style very short, as wide as or wider than long. Fruit orbicular, rounded to retuse above, 0.2 mm. long. Seed discoid, c. 1.25 mm. in diam., brown, shining, faintly reticulate.

Botswana. N: Maun, 30.xi.1974, *Smith* 1211 (K; SRGH). SE. Mochudi, Phutodikobo Hill, 910—1060 m., 6.iii.1967, *Mitchison* 29 (K). **Zambia**. C: Luangwa R., 30.v.1958, *Fanshawe* 4466 (K; NDO). S: Mazabuka, 6.iii.1963, *van Rensburg* 1600 (K; SRGH). **Zimbabwe**. N: Hurungwe Distr., Mana Pools Nat. Park, vii.1970, *Guy* 893 (K; SRGH). C: Harare, v.1914, *Craster* 156 (K). E: Inyanga, ad villam Cheshire, c. 300 m., 15.i.1931, *Norlindh & Weimarck* (K; L). S: Great Zimbabwe, 29.iii.1973, *Chiparawasha* 644 (K; SRGH). **Malawi**. S: Chikwawa Distr., W. bank of Shire R. at Kasisi, c. 6 km. N. of Chikwawa, c. 95 m., 21.iv.1970, *Brummitt* 10027 (K; MAL; SRGH). **Mozambique**. T: Between Marueira and Songo, c. 1 km. towards Songo, 24.iii.1972, *Macêdo* 5083 (K). MS. Gorongosa Nat. Park, SE. Urema Plains, iii.1972, *Tinley* 2457 (K; SRGH). GI: Dumela, 30.iv.1961, *Drummond & Rutherford-Smith* 7608 (K; LISC; SRGH). M: Quinta to Umbelúzi, 27 km. W. of Maputo, 9.v.1946, *Gomes e Sousa* 3452 (K).

A native of tropical America now widespread as a weed in tropical Africa. Found along roadsides, on waste ground, on overgrazed areas, in grassland, along watercourses; appears to thrive on being trodden, being generally on compacted sandy soil.

2. **Alternanthera caracasana** Kunth in H.B.K., Nov. Gen. Sp. **2**: 205 (1817). TAB. **30** fig. A. Type from Venezuela.

Illecebrum peploides Humboldt & Bonpland ex Schultes, Syst. Veg. ed. 15, **5**: 517 (1819). Type, Santo Domingo.

Telanthera caracasana (Kunth) Moq. in DC., Prodr. **13**, 2: 370 (1849). Type as for *Alternanthera caracasana*.

Alternanthera peploides (Schultes) Britton in Britton & Brown, Bot. Puerto Rico

& Virgin Is. 2: 279 (1924).—Schinz in Engl. & Prantl Pflanzenfam. ed. 2, **16 C**: 74 (1934); in J.H. Ross Fl. Natal: 159 (1973). Type as for *Illecebrum peploides*.

Prostrate, mat-forming perennial herb with a stout vertical rootstock, also rooting at the lower nodes, much-branched from the base outwards, mats up to c. 1 m. across. Stem and branches terete, striate, stout to slender, more or less densely villous with long, white hairs. Leaves broadly ovate to broadly elliptic or obovate, 0.8—3.2 × 0.4—1.5 cm., rounded to subacute at the apex, sharply mucronate, narrowed below to a distinct petiole up to c. 1 cm. long, glabrous to thinly long-pilose on both surfaces, especially about the base. Inflorescences sessile, axillary, solitary or 2—3 together, usually shortly cylindrical, 0.5—1.5 cm. long and 0.4—0.7 cm. wide; bracts membranous, white or stramineous, 3—3.5 mm. long, deltoid-ovate, glabrous or slightly pilose at the basal margins, distinctly aristate with the excurrent midrib, entire; bracteoles lanceolate, sometimes falcate, commonly somewhat wider on one side of the midrib than the other, 3—3.5 mm. long, glabrous or slightly pilose along the keel, subentire. Tepals extremely dissimilar; the outer 2 (abaxial) deltoid-lanceolate, 4—4.5 mm., very rigid, 3-nerved below with the laterals meeting the midrib c. two thirds of the way up, midrib excurrent to form an arista less than one quarter the length of the tepal; inner (adaxial) tepal oblong, flat, 3—4 mm., entire or faintly denticulate at the apex, 3-nerved with the laterals meeting the midrib c. two thirds up, finely mucronate; lateral tepals 2.75—3 mm. long, sinuate in lateral view with the two sides connivent above, entire or almost so, sharply mucronate; abaxial and adaxial tepals with glochidiate and barbed whitish hairs to about halfway or more, the laterals with a long tuft in c. the centre half of the midrib. Stamens 5, all with anthers, at anthesis subequalling or slightly exceeding the ovary and style, the alternating pseudostaminodes narrowly triangular-subulate, subequalling or slightly shorter than the filaments. Ovary compressed, squat, style very short, about as long as wide. Fruit orbicular, rounded to retuse above, c. 2 mm. long. Seed discoid, c. 1.25 mm. in diam., brown, shining, faintly reticulate.

Botswana. SE: "Railway siding, Bechuanaland", no further details, viii.1929, *Sandwith* s.n. (K). **Zambia**. B. near Luena R. 14 km. ESE. of Kaoma, 21.xi.1959, *Drummond & Cookson* 6703 (K; SRGH). N: Mbala Aerodrome, 1520 m., 2.v.1955, *Richards* 5471 (K). W: Ndola, 1220 m., iv.1961, *Wilberforce* A/57 (K). C: Great E. Rd. 22 km. E. of Lusaka, 1100 m., 5.xii.1971, *Kornas* 0601 (K). S: Mazabuka, 1000 m., 9.x.1931, *Trapnell* CRS 441 (K). **Zimbabwe**. N: Mazowe Citrus Estate, 1210 m., 18.iii.1971, *Searle* 221A (K; SRGH). C: Harare, 1485 m., 23.xi.1968, *Biegel* 2691 (K; LISC; SRGH). **Malawi**. N: Chitipa Distr., by Rest House, Chisenga, 1540 m., 12.vii.1970, *Brummitt* 12026 (K; MAL; SRGH; LISC; PRE; UPS). C: Dedza Distr., Forest Institute Building, Chargari Forest, 10.i.1972, *Salubeni* 1742 (K; SRGH).

A native of tropical America now widespread in tropical Africa. By roadsides and railway tracks, in lawns, on waste ground, usually on compacted sandy soil where, like *A. pungens*, it seems to thrive on being trodden.

3. **Alternanthera tenella** Colla in Mem. R. Accad. Sci. Torino **33**: 131 t.9 (1828). Represented in Africa only by var. *bettzickiana* (Regel) Veldk. in Taxon **27**: 313 (1978). Type, cultivated material from St. Petersburg Botanic Garden (LE, holotype).

Telanthera bettzickiana Regel in Ind. Sem. Hort. Petrop. **1862**: 28 (1862); in Gartenflora **11**: 178 (1862). Type as above.

Alternanthera bettzickiana Nichols., Ill. Dict. Gard. **1**: 59 (1884) nomen subnudum.

Alternanthera bettzickiana (Regel) Voss in Vilm., Blumengart. ed. 3, **1**: 689 (1895) sphalm. *bettzichiana*.—Schinz in Engl. & Prantl Pflanzenfam. ed. 2, **16 C**: 75 (1934).—Hauman in F.C.B. 2: 79 (1951).

Alternanthera ficoidea var. *bettzickiana* (Nichols.) Backer in Fl. Males. Ser. **1**, 4: 93 (1949).

Erect or ascending, bushy perennial herb (commonly cultivated as an annual), c. 5—45 cm. tall, stem and branches villous when young but soon glabrescent, older parts terete, younger bluntly quadrangular. Leaves narrowly or more broadly elliptical to oblanceolate or rhomboid-ovate, acute to acuminate at the apex, attenuate into a slender, indistinctly demarcated petiole below, thinly furnished with fine, whitish hairs to subglabrous, often reddish or purple suffused and not rarely variegated. Heads axillary, sessile, usually solitary,

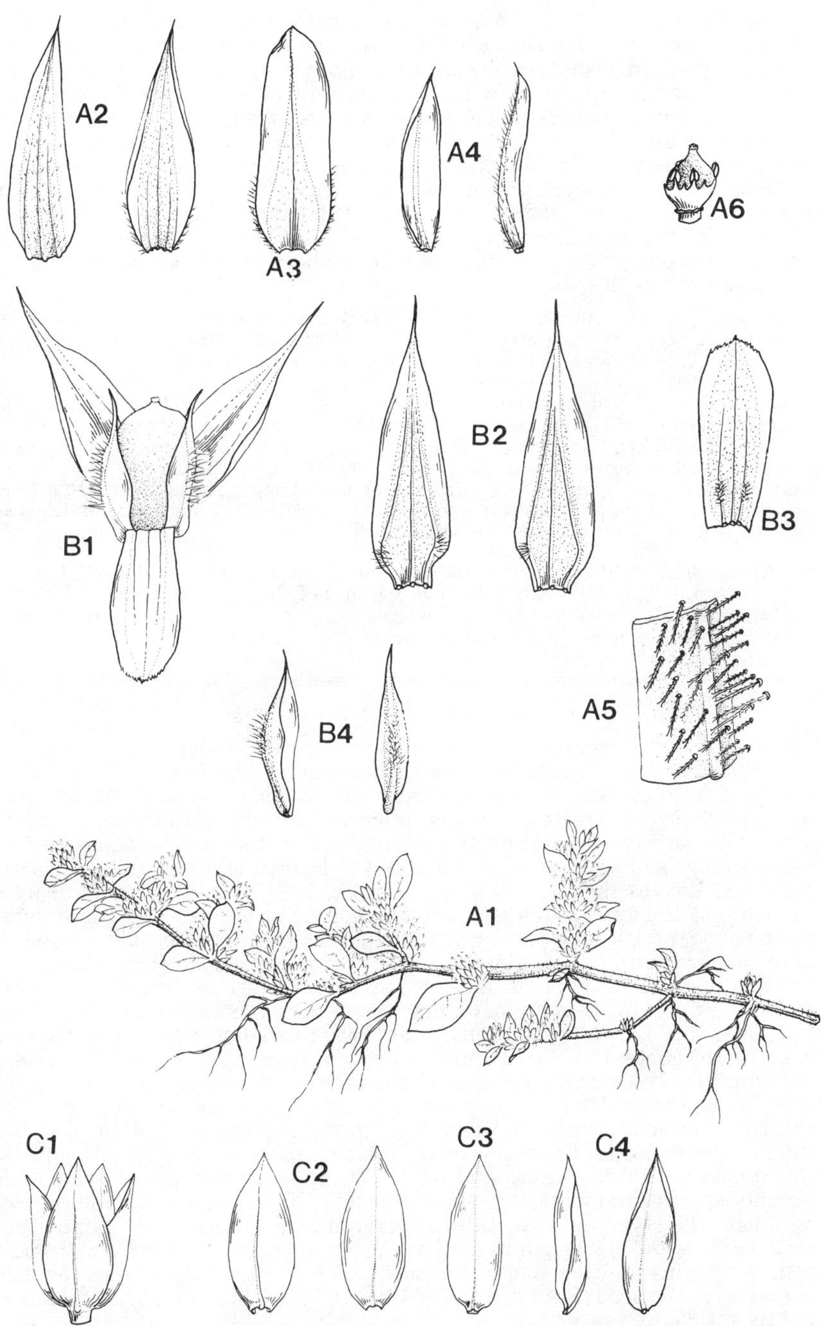

Tab. 30. A.—ALTERNANTHERA CARACASANA. A1, flowering branch (×1); A2, outer tepals (×8); A3, intermediate (adaxial) tepal (×8); A4, inner tepals (×8); A5, hairs of inner tepal (×24); A6, gynoecium and androecium (×8), A1–6 from *Hooper & Townsend* 1649. B.—ALTERNANTHERA PUNGENS. B1, flower with perianth segments spread (×8); B2, outer tepals (×8); B3, adaxial tepal (×8); B4, inner tepals (×8), B1–4 from *Archbold* 7. C.—ALTERNANTHERA SESSILIS. C1, single flower (×8); C2, outer tepals (×8); C3, intermediate tepal (×8); C4, inner tepals (×8), C1–4 from *Bally & Smith* 14836.

globose to ovoid, 4–6 mm. in diam.; bracts pale, deltoid-ovate, c. 2 mm. long, glabrous, lacerate-margined, aristate with the excurrent midrib; bracteoles similar but slightly shorter. Tepals white, lanceolate to oblong-elliptic, 3.5–4 mm. long, acute, mucronate with the excurrent midrib; outer 2 prominently 3-nerved below and darker in the nerved area, with a line of whitish, minutely barbellate hairs on each side of this area, the hairs becoming denser towards the base of the tepal; inner 2 slightly shorter, narrower and less rigid, mostly 1–2-nerved; central tepal intermediate. Stamens 5, at anthesis much exceeding the ovary and style, the alternating pseudostaminodes subequalling the filaments plus anthers, narrowly oblong, laciniate at the apex. Ovary strongly compressed, obpyriform, 0.6 mm. long, style about the same length. Ripe fruit and seeds not seen.

Zambia. C: Ornamental, Mt. Makulu Research Station, 16 km. S. of Lusaka, 20.ix.1960, *Coxe* 15 (K; SRGH). **Zimbabwe**. C: Grown as a dense, clipped edging to flower bed, Harare, 1460 m., 2.viii.1970, *Biegel* 3387 (K; SRGH). **Malawi**. S: Makwapala Expt. St., near Zomba, 713 m., 7.v.1937, *Lawrence* 369 (K).

Said to be a native of S. America, probably Brazil, but no certainly wild material has been seen. The plant is widespread as a decorative border plant because of its more or less variegated foliage, and is believed to be merely a cultigen of *A. tenella* (= *A. ficoidea* auctt.). As will be seen above, ripe seeds (normally abundant in the genus) have not been seen, and *A. bettzickiana* is propagated readily by cuttings. The collector of *Lawrence* 369 observes that the plant is "common all over Malawi at elevations between 1,500 and 4,000 ft .... used on tea plantations hold soil".

4. **Alternanthera nodiflora** R. Br., Prodr. Fl. Nov. Holl.: 417 (1810).—Baker & Clarke in F.T.A. **6**, 1: 73 (1909).—Schinz in Engl. & Prantl Pflanzenfam. ed. 2, **16 C**: 72 (1934).—Hauman in F.C.B. 2: 74 (1951).—Cavaco in Mém. Mus. Hist. Nat. Paris Sér. B, **13**: 160 (1962).—Podlech & Meeuse in Merxm. Prodr. Pl. SW. Afr. **33**: 6 (1966). Type from Australia.

Alternanthera sessilis var. *nodiflora* (R. Br.) Kuntze, Rev. Gen. Pl. 2: 540 (1891). Type as above.

(Annual? or) perennial herb with a stout vertical rootstock, prostrate to ascending or erect, frequently rooting at the lower nodes, slightly to considerably branched from the base outwards, mostly 12–40 cm. Stem and branches green to purplish-brown, terete below and more or less tetragonous above, with a narrow line of whitish hairs down each side of the stem and branches (at least when young) and tufts of white hairs in the branch and leaf axils, otherwise glabrous. Leaves linear to narrowly elliptic, $2-7 \times 0.3-1.2$ cm., acute to subacute at the apex, attenuate below, glabrous or occasionally thinly pilose when young; petiole indistinct or absent. Inflorescences sessile, axillary, solitary or in clusters of up to c. 5, globose, c. 7–10 mm. in diam., often meshing with one another and those of the opposite leaf axil to form a dense mass; bracts scarious, white, ovate-acuminate, mucronate with the excurrent pale midrib, glabrous, (2) 2.25–2.75 mm. long; bracteoles ovate-lanceolate, slightly more longly mucronate, (2)2.25–2.5 mm. long, glabrous. Tepals ovate-acuminate, tapering from below the middle (in smaller forms less long-pointed and tapering from about the middle), (2.75)3–4 mm. long, white to pink-tinged, glabrous, shortly mucronate with the excurrent nerve, the margins often obscurely lacerate-denticulate. Stamens 5 (2 filaments anantherous), at anthesis subequalling the filaments and style, the alternating pseudostaminodes resembling the filaments but a little shorter. Ovary strongly compressed, roundish, the style very short. Fruit glabrous, obcordate, 1.75–2 mm. long, 2–2.5 mm. wide, about half as long as the perianth or rather more with tumid yellow margins on each side of the seed much thicker than the narrow keel, darker over the seed. Seed discoid, c. 1.25–1.5 mm. in diam., brown, shining, faintly reticulate.

Botswana. N: 14 km. SE. of Nata River bridge near Makgadikgadi Pans, 895 m., 23.iv.1957, *Drummond & Seagrief* 5194 (K; LISC; SRGH). SE: Content Farm, Gaberone Distr. 5.iv.1972, *Kelaole* A7 (PRE). **Zambia**. C: Luangwa Game Reserve, flood plain of Luangwa R., 29.iv.1965, *Mitchell* 2734 (K). E: Nsefu Game Camp, edge of pool left by Luangwa R., 750 m., 15 x.1958, *Robson* 121 (BM; K; LISC; SRGH). S: Mazabuka, roadside at Lake Chalimbana, E. of Magoye R., 24.v.1963, *van Rensburg* 2214 (K; SRGH). **Zimbabwe**. N: Binga Distr., 11 km. N. of Kalinda Mine - Binga road, 29.iv.1959, *Noel*

3731 (K; SRGH). W: Shangani Distr., Gwampa vlei, vi.1956, *Goldsmith* 119/56 (K; SRGH). S: Bikita Distr., Devure Ranch, N. bank of Dopsoa Dam, 22.vii.1958, *Chase* 6953 (K; LISC; SRGH). **Malawi**. :N. Kondowe to Karonga, vii.1896, *Whyte* s.n. (K). S: Mangochi Distr., near Monkey Bay, 14.ii.1969, *Williams* 26 (K; SRGH). **Mozambique**. Z: Sena, vii.1859, *Kirk* s.n. (K). T: Mazowe R. 8 km. from Zimbabwe border, 304m., 22.ix.1948, *Wild* 2597 (K; SRGH). MS: Beira, 6 km. from Tambabra para near Mitondo, c. 170 m., 15.v.1971, *Torre & Correia* 18453 (LISC). M: Massangena, vii.1932, *Smuts* P.378 (K).

A native of Australia now widespread in tropical Africa. Occurs chiefly in sandy river beds, along banks of watercourses, in pans, swamps & c., but also along roadsides; on sandy or gravelly soil or black clay.

When well-grown there is no difficulty in separating this plant from *A. sessilis*, but small plants without ripe fruit can be difficult.

5. **Alternanthera sessilis** (L.) DC., Cat. Hort. Monsp.: 77 (1813).—Hauman in F.C.B. **2**: 73 (1951).—Cavaco in Mém. Mus. Hist. Nat. Paris Sér. B, **13**: 158 (1962).—Podlech & Meeuse in Merxm. Prodr. Fl. SW. Afr. **33**: 6 (1966).—J.H. Ross Fl. Natal: 159 (1973). TAB. **30** fig. C. Type from Sri Lanka.

Gomphrena sessilis L., Sp. Pl.: 225 (1753). Type as above.

Illecebrum sessile (L.) L., Sp. Pl. ed. 2: 300 (1762). Type as above.

Alternanthera "achyranth." Forssk., Fl. Aegypt.-Arab.: 1: ix (1775) et auctt. mult. subnom. *"achyranthoides"*, incl. Baker & Clarke in F.T.A. **6**, 1: 73 (1909).

Alternanthera repens Gmel., Syst. Nat. ed. **13**, 2, 1: 106 (1971). Type from Egypt.

Annual or usually perennial herb; in drier situations with slender, more solid stems, prostrate, decumbent or erect, more or less much-branched, to c. 30 cm.; in wetter places ascending or most commonly prostrate with stems c. 0.1—1 m. long, rooting at the nodes, more or less fistular, with numerous lateral branches; when floating very fistular, the stems attaining a metre or more in length and over 1 mm. thick, with long clusters of whitish rootlets at the nodes. Stem and branches green to pink or purplish, with a narrow line of whitish hairs down each side of the stem and branches (at least when young) and tufts of white hairs in the branch and leaf axils, otherwise glabrous, striate, terete below, more or less tetragonous above. Leaves extremely variable in shape and size, linear-lanceolate to oblong, oval, or obovate-spathulate, 1—9 (15) × 0.2—2(3) cm., blunt to shortly acuminate at the apex, cuneate to attenuate at the base, glabrous or thinly pilose, especially on the inferior surface of the midrib; petiole obsolete to c. 5 mm. long. Inflorescences sessile, axillary, solitary or in clusters of up to c. 5, subglobose (slightly elongate in fruit), c. 5 mm. in diam.; bracts scarious, white, deltoid-ovate, mucronate with the excurrent pale midrib, glabrous, c. 0.75—1 mm. long; bracteoles similar, 1—1.5 mm., also persistent. Tepals oval-elliptic to lanceolate-ovate, equal, 1.5—2.5 mm. long, acuminate to rather blunt, white to pink-tinged, glabrous, shortly but distinctly mucronate with the stout, excurrent midrib, the margins often obscurely lacerate-denticulate. Stamens 5 (2 filaments anantherous), at anthesis subequalling the ovary and style, the alternating pseudostaminodes resembling the filaments but usually somewhat shorter. Ovary strongly compressed, roundish, style extremely short. Fruit glabrous, obcordate or cordate-orbicular, 2—2.5 mm. long, dark at maturity, strongly compressed with the margins each side of the seed not much thicker than the narrow, pale, yellowish keel, slightly longer to slightly shorter than the perianth. Seed discoid, c. 0.75—1 mm. in diam., brown, shining, faintly reticulate.

Botswana. N: Karongana/Thagoe junction, 19.viii.1975, *Smith* 1425 (K; SRGH). SE: Gaberone Dam, 970 m., 11.iv.1977, *Hansen* 3120 (C; GAB; K; PRE; SRGH). **Zambia**. B: Near Senanga, 1060 m., 30.vii.1952, *Codd* 7244 (K; PRE). N: Mbala Distr., Kaka New Road, Saisi R., 1500 m., 17.ii.1962, *Richards* 16121 (K; SRGH). W: Mwinilunga Distr., by R. Lunga just below R. Mudjanyama, 25.xi.1937, *Milne-Redhead* 3393 (K). S: Namwala Distr., Baambewe Village on the Shanakabinza dambo, 12.xii.1962, *van Rensburg* 1072 (K; SRGH). **Zimbabwe**. N: Binga, D.C.'s Harbour, at water's edge, 10.iii.1966, *Jarman* 383 (K; SRGH). C: Chegutu, Distr., Nkuti 24.i. Farm, i.1969, *Mavi* 944 (K; SRGH). E: Mutare Distr., Eastlands Farm,, c.910 m., 7.iii.1955, *Chase* 5713 (BM; SRGH). S: Chiredzi Distr., Gonarezhou, between Chitsa's store and Sabi/Lundi rivers junction, in Tamboharta Pan, 31.v.1971, *Grosvenor* 585 (K; SRGH). **Malawi**. N: Nkhata Bay Distr., Chinthehe, near Luweya R., 570 m., 21.ii.1961, *Richards* 14439 (K; SRGH). C: N. bank of Bua R. by Rd., bridge 19 km. N. of Nkhota Kota, 475 m., 16.vi.1970, *Brummitt* 11463 (K; MAL). S: W. bank of Shire R. at Kasisi, 6 km. N. of Chikwawa, 95 m., 21.iv.1970, *Brummitt* 10022 (K; SRGH). **Mozambique**. N: Cuamba, at the permanently wet margins of the R.

Cúrio, 2.viii.1934, *Torre* 517 (COI; LISC). GI: Inhambane, Panda, 7.iv.1959, *Barbosa & Lemos* 8531 (COI; K; LISC; LMJ; SRGH). T: Tete, Isl. de Micune, 19.vi. 1949, *Barbosa & Carvalho* 3171 (LISC; LU). Z. Quelimane Distr., Namagoa, 25.vi.1949, *Faulkner* K455 (K).

A widespread plant in the tropics and subtropics of both Old and New Worlds, growing in water (to c. 1 m. deep or perhaps more - and the plant then floating or emergent) or along the margins of rivers, pools or irrigation ditches, or in sandy or clay flood plains, or in depressions along roadsides, which were waterlogged seasonally.

23. GOMPHRENA L.

Gomphrena L., Sp. Pl. 1: 224 (1753); in Gen. Pl. ed. 5: 105 (1754).

Annual or occasionally perennial herbs with entire, opposite leaves. Inflorescences terminal or axillary, capitate or spicate, solitary or glomerate, often subtended by a pair of sessile leaves, bracteate with the bracts persistent in fruit, the axis frequently thickened. Flowers hermaphrodite, each solitary in the axil of a bract, bibracteolate; bracteoles laterally compressed, carinate, often more or less winged or cristate along the dorsal surface of the midrib, deciduous with the fruit. Tepals 5, erect, free or almost so, more or less lanate dorsally, at least the inner 2 usually more or less indurate at the base in fruit. Stamens 5, monadelphous, the tube shortly 5-dentate with entire to very deeply bilobed teeth. Style short or long, stigmas 2, suberect or more or less divergent to very short. Ovary with a single pendulous ovule. Fruit a thin-walled, irregularly rupturing utricle. Seed ovoid, compressed.

About 120 species, chiefly in the Americas but a few representatives in Australia.

1. Bracteoles with a distinct crest on the dorsal surface, at least near the apex . . 2
– Bracteoles with no trace of a crest . . - - - - - - - 3. *martiana*
2. Bracteoles with the dorsal crest large and conspicuous, extending from the apex almost to the base of the midrib - - - - - - - - 2. *globosa*
– Bracteoles with the dorsal crest small, confined to about the upper third of the dorsal surface - - - - - - - - - - - - - - - 1. *celosioides*

1. **Gomphrena celosioides** Mart. in Nov. Act. Acad. Caesar-Leop. Carol. **13**: 301 (1826).—Cavaco in Mém. Mus. Hist. Nat. Paris Sér. **B**, 13: 165 (1962).—Podlech & Meeuse in Merxm. Prodr. Fl. SW. Afr. **33**: 16 (1966).—J.H. Ross Fl. Natal: 158 (1973). TAB. **31**. Type from Brazil.

Gomphrena globosa subsp. *africana* Stuchlik in Beih. Bot. Centralbl. **30**, 2: 396 (1913). Type from Transvaal.

Gomphrena decumbens (non Jacq.) auctt. incl. Stuchlik in Burtt Davy, F.T.F.T. **1**: 185, f.20 (1926).—Mears in Taxon **29**: 87 (1980).

Gomphrena alba Peter in Fedde, Repert. Beih. **40**, 2, Descriptiones: 24 (1932). Type from Tanzania.

Gomphrena celosioides f. villosa Suesseng. in Fedde, Repert. **42**: 57 (1937). Type from Namibia.

Perennial herb, prostrate and mat-forming to ascending or erect, c. 7–30 cm., much-branched from the base and also above; stem and branches striate, often sulcate, green to reddish, when young usually more or less densely furnished with long, white, lanate hairs, more or less glabrescent with age. Leaves narrowly oblong to oblong-elliptic or oblanceolate, c. 1.5–4.5 (8) × 0.5–1.3 (1.8) cm., obtuse to subacute at the apex, mucronate, narrowed to a poorly demarcated petiole below, the pair of leaves subtending the terminal inflorescence more abruptly narrowed and sessile, oblong or lanceolate-oblong, all leaves glabrous or thinly pilose above, thinly to densely furnished with long whitish hairs on the margins and lower surface, sometimes more or less lanate on both surfaces. Inflorescences sessile above the uppermost pair of leaves, white, at first subglobose and c. 1.25 cm. in diam., finally elongate and cylindrical, c. 4–7 cm. long with the lower flowers falling, axis more or less lanate; bracts deltoid-ovate, 2.5–4 mm. long, shortly mucronate with the excurrent midrib; bracteoles strongly laterally compressed, navicular, c. 5–6 mm. long, mucronate with the excurrent midrib, furnished along about the upper one-third of the dorsal surface of the midrib with an irregularly dentate or subentire wing. Tepals c. 4.5–5 mm. long, narrowly lanceolate, 1-nerved, the outer 3 more or less flat, lanate only at the base, nerve thick and greenish below, thinner and excurrent in a short mucro at the apex; inner 2 sigmoid in lateral view at the strongly indurate base, densely lanate almost to the apex, slightly longer. Staminal tube subequalling the perianth, the 5 teeth deeply bilobed with obtuse lobes subequalling the c. 0.75 mm. long anthers which are set between them; pseudostaminodes absent. Style and stigmas together c. 1 mm. long, style very short, stigmas divergent. Capsule shortly compressed-pyriform, c. 1.75 mm. long. Seeds compressed. ovoid, c. 1.5 mm. in diam., brown, faintly reticulate, shining.

Botswana. N: Maun, Camp No. 9, 15.i.1974, *Smith* 782 (K; SRGH). SE: Farm, Springfield, 3 km. S. of Lobatsi, E. of railway, 17.i.1960, *Leach & Noel* 144 (SRGH). **Zambia**. B: Sesheke, by Zambezi R., 28.xii.1952, *Angus* 1063 (FHO; K). N: Mbala Distr., Chilongowelo Garden, 4.ii.1952, *Richards* 684 (K). W: Mwinilunga Distr., Matonchi Farm, 16.xi.1937, *Milne-Redhead* 3271. (BM; K). C: Kabwe Distr., Kamaila Forest St., 36 km. N. of Lusaka, 9.ii.1975, *Brummitt, Hooper & Townsend* 14289 (K; NDO). E: Chipata 1030 m., 1.i.1936, *Winterbottom* 21 (K). S: 3 km. from Namwala on Kafue Nat. Park Rd., 11.xii.1962, *van Rensburg* 1068 (K; SRGH). **Zimbabwe**. N: Hurungwe Safari Area, c. 349.5 km. Harare to Chirundu, 325 m., 14.ii.1981, *Philcox, Leppard & Dini* 8522 (K; SRGH). W: Hwange Nat. Park. Guvalala Pan, 11.xi.1968, *Rushworth* 1255 (K; LISC; SRGH). C: Harare, Mukavisi Vlei, 1425 m., i.1917, *Eyles* 590 (K; SRGH). E: Roadside c. 5 km. N. of Mutare, 1400 m., *Fries, Norlindh & Weimarck* 3992 (K; L). S: Chiredzi Distr., Nyajena common land, Mabagweshe, 18.v.1971, *Taylor* 178 (K; SRGH). **Malawi**. N: Mzimba Distr., 5 km. W. of Mzuzu, Katoto, 1365 m., 13.iii.1974, *Pawek* 8218 (K; MAL; MO; SRGH; UC). S: Blantyre Distr., by Matope Bridge, 470 m., 12.ii.1970, *Brummitt & Banda* 8542 (K; LISC; MAL; PRE; SRGH). **Mozambique**. N: Mocambique Distr., Erati, Namapa, Estação Experimental do C.I.C.A., 20.iii.1961, *Balsinhas & Marrime* 300 (COI; K; L; LISC; LMJ; SRGH). Z: Quelimane Distr., Namagoa, s.d., *Faulkner* K120 (K; SRGH). T: Tete, 25.x.1965, *Myre & Rosa* 4733 (LISC). MS: Beira, iv.1921, *Dummer* 4668 (K). M: Maputo, 6.ix.1945, *Pedro* 37 (K; PRE).

A native of S. America (S. Brazil, Paraguay, Uruguay and Argentina) which has spread with great rapidity during the present century to become a widely distributed weed in the tropics and subtropics. It occurs along roadsides, in waste and disturbed ground, as a weed of cultivation, in open places in forests and grassland, on anthills, and on dried-out pans and flood-plain; chiefly on sandy soil, but also on black clay. According to Mears ("The Linnean species of *Gomphrena* L." in Taxon **29**: 86–7 (1981)) this widespread weed has been misidentified as *G. celosioides* Mart. and is really *G. decumbens* Jacq., which in turn should now be called *G. serrata* L. He believes that true *celosioides* is restricted to South America except in cultivation. However, the original description and (very good) plate of *G. decumbens* indicate a plant with (1) an annual root (2) bracteoles with a serrate keel reaching almost to the base and (3) heads sessile in groups of three on the stems and principal branches - none of which characters can possibly apply to the African weedy species. Nor does it agree with Linnaeus's description of *"capitulis solitariis terminalibus sessilibus"* - which could apply to the African species - or various others. Mears's neotypification of *G. serrata* in the above paper seems to me conjectural, and as I cannot separate the African plant from the isotypes of *G. celosioides* which I have seen, I prefer to retain this name. *G. serrata* L. may indeed be an earlier name for *G. celosioides* but (cf. my remarks on *Achyranthes nodosa* Schum. under *Pandiaka involucrata*) in the absence of an unquestionable type I feel it very desirable to adhere to the established name.

2. **Gomphrena globosa** L., Sp. Pl. **1**: 224 (1753). Type, Hort. Sicc. Cliffort. p. 86 (BM, lectotype, Mears in Taxon **29**: 86 (1980)).

Annual herb, decumbent or erect, branched from the base and also above, c. 15–60 cm.; stem and branches striate or sulcate, more or less densely clothed with appressed white hairs at least when young. Leaves broadly lanceolate to oblong or elliptic-oblong, 2.5–12 (15) × 2–4 (6) cm., narrowed to an ill-defined petiole below, thinly pilose on both surfaces, the pair of leaves subtending the terminal inflorescence sessile or almost so, broadly to subcordate-ovate. Inflorescences sessile above the uppermost pair of leaves, usually solitary, globose or depressed-globose, c. 2 cm. in diam., pinkish to deep red; bracts deltoid-ovate, 3–5 mm. long, mucronate with the shortly excurrent midrib; bracteoles strongly laterally compressed, navicular, c.8–12 mm., mucronate with the excurrent midrib, furnished from the apex almost to the base of the dorsal surface of the midrib with a broad, irregularly dentate crest. Tepals similar to those of *G. celosioides* but longer (6–6.5 mm.), the outer more lanate and the inner less markedly indurate at the base. Staminal tube subequalling the perianth, the 5 teeth deeply bilobed with obtuse lobes subequalling the anthers; pseudostaminodes absent. Style and stigmas together c. 2.5 mm. long,

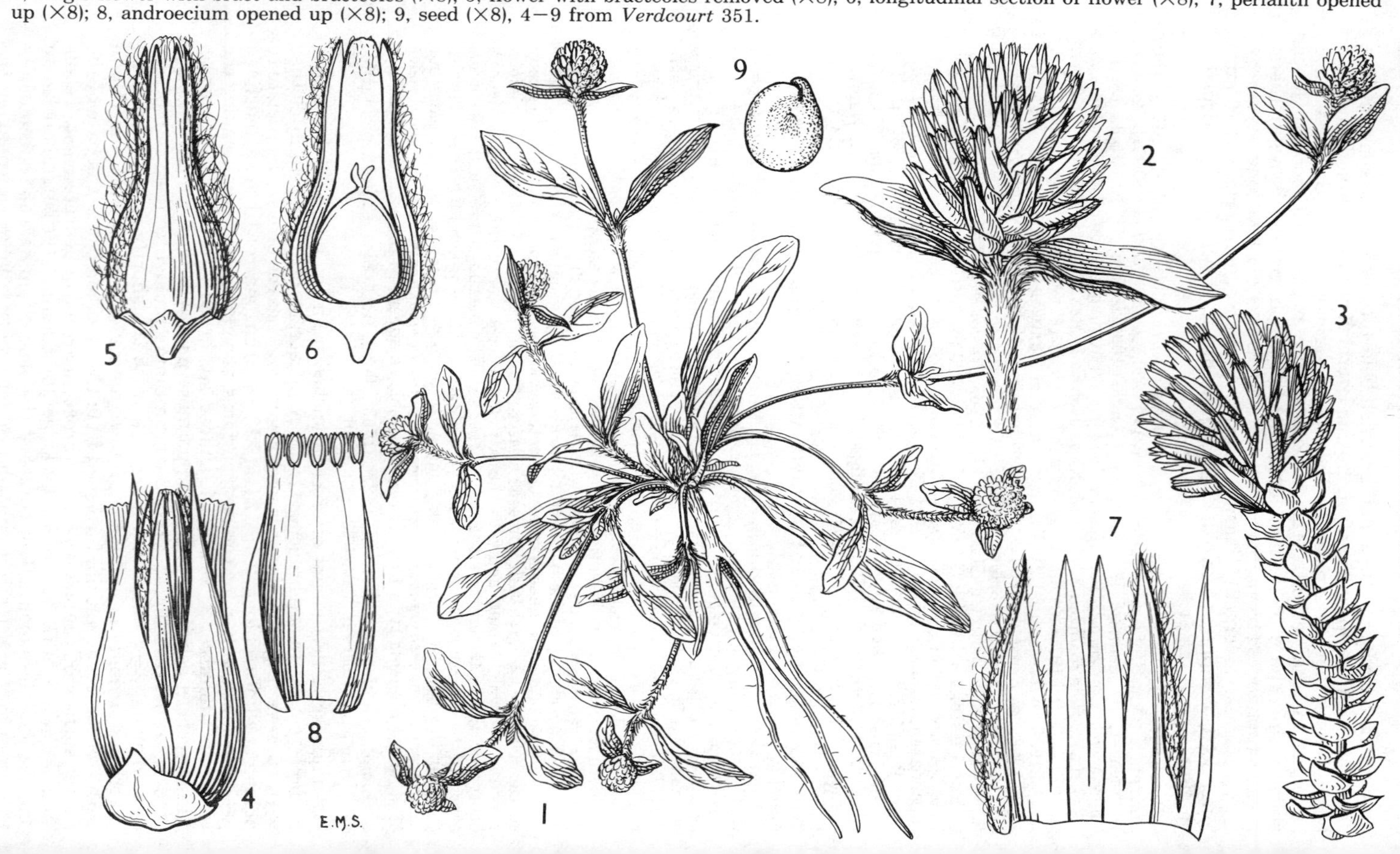

Tab. 31. GOMPHRENA CELOSIOIDES. 1, flowering plant ($\times \frac{2}{3}$) *Hancock* 41; 2, flowering head ($\times$3) *Verdcourt* 351; 3, fruiting head ($\times$3) *Hancock* 41; 4, single flower with bract and bracteoles ($\times$8); 5, flower with bracteoles removed ($\times$8); 6, longitudinal section of flower ($\times$8); 7, perianth opened up ($\times$8); 8, androecium opened up ($\times$8); 9, seed ($\times$8), 4–9 from *Verdcourt* 351.

stigmas divergent, subequalling or slightly longer than the style. Capsule oblong-ovoid, compressed, c. 2.5 mm. long Seeds compressed-ovoid, c. 2 mm. in diam., brown, almost smooth, shining.

Zambia. E: Luangwa Valley, Munkanya, 2.iii.1968, *Rabson Phiri* 90 (K). **Malawi**. S: Kundwelo village, Palombe Plain, 636 m., 29.vii.1956, *Newman & Whitmore* 277 (BM; SRGH). "cult. for flowers?". **Mozambique**. N: Litunde, expontânea?, xi.1933, *Torre* 5 (COI; LISC). MS: Maringua, 15 km. N. of Sabi R., 27.vi.1950, *Chase* 2579 (BM; SRGH), "possibly garden escape".

A native of tropical America, long cultivated in the warmer parts of the world and its precise native area obscure. The Zambian specimen cited above bears the note: "Pink to reddish flowers. Grows on sand or clay. Good for yarding".

3. **Gomphrena martiana** Gillies ex Moq. in DC., Prodr. **13**, 2: 400 (1849).

Prostrate or sprawling to ascending herb, to at least 50 cm., branched from the base upwards; stem and branches feebly striate, more or less densely furnished in the upper parts with upwardly appressed, rather long white hairs, becoming glabrescent with age. Leaves ovate to oblong-ovate, elliptic or subspathulate, lamina of the larger leaves c. 2—10 × 1—5 cm. (reducing above on both stem and branches), acute to rather blunt at the apex, cuneate or attenuate to a distinct petiole c. 0.5—3.5 cm. long, subglabrous or with scattered hairs on the superior surface, with appressed hairs chiefly along the primary venation below. Inflorescences of mostly c. 4 agglomerated clusters forming a head c. 1—2.5 × 0.5—3 cm., frequently broader than long, subtended by 4—5 unequal leaves, on a slender appressed-pilose peduncle c. 1.5—5.5 cm. long; bracts small, delicate and hyaline, c.1.5—2 mm. long, deltoid-ovate, shortly aristate; bracteoles concave, c. 4.5—5 mm. long, deltoid-ovate with a short mucro formed by the excurrent nerve, firm and yellowish above, hyaline towards the base, with no dorsal crest. Tepals 4—5 mm. long; outer 2 linear-oblong, rather abruptly narrowed to the slightly denticulate, mucronate apex, 1-nerved, the firm central part near the base finely striate and more or less tomentose; inner 3 progressively slightly shorter, narrower and more narrowly pointed; all very indurate at the base in fruit. Staminal tube c. 3 mm. long, divided to c. half way, with only a short blunt tooth or knob on each side of the antheriferous tooth. Ovary at anthesis c. 1.5 mm. long; style very short, stigmas filiform. Capsule shortly ellipsoid, c. 1.5 mm. long. Seeds ellipsoid, c. 1.25 mm. in diam., brown, faintly reticulate.

Mozambique. M: Near Polana, 4.i.1941, *Torre* 2452, (C; COI; K; LISC; LMU; MO; WAG).

A native of southern S. America; presumably introduced into Mozambique as a garden plant, or with garden plants, and now occurring as a ruderal. It is recorded by one collector (*Balsinhas* 2211, K; LISC) from "Jardin Vasco da Gama, infestante".

133. CHENOPODIACEAE

By J.P.M. Brenan †

Mostly annual or perennial herbs, sometimes shrubs rarely small trees, often halophytic or nitrophilous and with tendency to fleshiness. Leaves alternate, rarely opposite, simple, without stipules, entire or not; sometimes leaves reduced to scales or green ring-like joints along the stem. Flowers small to minute, mostly green or grey, solitary and axillary, or more often variously clustered or cymose, usually regular, unisexual or male and female. Perianth (1) 2—5 (or more, but not in the Flora Zambesiaca area) lobed; lobes united below, imbricate or almost valvate, presistent, after flowering, or variously modified in fruit. Petals and sepals not differentiated. Stamens as many as or fewer than the perianth segments, and opposite them. Ovary normally free and superior, unicellular. Stigmas 2 (5). Ovule solitary, campylotropous on a long or short almost always basal funicle. Fruit usually utricular, indehiscent, rarely circumscissile, often included in and falling with the perianth. Embryo peripheral, curved, annular or spiral surrounding the endosperm, or endosperm absent.

A family of some 1500 species, widely distributed, predominantly halophytic, weakly represented in the wet tropics, most richly represented in drier parts of the temperate and subtropical zones.

In addition to the genera mentioned below, *Beta vulgaris* L. (Beet) is recorded as cultivated in Mozambique (Gomes e Sousa. Pl. Menyharth., in Bot. Soc. Estud. Col. Mocamb. 32: 65 (1936)). It is probably cultivated elsewhere in the Flora Zambesiaca area.

Eyles in Trans. Roy. Soc. S. Afr. 5: 347 (1916) records (under *Chenopodiaceae*) "*Kochia decumbens* (Hoscht.) Torre & Harms", with *Pentodon decumbens* Hochst. as a synonym. The record was based on *Monro* 925 from Victoria, Zimbabwe. *Pentodon* is *Rubiaceae*, *Kochia Chenopodiaceae*, but no *Kochia* is known from our area, and the record must represent some inexplicable muddle between families.

1. Stems seemingly leafless, apparently built up of numerous superposed tubular or ring-like succulent green segments which ultimately shrivel and fall away from the stem proper; flowers immersed in a fleshy spike; plant glabrous - - - - 2
— Stems obviously leafy, though leaves sometimes very small and scale-like, not appearing to be built up of green fleshy segments; flowers not immersed in a fleshy spike; plants mostly hairy, mealy or glandular, sometimes glabrous - - 4
2. Perennials: flowers exposed or not, with or without a truncate lateral shield, seeds with a soft or hard testa, endosperm present or absent - - - **6. Salicornia**
— Annual herbs; flowers always exposed, each with a truncate lateral shield; seeds with a thin membranous testa, minutely hairy; endosperm absent - - - - - 3
3. Fruiting spikes disarticulating into separate ring-like segments, cylindrical, corky (in the Flora Zambesiaca area); flowers concealed in spike, seed with hard crustaceous testa; endosperm present and obvious- - - - - - - - -**4. Halosarcia**
— Fruiting spikes persistent or breaking up irregularly; flowers exposed (except in *S. natalensis* which has tapering more slender spikes); seed with soft membranous testa; endosperm absent - - - - - - - - - - - - - **5. Sarcocornia**
4. Calyx (in fruit) horizontally winged; leaves (in the Flora Zambesiaca area) scale-like and imbricate - - - - - - - - - - - - - **8. Salsola**
— Calyx (in fruit) not horizontally winged; leaves various - - - - - - 5
5. Plants (in the Flora Zambesiaca area) with obvious mealy pubescent silky or glandular indumentum, at least on young parts; flowers without bracteoles (except the female in *Atriplex*); leaves various but often flattened and more or less toothed or lobed (except in *Chenolea*) - - - - - - - - - - - - - - - 6
— Plants (in the Flora Zambesiaca area) glaborous or nearly so, sometimes slightly all subtended by small scarious bracteoles; leaves linear to oblong, fleshy, entire; flowers all sessile and axillary - - - - - - - - - - - - - **7. Suaeda**
6. Leaves densely appressed silky pubescent, entire, linear to lanceolate; flowers sessils and axillary, 1 (3) together, hermaphrodite - - - - - - - **3. Chenolea**
— Leaves (at least when young) more or less mealy, glandular or with spreading pubescence, never silky, in our species almost always wider and more or less toothed or lobed; flowers various, sessile and axillary in a few species of *Chenopodium* and *Atriplex*, but then in dense several-to many flowered clusters, if hermaphrodite then with female mixed - - - - - - - - - - - - - - - - - 7
7. Flowers hermaphrodite and female, all without bracteoles and with a calyx - - - - - - - - - - - - - - - - - - **1. Chenopodium**
— Flowers male and female, male without bracteoles and with calyx, with 2 bracteoles but no calyx, the bracteoles enlarging and clasping the fruit - - **2. Atriplex**

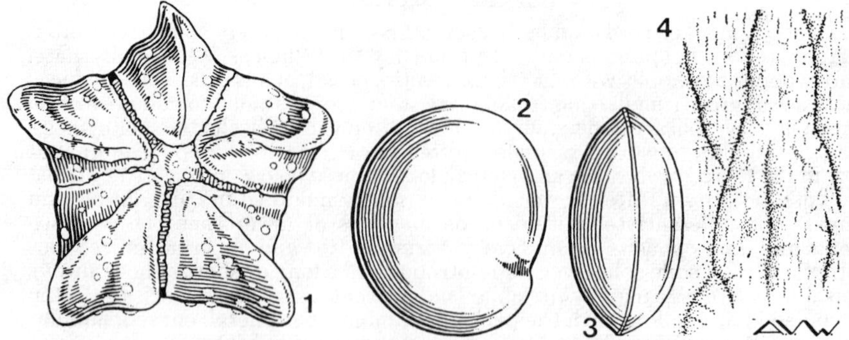

Tab. 32. CHENOPODIUM ALBUM. 1, perianth containing fruit, seen from above (×20); 2, seed, front view (×20); 3, seed, lateral view (×20); 4, portion of side of seed-testa (×200), 1—4 from F.T.E.A.

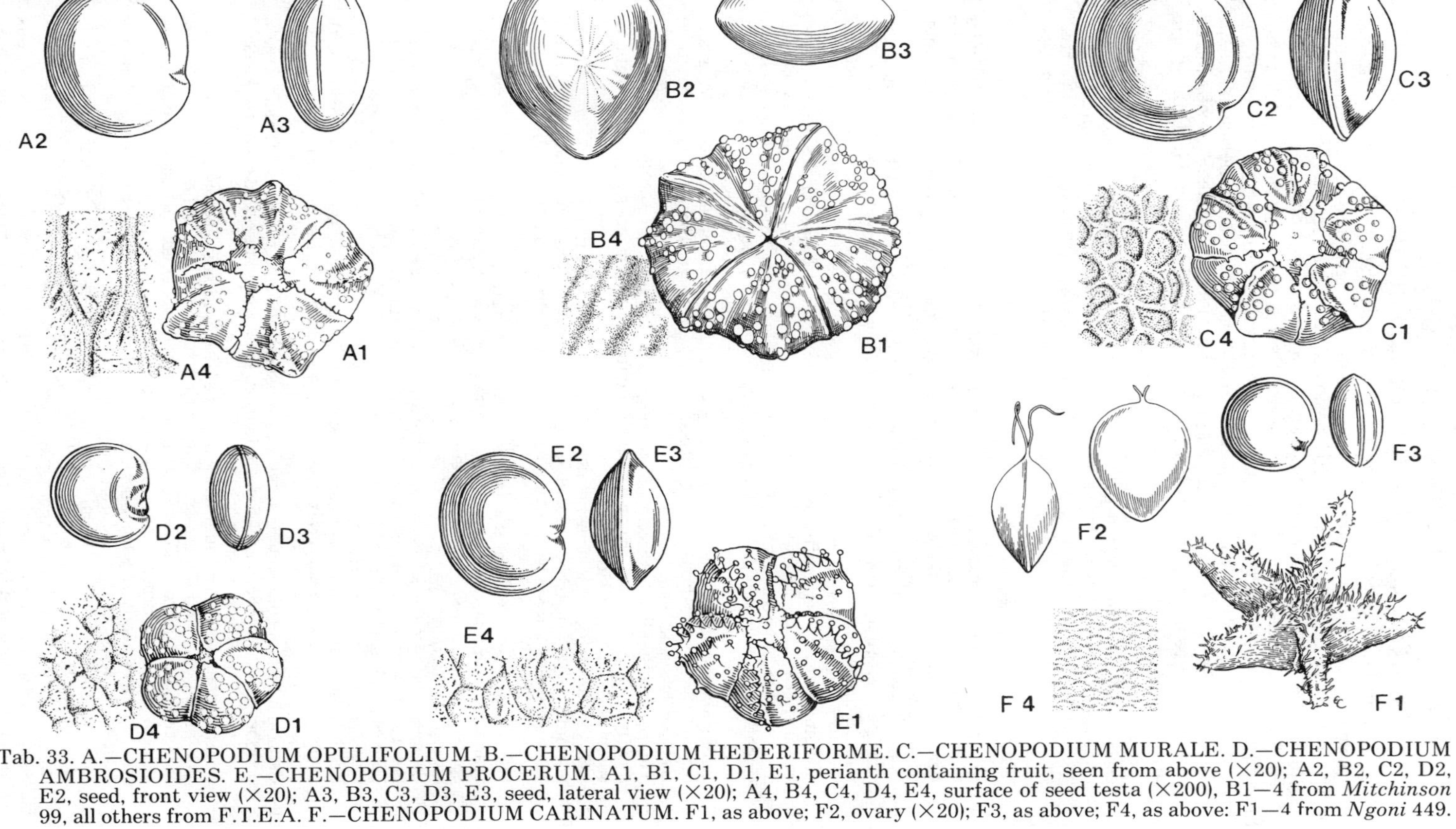

Tab. 33. A.—CHENOPODIUM OPULIFOLIUM. B.—CHENOPODIUM HEDERIFORME. C.—CHENOPODIUM MURALE. D.—CHENOPODIUM AMBROSIOIDES. E.—CHENOPODIUM PROCERUM. A1, B1, C1, D1, E1, perianth containing fruit, seen from above (×20); A2, B2, C2, D2, E2, seed, front view (×20); A3, B3, C3, D3, E3, seed, lateral view (×20); A4, B4, C4, D4, E4, surface of seed testa (×200), B1—4 from *Mitchinson* 99, all others from F.T.E.A. F.—CHENOPODIUM CARINATUM. F1, as above; F2, ovary (×20); F3, as above; F4, as above: F1—4 from *Ngoni* 449.

1. CHENOPODIUM L.

Chenopodium L., Gen. Pl., ed. 5: 103 (1754).

Mostly annual or perennial herbs, glabrous, pubescent, glandular or mealy with vesicular hairs. Leaves membranous to more or less fleshy, entire, toothed, or pinnately divided, alternate, mostly petiolate, normally broad. Flowers mostly in cymose clusters ("glomerules") variously arranged but often paniculate and mixed, without bracteoles. Perianth of both sorts of flower normally (3) 4–5 lobed, unaltered or nearly so in fruit, or sometimes becoming fleshy. Stamens 1–5. Stigmas 2 (5). Fruit with a membranous indehiscent pericarp. Seeds "horizontal" (vertically compressed) or, less commonly, "vertical" (horizontally compressed), testa normally thin, hard and brittle. Embryo annular. Endosperm present.

Species about 100–150, cosmospolitan, ranging from temperate to tropical regions, often weedy; relatively few in Africa compared with America and Australia.

I have been unable to assign the record of *C. vulvaria* by Macnae & Kalk, Nat. Hist. Inhaca I., Moçamb. 144 (1958). *C. vulvaria* L. is a distinctive European species, at present unknown from tropical Africa.

1. Plant more or less mealy, at least on young parts, with grey or whitish vesicular hairs, often malodorous; other sorts of hair and also glands absent; stamens (of hermaphrodite flowers) always 5; seeds usually black when ripe, and usually 1 mm. or more in diam. - - - - - - - - - - - - - - - - 2
– Plant pubescent, and with yellow to amber glands, aromatic, without vesicular hairs; stamens (of hermaphrodite flowers) 1–5; seeds black to red-brown when ripe, 0.5–1.25 mm. in diam. - - - - - - - - - - - - - - - - - - 7
2. Testa of seed marked with very close minute rounded pits; seeds sharply keeled on margin; pericarp very difficult to detach from seed; inflorescences always cymose and leafy - - - - - - - - - - - - - - - - 6. *murale*
– Testa of seed more or less furrowed and irregulary roughened, sometimes almost smooth, never closely pitted; seeds bluntly keeled on margin (rather acutely only in *C. olukondae*); pericarp readily rubbed or scraped off seed . - - - - - 3
3. Young shoots suffused with conspicuous vivid amaranth-purple; leaves (median and inferior) large, mostly 10–18 × 10–16 cm., with about 15–30 shallow irregular teeth or lobes on each margin - - - - - - - - - - 2. *giganteum*
– Young shoots not or scarcely purple-tinged; leaves (median and inferior) smaller, mostly 1–8 × 0.5–5.5 cm., with up to about 10 teeth or lobes on each margin, or sometimes entire or subentir - - - - - - - - - - - - - - 4
4. Leaves (except juvenile ones following the cotyledons) distinctly larger than broad, normally by at least 1½ times; stems often more or less red; branching commonly erect or suberect; testa with spaced irregular radial furrows, otherwise almost smooth - - - - - - - - - - - - - - - - - - - 1. *album*
– Leaves (at least median and inferior cauline) nearly or quite as broad as long; branching variable, divergent or erect; testa usually with radial furrows more closely spaced than in *C. album* and often also with roughening in between (smooth or nearly so in *C. hederiforme*) - - - - - - - - - - - - - - 5
5. Seeds subelliptical, distinctly though slightly longer than wide, 1–1.2 × 0.9 mm. testa almost smooth except for a few faint radial furrows and slight roughening in between; foliage very broadly cuneate or even concave-subtruncate at base - - - - - - - - - - - - - - - - - 5. *hederiforme*
– Seeds subcircular, about as long as wide, 1.0–1.5 mm. in diam.; testa marked with numerous close to moderately spaced furrows with roughening in between - - - - - - - - - - - - - - - - - - - 6
6. Central lobe of median and inferior leaves usually with up to about 10 shallow teeth or lobes on each side; basal lobes on either side ascending; seeds bluntly keeled - - - - - - - - - - - - - - - - - 3. *opulifolium*
– Central lobe of median and inferior leaves entire or almost so; basal lobes on either side promiment and divergent; seeds with a rather sharp keel - 4. *olukondae*
7. Inflorescence built up of distinct though sometimes small dichasial cymes in the axils of leaves or bracts, these cymes usually aggregated as though into a spike; seeds black or nearly so when ripe; stamens 1–2; inferior and median leaves pinnately divided, at least in their inferior par - - - - - - - - - - - 8
– Inflorescence built up of small sessile or subsessile clusters of flowers in the axils of leaves or bracts, flowers not in dichasial cymes; seeds red-brown to blackish when ripe, stamens 1–5; leaves and sepals various - - - - 10. *carinatum*
8. Seeds 0.7–0.8 mm. in diam.; testa marked with very minute shallow contiguous rounded or angular pits; glands between veins on inferior surface of leaf, also those

on outside of sepals, all sessile (seen under × 20 lens); leaves pinnately divided throughout each side usually to within 2–3 mm. of midrib - - - - - - - - - - - - - - - - - 9. *schraderianum*

– Seeds 0.9–1.1 mm. in diam., testa marked with slightly impressed sinuose lines and minor roughness; glands between veins on inferior surface of leaf, also many of those on outside of sepals, shortly but distinctly stalked (seen under × 20 lens); lower part of leaf pinnately divided, upper part toothed but scarcely lobed - - - - - - - - - - - - - - - - - - 8. *procerum*

9. Sepals rounded, dorsally not at all keeled; leaves 3–14 cm. long; inflorescence an ample panicle, the upper flower-clusters bracteolate or not; stamens 4–5; seeds in each cluster some "vertical", others "horizontal" (see generic description) - - - - - - - - - - - - - - - - - 7. *ambrosioides*

– Sepals each dorsally with a conspicuous wing-like keel broadening upwards; leaves to 3 cm. long. Flowers all in least axils. Stamens 1; seeds all "vertical" (see generic description): red-brown, 0.5–0.75 mm. in diam. - - - - 10. *carinatum*

1. **Chenopodium album** L., Sp. Pl. 219 (1753).—Adamson in Adamson & Salter, Fl. Cape Penins.: 352 (1950).—Brenan in F.T.E.A., Chenopodiaceae: 6, fig. 1 (1954).—Wild, Common. Rhod. Weeds, fig. 20 (1955).—Aellen in Hegi, Ill. Fl. Mitteleur. ed. 2, 3, 2: 648 (1960). TAB. **32**. Type presumably from Europe.

Annual herb usually 10–150 cm. high, normally, much branched, but sometimes stems simple or subsimple especially in small plants; plant green or tinged red especially on stem (which is often red or pink striped), more or less clothed with grey-mealy hairs especially on young parts. Leaves very variable even on a single plant, rhombic-ovate to lanceolate, usually distinctly longer than broad by at least 1½ times (but the juvenile leaves following the cotyledons may be almost as broad as long), 1.2–8.2 × 0.3–5.5 cm.; leaf-margins entire or more commonly with up to about 10 shallow irregular teeth on each side, the lowermost tooth or lobe on each margin sometimes more prominent than the rest; apex of each margin sometimes more prominent than the rest; apex of leaf acute, or particularly in the lower cauline leaves subacute to rounded; superior leaves and bracts progressively smaller. Inflorescence a usually ample pancile of very numerous small or medium-sized (2–6 mm. in diam.) densely or laxly spicately or cymosely arranged dense rounded clusters ("glomerules") of minute grey to green flowers, the latter 1–1.5 mm. in diam. Perianth segments 5, papillose with grey mealy hairs on margins and outside, each with a prominent green keel in upper part. Stamens 5. Pericarp somewhat persistent, but easily rubbed or scraped off seed. Seeds black, shining, 1.2–1.6 (1.85) mm. in diam., bluntly keeled; testa (see under microscope) marked with faint irregular radial furrows, but otherwise almost smooth.

Botswana. SE: Gaborone, fl. 25.ii.1974, *Mott* 167a (K). **Zambia**. C: Mt. Makulu Research Station, fr. 29.iv.1960, *Simwanda* 129 (K; SRGH). S: Muckle Neuk, fl. & fr. 11.x.1954, *Robinson* 910 (K; SRGH). **Zimbabwe**. W: Bulawayo Distr., Aisleby Dam., fl. & fr. 16. iii.1968, *Best* 767 (K; SRGH). C: Harare, Chishawasha Seminary and Mission, fl. & fr. 7.ix.1957, *Phipps* 728 (K; SRGH). E: Chipinge Distr., Sabi Dam Farm, fr. ix.1960, *Suane* 329 (K; SRGH). S: Victoria Distr., Zimbabwe Nat. Park, fr. 27.iii. 1973, *Chiparawasha* 611 (K; SRGH). **Mozambique**. M: Catembe, fl. & fr. xii.1931, *Gomes e Sousa* 208 (K).

Cosmopolitan, especially in the Northern Hemisphere; in Africa widely scattered but mainly eastern. A weed of cultivated and disturbed ground; 1100–1600 m.

Mavi 1088 (K; LISC; SRGH) from Zimbabwe, N. Lomagundi Distr., Rothwell Farm, Kutama Mission, fl. & fr. 1.vi.1970, is close to *C. album* but has oblong leaves slightly toothed to subentire and obtuse to subacute at apex. It may well be the closely related European taxon known as *C. strictum* Roth or *C. album* L. subsp. *striatum* (Krasan.) J. Murr, from which some material from Namibia named *C. amboanum* (Murr) Aellen seems hardly distinguishable.

2. **Chenopodium giganteum** D. Don, Prodr. Fl. Nepal.: 75 (1825).—Adamson in Adamson & Salter, Fl. Cape Penins.: 352 (1950).—Aellen in Hegi, III. Fl. Mitteleur. ed. 2, 3, 2: 639 (1960). Type from Nepal but not so far found.

Chenopodium album subsp. *amaranticolor* Coste & Reynier in Reynier in Bull. Herb. Boiss., Sér. 2. 5: 979 (1905).—Mullin in Hara, Chater & Williams, Enum. Fl. Pl. Nepal **3**: 170 (1982). Type from France.

Chenopodium amaranticolor (Coste & Reynier) Coste & Reynier in Reynier in Bull. Soc. Fr. **54**: 181 (1907).

Annual herb closely related to *C. album*, c. 1–3 m. high, erect, much branched, with young shoots suffused with a conspicuous vivid amaranth-purple, and densely clothed with mealy vesicular hairs; stems red-striped. Leaves (of inferior and middle parts of stem and branches) broadly rhombic-ovate, large, (5) 10–18 × (4.5) 10–16 cm., broadly cuneate-attenuate at base, rounded at apex; margins with about 15–30 shallow irregular teeth or lobes on each side, lowermost teeth or lobes not especially prominent; upper leaves and bracts becoming smaller narrower and subacute to acute at apex. Inflorescence an ample panicle of very numerous small spicately arranged flower-clusters ("glomerules") 2–3 mm. in diam.; flowers green or reddish, 0.75–1.5 mm. in diam. Perianth segments 5, papillose with grey-mealy hairs outside, each with a prominent green keel in upper part. Stamens 5. Pericarp somewhat persistent, but easily rubbed or scraped off. Seeds black, shining, about 1–1.2 mm. in diam., bluntly keeled; testa (seen under microscope) marked with numerous irregular spaced radial furrows and with minor irregularities on surface between furrows, but giving a generally smoothish appearance.

Zimbabwe. C: Harare, fl. & fr. 6.xii.1971, *Biegel* 3674 (K; LISC; SRGH). E: Inyanga Distr., 6 km. N. of St. Triashill Mission, fl. 25.i.1978, *Best* 1326 (SRGH).

Native of India and possibly China. A garden weed; 1650 m.

I have also collected *C. giganteum* from cultivation in Zambia, N: Luwingu Distr., St. Maria Mission. Chiluwi Isl., Lake Bangweulu, fl. & fr. 14.x.1947, *Brenan & Greenway* 8108 (K), where it was grown for the young shoots that were eaten as "spinach".

The taxonomy of *C. giganteum* is far from clear or certain, and it is scarcely possible to solve the problem without better material and field-study from India. In particular, the name *C. giganteum* has been applied also to a plant cultivated for grain in the Himalayas for grain; the seeds here are about 2 mm. in diam., about 1 mm. thick (not 0.5 mm.) and with a very broad marginal keel; the testa is also red-brown rather than black and with a minutely roughened and reticulate surface, giving a matter not shining appearance. In other respects the plant is similar to ours. These differences may be due simply to genetic selection through long cultivation, but may be specific. The position is not made easier by the apparent absence of any type-specimen of *C. giganteum*.

A close relative, *Chenopodium quinoa* Willd., Sp. Pl. **1**: 1301 (1798), has been cultivated at Salisbury Experimental Station, fr. 1.iv.1936, *Arnold* 6339 (K). *C. quinoa* is likewise grown for grain in the Andes of S. America. For a fuller account see the article on Quinua in Underexploited Plants with Promising Economic Value: 20–23 (Nat. Acad. Sci. Washington, 1975). It has normally white, sometimes brownish or reddish seeds about 1.5–2.5 mm. in diam., with a broad keel like the Himalayan *C. giganteum*. The leaves have generally rather few marginal teeth (2–12 on each margin) and often prominent basal lobes.

3. **Chenopodium opulifolium** Schrad. ex Koch & Ziz. Cat. Pl. Palat.: 6 (1814).—Brenan in F.T.E.A., Chenopodiaceae: 6, t. 2 fig. 2 (1954).—Wild, Common Rhod. Weeds, t. 23 (1955).—Aellen in Hegi, III. Fl. Mitteleur. ed. 2, 3, 2: 648 (1960). TAB. **33** fig. A. Type from Germany.

Annual herb (or ? sometimes a short-lived perennial with stems becoming woody below), closely related to *C. album*, up to 0.6–1.5 (3) m. high, erect, normally much branched, green to almost white, rarely red-tinged, more or less clothed with mealy vesicular hairs, sometimes densely so on young shoots inflorescences and undersides of leaves. Leaves variable, mostly broadly and shortly rhombic-ovate, the median and lower almost as broad as long, c. 0.7–5.4 × 0.4–5.4 cm., leaf margins usually with up to about 10 mostly shallow teeth or lobes on each side, often the lowermost teeth or lobes more prominent divergent and often bilobed; sometimes margins entire or nearly so and leaves elliptic especially (it seems) if plants perennate; apex of leaf rounded to acute. Inflorescence a usually ample panicle of very numerous small densely or laxly spicately or rarely cymosely arranged dense rounded clusters ("glomerules") of minute grey to greenish flowers, the latter 1–1.5 mm. in diam. Perianth-segments 5, papillose with grey-mealy hairs on margins and outside, each segment with a prominent green keel in upper part. Stamens 5. Pericarp somewhat persistent, but readily scraped off seed. Seeds black, shining, subcircular, 1.0–1.5 mm. in diam. bluntly keeled; testa (seen under microscope) marked with numerous spaced radial furrows and minute papillose roughening in between.

Zambia. N: Mporokoso Distr., Mweru-Wa-Ntipa, Kangiri, fl. & fr. 7.vi.1957, *Richards* 9074 (K). **Zimbabwe**. W: Matobo Distr., fl. & fr. 18.iii.1947, *West* 2245 A (K). E: Chirinda, fr. 20.v.1906, *Swynnerton* 465 (BM; K). **Mozambique**. M: Maputo, Goba Fronteira, fl. & fr. 11.i.1980, *de Koning* 8001 (BM).

Europe and the Mediterranean Region, eastwards to India and (?) Mongolia, southwards through tropical Africa to our area and Angola, probably extending into S. Africa. A weed of cultivated ground; 1000–1300 m.

C. opolifolium is very close to *C. album*, differing in most of the leaves being about as long as wide, the usually more glaucous mealy inflorescence, and the rougher surface of the testa when viewed under the microscope. The red colouring, so often seen in *C. album*, is rarely present in *C. opulifolium*.

4. **Chenopodium olukondae** (Murr) Murr in Bull. Herb. Boiss., Sér. 4: 992 (1904).—Aellen in Merxm., Prodr. Fl. SW. Afr. 32: 10 (1967). Type from Namibia.

Chenopodium opulifolium subsp. *olunkondae* ("*oluhondae*") Murr in Magyar Bot. Lapok 1: 342 (1902).

Chenopodium album subsp. *olukondae* (Murr) Murr in Bull. Herb. Boiss., Sér. 2, **4**: 992, t. 4 fig. 8 a, b (1904).

Annual herb, (?) sometimes becoming thinly woody below, 0.6–1.5 m. high, much branched especially below, grey to green, uncertain whether red-tinged, more or less clothed with mealy vesicular hairs especially on young parts and inflorescences. Leaves variable, the median and inferior rhombic-triangular to elliptic or ovate in outline, mostly about $1-1\frac{1}{2}$ times as long as broad, 1.3–3.5 × 1.1–3 cm. cuneate at base, with a pronounced diverricate or somewhat deflexed often somewhat bifid lobe in inferior part, otherwise entire or nearly so in superior part, apex of leaves obtuse to subacute; superior leaves smaller, entire or subentire, elliptic to elliptic-lanceolate. Inflorescence an ample profusely branched panicle of very numerous small laxly spicately arranged dense rounded clusters ("glomerules") of minute greyish or grey-green flowers 1–1.5 mm. diam. Perianth segments 5, papillose with grey mealy hairs on margins and outside, each with a prominent green keel in upper part. Stamens 5. Pericarp readily rubbed or scraped off. Seeds black, shining, subcircular, 1.1–1.5 mm. in diam., rather sharply keeled; testa (seen under microscope) with numerous rather close radial impressed furrows.

Botswana. SE: Tlalambele-Moru area, near Soa Pan, fl. 8.i. 1974, *Ngoni* 272 (K; SRGH).

Also in S. and Namibia. Habitat insufficiently known: *Ngoni* 272 from a limestone outcrop.

The material of *C. olukondae* from the Flora Zambesiaca area is inadequate for certainty, especially in that it lacks ripe seeds. The relationship of *C. olukondae* with *C. opulifolum* is clearly very close. However, the foliage appears distinctive (see key) and also the rather sharp keel to the seeds. More material and observation is much needed.

Three specimens from Mozambique are close to *C. olukandae*, though evidently from very different habitats: GI: Maputo st. ix. 1944, *Pimenta* 4901 (LISC); Gaza, Chibuto, Maniquenique, Experimental Station, near houses, fr. 13.x.1957, *Barbosa & Lemos* 8023 (K); M: Inhaca Isl., basic sandy soil; 0–200 m., fl. 30.ix.1958, *Mogg* 28432 (K). Again, more material is desired.

5. **Chenopodium hederiforme** (Murr) Aellen in Fedde, Repert. **24**: 339 (1928); in Merx., Prodr. Fl. SW. Afr. **32**: 9 (1967). TAB. **33** fig. B. Type: Botswana (?), "Chansis", Fleck 1892.

Chenopodium opulifolium subsp. *hederiforme* Murr in Bull. Herb. Boiss., Sér. 2, **4**: 993 (1904).

Annual herb 0.4–2.5 m. high, erect, much branched, grey to glaucous green, uncertain whether red-tinged, more or less densely clothed with mealy vesicular hairs especially on young parts and inflorescences. Leaves variable; the median and inferior mostly triangular-ovate to triangular-subreniform, mostly as broad as or broader than long, 1.5–4.5 × 1.5–4.5 cm., broadly cuneate to concave-subtruncate at base; leaf margin with 4–11 coarse teeth or lobes on each side;

apex of leaves rounded to acute; upper leaves becoming narrower, more narrowly cuneate at base, with fewer (to almost no) marginal lobes or teeth, often markedly trilobed with a prominent ascending or divergent lobe on either side near base. Inflorescence an ample profusely branched panicle of very numerous small laxly arranged dense rounded clusters ("glomerules") of minute grey flowers 1–1.5 mm. in diam. Perianth segments 5, papillose with grey mealy hairs on margins and outside, each with a narrow prominent green keel in upper part. Stamens 5. Pericarp readily rubbed or scraped off seed. Seeds black or brownish-black, shining, broadly ellipsoid (slightly larger than wide), 1–1.2 × 0.8–0.9 mm., bluntly keeled; testa (seen under microscope) much smoother than in *C. opulifolium*, with a few faint radial furrows and some slight irregular roughenings elsewhere.

Caprivi Strip. Okavango R., 19 km. N. of Shakawe on Botswana border, fr. 16.v.1965, *Wild & Drummond* 7092 (K; LISC; SRGH). **Botswana**. N: Toromoja, Botletle R., fl. & fr. 27.iv.1975, *Ngoni* 449 (K; SRGH). SW: Ghanzi, fr. 9.v.1969, *Brown* 6035 (K; SRGH). SE: 8 km. N. of Gaberones, fl. 19.i.1960, *Leach & Noel* 229 (K; SRGH). **Zimbabwe**. S: Beitbridge Distr., Shashi R., 10 km. NW. of junction with Limpopo, fr. 6.v.1959, *Drummond* 6086 (K; SRGH).

Also in S. Africa and Namibia. On riverbanks and in savanna and wooded grassland.

This species has often been misidentified with *C. opulifolium*. Probably the most reliable distinction is in the seeds which are smaller; up to 1.2 × 0.9 mm., characteristically rather longer than wide, with a much smoother testa and a more readily detachable pericarp. The generally more laxly and divaricately branched inflorescence and the more coarsely toothed lower foliage very broadly cuneate or even concave-subtruncate at base generally enable recognition at sight. Although the original illustrations of *C. hederiforme* show leaves more deeply cut than in most specimens from the Flora Zambesiaca area, there appears to be a good deal of variation. I am unable to see any clear distinction between *C. hederiforme* and *C. petiolariforme* (Aellen) Aellen from S. Africa.

6. **Chenopodium murale** L., Sp. Pl.: 219 (1753).—Adamson in Adamson & Salter Fl. Cape Penins.: 353 (1950).—Brenan in F.T.E.A., Chenopodiaceae: 7, t. 2 fig. 3 (1954).—Wild, Common Rhod. Weeds, t. 22 (1955).—Aellen in Hegi, Ill. Fl. Mitteleur. 3, 2: 621 (1960). TAB. **33** fig. C. Type presumably from Europe.

Annual up to 90 cm. high, upright or spreading, normally much branched, green, rarely red-tinged, with mealy hairs especially on young parts but rarely densely clothed. Leaves variable, commonly rhombic-ovate in outline, rarely narrower, (0.7) 1.5–9 × (0.4) 0.8–5 (7) cm., without any tendency for especially prominent basal lobes, but with several coarse irregular ascending usually sharp teeth (about 5–15 on each margin, rarely fewer). Inflorescences leafy, composed of divericately branching cymes up to 5 cm. long, terminal and from upper and sometimes median axils. Flowers greenish, minute, about 1–1.5 mm. in diam. Perianth segments 5, papillose on margins and outside, each with a blunt raised green keel towards apex only. Stamens 5. Pericarp very difficult to detach from seed. Seeds black, somewhat shining, 1.2–1.5 mm. in diam., acutely keeled; testa (seen under microscope) marked with very close minute rounded pits.

Botswana. N: Nata village, fl. & fr. 20.iv.1976, *Ngoni* 519 (K; SRGH). **Zambia**. S: Muckle Neuk, fl. & fr. 26.x.1958, *Robinson* 2908 (K; SRGH). **Zimbabwe**. W: Plumtree Station, fl. & fr. 3.vii.1962, *Wild* 5836 (K; SRGH). C: Gweru, fl. & fr. 3.ix.1957, *Whellan* 1425 (K; SRGH). S: Mushandike Dam, fl. & fr. 18.viii.1969, *Blignaut* 196, 505 (K; SRGH).

Cosmopolitan. A weed of cultivated and disturbed ground.

C. murale is close to *C. album*, but can usually be distinguished at sight, though the general differences are hard to convey in words. The combination of rhombic-ovate leaves with more or less numerous jagged teeth, and of clearly cymose inflorescences usually suffices. In cases of doubt, the sharply keeled seeds with the testa surface closely pitted are constant and diagnostic.

7. **Chenopodium ambrosioides** L., Sp. Pl.: 219 (1753).—Gomes e Sousa, Pl. Menyharth, in Bol. Soc. Estud. Col. Moçamb. **32**: 65 (1936).—Adamson in Adamson & Salter, Fl. Cap. Penins.: 353 (1950).—Brenan in F.T.E.A., Chenopodiaceae: 10, t. 2 fig. 5 (1954).—Wild, Comm. Rhod. Weeds: t. 21 (1955).—Binns, H.C.L.M.: 27 (1968). TAB. **33** fig. D. Type from Spain.

Tab. 34. CHENOPODIUM SCHRADERIANUM. 1, habit ($\times\frac{1}{6}$) *Hislop* 48; 2, flowering branch ($\times\frac{2}{3}$); 3, part of lower surface of leaf showing sessile glands ($\times$20); 4, flower ($\times$30); 5, stamen ($\times$30); 6, ovary ($\times$30), 2–6 from *Teague* 199; 7, fruiting perianth ($\times$30) *Hislop* 48; 8, seed, front view ($\times$20); 9, seed, lateral view ($\times$20), 8–9 from F.T.E.A.

Herb, usually annual, rarely a short-lived perennial, up to 180 cm. high, upright, much branched, green (? occasionally red-tinged), shortly and inconspicuously pubescent or puberulous, often with some longer hairs on stem, also with numerous yellowish sessile glands particularly on inferior side of leaves, strongly aromatic. Leaves mostly lanceolate, the inferior with several (to c.10) coarse irregular ascending teeth on each margin, 3−14 × 0.5−4.5 cm.; superior leaves becoming smaller narrower and linear-entire, bracts even smaller, down to 23 × 0.5 mm. Inflorescence an ample much branched pancile of small sessile clusters of flowers arranged spicately along the ultimate branches, all or some clusters bracteate. Perianth-segments 3−5 (female flowers), 4−5 (unisexual flowers), pubescent to glabrous, glandular, variable connate, smooth and not at all dorsally keeled keeled. Stamens 4−5. Pericarp easily removed. Seeds deep red-brown, about 0.5−0.8 mm. in diam.; testa (see under microscope) almost smooth and with faintly impressed sinuouse lines.

Botswana. N: Thamalakane R., near Maun, fl. & fr. 13.iii.1961, *Richards* 14688 (K). SW: Ghanzi pan, fl. 8.iv.1969, *de Hoogh* 224 (K). SE: Mahalapye Distr., Sephare, fl. 25.ii.1958, *de Beer* 700 (SRGH). **Zambia.** B: Kalabo, fl. & fr. 13.xi.1959, *Drummond & Cookson* 6405 (K; LISC; SRGH). N: Mbala Distr., Sunzu, fl. & fr. 23.ii.1952, *Richards* 943 (K). W: Matonchi Farm, fl. 29.i.1938, *Milne-Redhead* 4401 (K). C: Mt. Makulu Research Station, fl. 13.i.1958, *Angus* 1813 (K; LISC). E: Mvuvje R., near Petauke, Great East Road, fl. & fr. 5.xii.1958, *Robson* 842; (K; LISC; SRGH). S: Machili, fl. 21.ix.1969, *Mutimushi* 3762 (K).**Zimbabwe.** N: Darwin Distr., R. Mazoe near Winda Pools, fl. 5.ix.1958, *Phipps* 1304 (K; SRGH). W: Bulawayo Distr., Hillside Dam, fl. & fr. x.1958, *Miller* 5483 (K). C: Harare, fl. & fr. 20.ix.1958, *Drummond* 5825. (K; LISC; SRGH). E: Sabi Valley, Nyanyadzi, fl. & fr. 3.ii.1948, *Wild* 2490 (K; SRGH). S: Chibi-Belingwe Distr., Lundi R. near Shabani, fl. & fr. 4.v.1958, *West* 3597 (K; SRGH). **Malawi.** N: Mzimba Distr., far Rd. to Mzuzu, fl. 5.iii.1975, *Pawek* 9118 (K; SRGH). C: Dedza Secondary School, fl. 24.v.1968, *Jeke* 193 (K; LISC; SRGH). S: Blantyre Distr., Jacarandas Hostel, fl. 13.iii.1969, *Msinkhu* 14 (K). **Mozambique.** T: Maroeira-Bucha-Rio Zambeze, fl. & fr. 3.xi.1973, *Aguiar Macêdo* 5351 (LISC). MS: Chimoio, Garwzo, fl. 28.iii.1948, *Barbosa* 1259 (LISC). GI: Gaza, Aldeia da Barragem, fl. & fr. 20.xi.1957, *Barbosa & Lemos* 8226 (K; LISC). M: Maputo, fl. xii.1945, *Pimenta* 4000 (LISC).

Found throughout the tropical and subtropical regions of the world, but especially polymorphic in S. America. Occurs in a variety of disturbed habitats, gardens, cultivated fields, waste ground etc., but especially often on sand by rivers; 550−1620 m.

The axillary, sessile flower clusters separate *C. ambrosioides* from all the other species of the genus in the Flora Zambesiaca area, except for *C. carinatum*. See under that species for the distinction.

The taxonomic treatments of *C. ambrosioides* have been diverse, and there is no finality as yet. However, in Africa the species is uniform and typical. The description above is drawn up from *C. ambrosioides* in a narrow sense, as it occurs in Africa. without citing a specimen, Aellen (in Hegi, Ill. Fl. Mitteleur. 3, 2: 596 (1960)) records from "Rhodesia" *C. spathulatum* Sieb. ex Moq. in DC., Prodr. **13**, 2: 73 (1849). This is almost certainly to be referred to *C. ambrosioides*, from which *C. spathulatum* does not appear specifically distinct.

8. **Chenopodium procerum** Hochst. ex Moq. in DC., Prodr. **13**, 2: 75 (1849).—Brennan in F.T.E.A., Chenopodiaceae: 11, t. fig. 6 (1954).—Binns, H.C.L.M.: 27 (1968), excl. syn. *C. botrys*.—Richards & Morony, Check List Fl. Mbala (Abercorn) & Distr.: 17 (1969). TAB. **33** fig. E. Type from Ethiopia.

Annual herb up to 2 m. or more high, upright, with few to many branches, green or often strongly red-tinged, glandular-pubescent all over, usually shortly so, strongly aromatic. inferior and median leaves elliptic or ovate-elliptic in outline, mostly 2.5−14 × 1.5−7 (9) cm., acute at apex, pinnately divided each side into 3−5 sharply toothed lobes, the lower lobes extending to near midrib, the superior much shallower and the top part of the leaf normally toothed but scarcely lobed; glands between veins on inferior surface shortly but distinctly stalked (seen under a × 20 lens); superior leaves smaller narrower and less divided; uppermost often oblong and obscurely pinnate-dentate. Inflorescences composed of dichotomously branched axillary cymes which are usually aggregated into leafy or leafless continuous cylindrical inflorescences 1.5−6 cm. wide and up to 60 cm. or more long. Flowers greenish or red-tinged, minute, 0.5−1.5 mm. in diam. Perianth segments 5, each with a green glandular keel towards apex, glandular outside with many glands distinctly stalked (× 20 lens necessary). Stamens 1–2. Pericarp easily scraped off seed. Seeds black or nearly so, glossy, 0.9–1.1 mm. in diam., with a rather prominent but blunt keel; testa (seen under microscope) very slightly rough with slightly impressed irregular sinuose lines and other minor roughnesses.

Zambia. N: Lake Chila, fl. & fr. 13.v.1955, *Richards* 5704 (K). **Zimbabwe**. E: Chimanimani fl. & fr. 27.iv.1907, *Swynnerton* 1508 (BM). **Malawi**. N: Musuku Plateau, fr. vii.1896, *Whyte* (K). C: Dedza, fl. 24.iii.1969, *Salubeni* 1280 (K; SRGH). S: Ncheu, fl. & fr. ii.v.1970, *Salubeni* 1471 (K; SRGH). **Mozambique**. N: Lichinga, fr. i.vi.1934, *Torre* 168 (LISC).
Also in eastern Africa northwards to Ethiopia and the Sudan. Upland grasslands, sometimes becoming a weed in waste or cultivated ground; 1000–2100 m.

C. procerum is related to *C. schraderianum* but is often taller, with the leaves acute not obtuse and with different more irregular lobing, stipitate not sessile glands, and with other differences in size and testa-marking of the seeds.

9. **Chenopodium schraderianum** Schultz., Syst. Veg. **6**: 260 (1820).—Brenan in F.T.E.A., Chenopodiaceae: 12, t. 2 fig. 7 (1954).—Wild, Comm. Rhod. Weeds: t. 24 (1955). TAB. **34**. Type a cultivated plant.
Chenopodium foetidum Schrad. in Magaz. Ges. Naturf. Freunde Berlin **2**: 76 (1808).—R.E. Fries, Wiss. Ergebn. Schwed. Rhod.- Kongo Exped.: 20 (1914) nom. illegit, non Lamarck (1778). Type as for *C. schraderianum*.
Chenopodium foetidum subsp. *resediforme* Murr in Bull. Herb. Boiss., Sér. 2, **4**: 990 (1904).
Chenopodium foetidum "var. *resediforme* (Murr) A & G." Aellen in Fedde, Repert. **24**: 346 (1928). It is not clear when this combination was made. Ascherson & Graebner, Syn. Mitteleur. Fl. **5**, 1: 24 (1913) apparently made it a subvariety.
Chenopodium botrys sensu Eyles in Trans. Roy. Soc. S. Afr. **5**: 347 (1916) non L.
Chenopodium sp. Eyles in Trans. Roy. Soc. S. Afr. **5**: 347 (1916).

Annual 0.1–1.5 m. high, upright; main stem simple or with few, rarely many lateral branches, especially near base; plant green; sometimes red, shortly glandular and pubescent all over, strongly aromatic. Inferior and median leaves elliptic to oblong in outline, mostly 1–5 (8) × 0.5–3 (5) cm., mostly obtuse at apex, pinnately divided throughout each side into 3–5 narrow blunt lobes extending usually to within 1–3 mm. of midrib; lobes entire or with a few blunt teeth; glands between veins on inferior surfaces of leaf all sessile (seen under × 20 lens), not accompanied by hairs; superior leaves progressively smaller and less divided. Inflorescences composed of dichotomously branched axillary cymes which are usually aggregated into more or less leafy or leafless continuous cylindrical inflorescences 1.5–6 cm. wide and up to about 60 cm. long. Flowers greenish or red-tinged, minute, 0.5–1 mm. in diam. Perianth segments 5, each with a prominent toothed keel outside from near apex to near base, glandular outside; glands all sessile (use × 20 lens). Stamens 1–2. Pericarp easily rubbed or scraped off seed. Seeds black or nearly so, somewhat glossy, 0.7–0.8 mm. in diam., bluntly and not prominently keeled; testa (seen under microscope) with very minute shallow contiguous rounded or angular pits.

Botswana. SW: Khutse, fl. & fr. iv.1972, *Baker* 22 (K; SRGH). **Zambia**. S: Choma, fl. & fr. v.1909, *Rogers* 8027 (K; SRGH). **Zimbabwe**. N: Mazoe, fl. & fr. viii.1905, *Eyles* 180 (BM; SRGH). W: Matopos Hills, fr. vii.1909, *Rogers* 5370 (K). C: Makoni, Dunedin. fl. & fr. 9.ii.1931, *Norlindh & Weimarck* 4953 (K; SRGH). E: Odzani R. valley, fl. & fr. 1914, *Teague* 199 (K).
Also widespread in eastern Africa from the Sudan, Ethiopia and Somalia southwards to S. Africa. Upland grasslands, but often a weed in disturbed or cultivated ground; 1200–2000 m.

Among the species occurring in the Flora Zambesiaca area closely related only to *C. procerum*. For the distinctions see under that species.

10. **Chenopodium carinatum** R. Br., Prodr. Fl. Nov. Holl. **1**: 407 (1810).—Brenan in F.T.E.A., Chenopodiaceae: 13, t. 2 fig. 8 (1954).—Aellen in Hegi, Ill. Fl. Mitteleur. ed. 2, **3**, 2: 598 (1960).—P.G. Wilson in Nuytsia **4**: 173 (1983). TAB. **33** fig. F. Type from Australia.

Annual herb, prostrate to erect, normally branched near base into simple to much branched stems up to 60 cm. long; plant green, rarely red-tinged, pubescent and glandular, aromatic. Leaves in outline ovate to elliptic, rarely narrow-elliptic, small, mostly 0.3−3 × 0.25−2 cm., sometimes as wide as long, with 2−4 (6) usually coarse sometimes obscure entire or denticulate teeth or lobes on each margin, rarely entire or almost so; glands between veins on inferior side of leaves sessile to subsessile, not accompanied by hairs unless on veins. Flowers greenish, minute, c. 0.4−0.75 mm. in diam., sessile or subsessile in small dense rounded axillary clusters at most of the nodes. Perianth segments normally 5, pubescent and glandular, each with a conspicuous wing-like hairy keel broadening upwards. Stamen 1. Seeds all "vertical" (laterally compressed), deep red-brown, shining, 0.5−0.75 mm. in diam., bluntly or sharply keeled; testa (seen under microscope) almost smooth.

Botswana. SE: Mochudi, Phutodikobo Hill, fl. & fr. 14.iii.1967, *Mitchison* 49 (K). **Zimbabwe**. C: Marondera fl. & fr. 12.xii.1971, *Clatworthy* 214452 (K; SRGH). S: Gwanda, fl. & fr. xi.1976, *Erasmus* 249997 (SRGH).

Native of Australia, introduced as a weed into other parts of the world including S. Africa and Europe. A weed of disturbed ground in the Flora Zambesiaca area; 1000−1600 m.

C. carinatum is very distinct from all the species in the Flora Zambesiaca area, except *C. ambrosioides*, by having the flowers in dense sessile axillary clusters. *C. ambrosioides* differs in being more robust with larger leaves and having the perianth segments dorsally rounded, not keeled.

2. ATRIPLEX L.

Atriplex L., Gen. Pl., ed. **5**: 472 (1754).

Annual or perennial herbs or shrubs, usually (always in the Flora Zambesiaca area) more or less mealy with vesicular hairs. Leaves alternate, rarely opposite, mostly petiolate, normally relatively broad and flat. Flowers in clusters, which are axillary or aggregated into terminal panicles or spikes, monoecious or dioecious. Male flowers without bracteoles, with a 3−5 lobed perianth and 3−5 stamens; female flowers with 2 relatively large bracteoles, without perianth or staminodes. Fruits enclosed by the persistent accrescent, and often modified bracteoles, with a membranous indehiscent pericarp. Seeds almost always "vertical" (i.e. laterally compressed), testa commonly thin and hard. Embryo annular, with inferior radicle pointing upwards or upwards and outwards. Endosperm present.

About 200−250 species, widespread especially in temperate and subtropical regions, many in sandy or saline habitats by sea or inland and sometimes ecologically important in inland steppe regions.

1. Leaves green when mature, grey vesicular mealy hairs scattered except when young; fruiting bracteoles 1.5−3 mm. wide, smooth in centre - - - 3. *suberecta*
— Leaves densely grey-mealy beneath, and often above as well; fruiting bracteoles 3−8 mm. wide, more or less roughened, tuberculate or with horn-like projections in centre - 2
2. Leaves entire or with (2) obscure teeth; bracteoles mostly muricate or tuberculate in centre - - - - - - - - - - - - - - - - - 1. *halimus*
— Inferior leaves with 2−5 shallow teeth or lobes; bracteoles with finger-like projections in centre - - - - - - - - - - - - - - - 2. *amboensis*

1. **Atriplex halimus** L., Sp. Pl., ed. **1**: 1052 (1753). TAB. **35**. Type from Europe.
Atriplex halimus var. *granulate* Chevall. in Bull. Herb. Boiss., Sér. 2, **5**: 444 (1905).—Brenan, F.T.E.A., Chenopodiaceae: 15 (1954). Type from Algeria.

Shrub or woody herb, procumbent or sprawling (probably always so in the Flora Zambesiaca area) to suberect, 0.5−3 m. high, densely mealy all over so that the whole plant is whitish to pale grey-green. Leaves ovate to oblong or elliptic in outline, 1−4 cm. long, 0.4−2.5 (3) cm. wide, rounded to acute at apex, cuneate to attenuate at base, entire or sometimes with 1 (2) obscure shallow lobes or teeth. Inflorescence a terminal panicle with the ultimate branches spiciform and mostly leafless; lower flower clusters in leaf axils. Fruiting bracteoles sessile, their basal connate part absent or very short, up to 1.5 mm.; free apical part of each bracteole reniform to broadly deltoid-ovate, as broad as or more commonly broader than long, 2.5–7 mm. long, 3–9 mm. wide, acute to rounded at apex, cordate or sometimes truncate or very broadly cuneate at base, smooth to conspiciously muricate or tuberculate in centre; margins on either side with 1–4 obscure to pronounced teeth.

Mozambique. MS: Chiloane Isl., fl. 18.xi.1958, *Mogg* 29224 (LISC). GI: between Chibuto and Magul, on the Caniçado Rd. by the R. Changane, fl. & fr. 9.x.1958, *Barbosa* 8345 (K). M: Chibuto, fl. & fr. 24.iii.1948, *Torre* 7544 (LISC).
Mediterranian Region from France and Spain to Syria and Jordan; in N. Africa, Kenya, Tanzania, Mozambique, Mauritius and, apparently, Madagascar.

Additional material, especially from East Africa, shows that the prescence of tubercles or murications in the centre of the fruiting bracteoles is too inconsistent to justify continued recognition of var. *granulata* Chevall. Indeed *Greenway & Myles Turner* 10020 from Tanzania shows both smooth and muricate bracteoles on one and the same plant.

2. **Atriplex amboensis** Schinz in Verh. Bot. Prov. Brand. **31**: 211 (1890).—Aellen in Bot. Jahrb. (Engl.) **70**: 391, t. 24, fig. 1–2 (1940); in Merxm., Prodr. Fl. SW. Afr. **32**: 5 (1967). Type from Namibia.

Annual or short-lived perennial herb with stems becoming thinly woody below, erect, 18–35 cm. high. Leaves elliptic to oblong or lanceolate in outline, 1–2.5 cm. long, 0.2–1.2 cm. wide, obtuse at apex, cuneate-attenuate at base, superior subentire, inferior with 2–4 spreading teeth or lobes on each margin, all densely grey-mealy on both surfaces. Flowers in dense axillary clusters, aggregated towards ends of stems and branches, upper clusters mostly without subtending leaves, clusters mostly bisexual, but upper predominantly male, lower predominantly female. Fruiting bracteoles subsessile, or with pedicels up to about 2 mm. long; basal connate part of bracteoles very short, 1–2 mm. long; free apical part of each bracteole broadly deltoid in outline, broader than long, 3–4 mm. long, 7–8 mm. wide, subacute at apex, with 1–3 pronounced teeth on each margin, in centre with about 6–10 spreading finger-like simple or branched projections 1–3 mm. long arranged in two groups.

Zambia. B: Liuuwa Plain, Paramount Chief's Game Reserve, c. 30 km. N. of Kalabo, fl. & fr. 14.xi.1959, *Drummond & Cookson* 6467 (K; LISC; SRGH).
Also in Namibia. In saline pan.

This rare species has untill now been known only from the type-gathering from Namibia (*Schinz* 468, Omadongo), with which the Zambian specimens agree well. The fruiting bracteoles with clusters of spreading finger-like projections in the centre are very characteristic. It appears to be an annual or short-lived perennial though becoming woody at base. In this way also it appears to differ from the more definitely shrubby *A. halimus* The description of *A. amboensis* in Merxm., Prodr. Fl. SW. Afr. 32: 4 (1967) as "stark verholzter.......Strauch" is misleading.

3. **Atriplex suberecta** Verdoorn in Bothalia **6**: 418, fig. 2 (p. 419) (1954). Type from S. Africa (Orange Free State).
Atriplex muelleri sensu Aellen in Bot. Jahrb. (Engl.) **70**: 390, t. 24, E (1940).—Adamson in Adamson & Salter, Fl. Cape Penins.: 356 (1950).—Brenan in F.T.E.A., Chenopodiaceae: 17 (1954) non Benth.

Annual herb, 12–60 cm., decumbent to suberect. Leaves variable in shape, in outline ovate to elliptic or oblong, 1–6 cm. long, 0.5–3 cm. wide, mostly obtuse to subtruncate at apex, cuneate to attenuate at base, normally with 1–5 coarse teeth on each margin; inferior leaves occasionally subentire or entire; grey vesicular hairs scattered and often sparse on both surfaces except on young leaves. Flowers clustered in leaf-axils, upper clusters male or mixed, lower female. Fruiting bracteoles almost sessile, their basal connate part about 2 mm. long, obconical or campanulate, becoming thick hard and pale; free apical part of each bracteole semicircular to triangular in outline, 2–3 mm. long, 1.5–3.5 mm, wide acute or subacute at apex, with 1–4 small teeth on each margin; bracteoles in centre smooth or venose, not at all tuberculate or muricate.

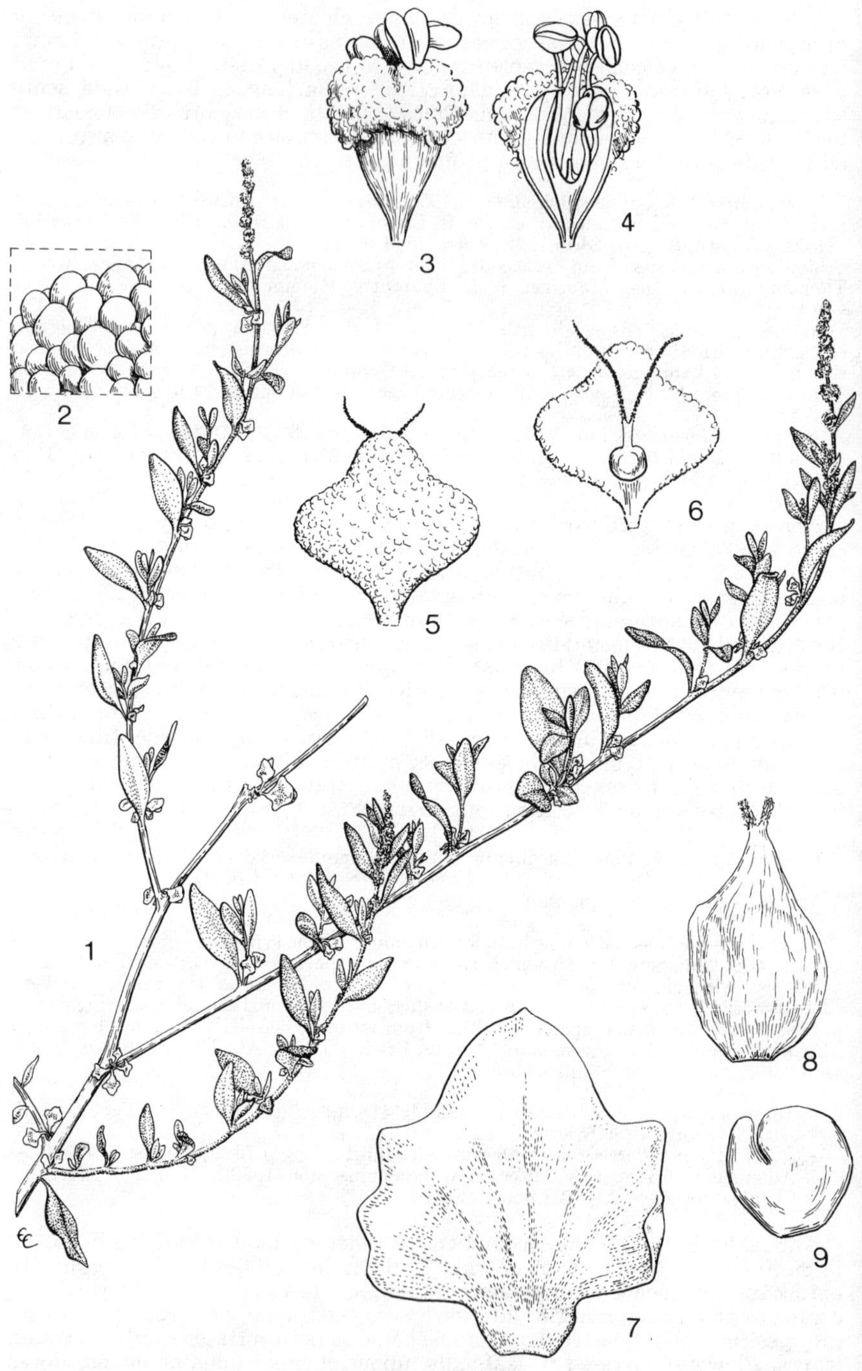

Tab. 35. ATRIPLEX HALIMUS. 1, habit (×$\frac{1}{3}$); 2, vesicular hairs on leaf surface (×60); 3, male flower (×15); 4, male flower, 2 perianth lobes removed (×15); 5, female flower (×15); 6, female flower, one bracteole removed (×15); 7, bracteole in fruiting stage (×5); 8, fruit (×10); 9, seed (×10), all from *Barbosa* 8345.

Zimbabwe. W: Bulawayo Sewage Farm, fl. & fr. 20.ix.1957, *Cronin* 76661 (K; SRGH). Also in S. Africa and Namibia; probably an introduction in Kenya.

3. CHENOLEA Thunb.

Chenolea Thunb. Nov. Gen. Pl.: 9 (1781).

Perennial herbs. sometimes thinly woody below more or less densely silky-hairy. Leaves more or less fleshy, entire, alternate, sessile, narrow but not scale-like. Flowers axillary, solitary or sometimes up to 3 together, sessile, subtended by a minute bidentate cupule but without bracteoles, hermaphrodite. Perianth 5-lobed, the lobes each developing in fruit dorsally a low blunt pyramidal horn without any wings. Stamens 5. Stigmas 2 (3). Fruit with membranous pericarp. Seeds "horizontal" (i.e. vertically compressed); testa apparently thinly crustaceous. Embryo annular. Endosperm absent.

A genus of 4 species in southern Africa and Mediterranean region.

Chenolea diffusa Thunb. Nov. Gen. Pl.: 10 (1781).—Wright in Dyer, Fl. Cap. **5**, 1: 447 (1910).—Adamson in Adamson & Salter, Fl. Cape Penins.: 354 (1950).—Macnae & Kalk, Nat. Hist. Inhaca I., Moçamb.: 144 (1958).—Aellen in Merxm., Prodr. Fl. SW. Afr. **32**: 7 (1967). TAB. **36**. Type from S. Africa (Saldanha Bay).

Low, decumbent, straggling plant about 10—25 cm. high. Stems thinly to densely silky-hairy. Leaves linear to lanceolate, ascending when young, later spreading, rather densely arranged along stems, 4—15 (17) × 1—3 mm., acute at apex more or less densely appressed-silky on both surfaces. Flowers yellowish-green, 1.5—2 mm. in diam., densely silky-hairy outside. Fruiting perianth c. 3—5 mm. in diam. Seed ellipsoid, c. 2—2.2 × 1.5—1.8 mm. brownish-black.

Mozambique. GI: Inhambane Bay, Isl., st. 26.xi.1958, *Mogg* 32175 (LISC; SRGH). LM: Marracuene, fl. i.vi.1959, *Barbosa & Lemos* 8541 (K; LISC).
Also in S. Africa. Mud in salt-marshes and near mangrove-swamps, at about high-tide limits.

4. HALOSARCIA P.G. Wilson

Halosarcia P.G. Wilson in Nuytsia **3**: 28 (1980).
Arthrocnemun sensu auct. mult. pro parte, e.g. Baker & Clarke in F.T.A. **6**, 1: 85 (1909).—Moss in Journ. S. Afr. Bot. **20**: 4 (1954).—Tölken in Bothalia **9**: 273 (1967).

Shrubs, usually thinly woody, low, glabrous, seemingly leafless, apparently built up of numerous, superposed, more or less tubular segments which are green or glaucous and ultimately shrivel and fall away from the stem; each segment at apex forming a little cup usually with two short teeth or lobes and embracing the base of the next higher segment. Fertile segments aggregated into spikes at ends of stem and lateral branches. Spikes (in our species) ultimately disarticulating into rings, in others sometimes persistent. Flowers minute, usually (always in our species) in clusters (cymules) of 3, a pair of opposite clusters to each fertile segment, which consists of two normally fused bracts. Perianth minutely and irregularly 3-lobed at apex (with us), never with a truncate lateral shield. Stamens 0—1 per flower. Seeds (in our species) with a hard crustaceous testa, smooth. Embryo curved. Endosperm present and obvious.

A genus of 23 species, all except *H. indica* confined to Australia.

Halosarcia indica (Willd.) P.G. Wilson in Nuytsia **3**: 63 (1980). TAB. **37**.
Salicornia indica Willd. in Ges. Naturf. Fr. New. Schr. **2**: 111, t. 4, fig. 2 (1799).
Arthrocnenum indicum (Willd.) Moq., Chenop. Monogr. Enum.: 113 (1840) pro parte.—Moss in Journ. S. Afr. Bot. **20**: 5 (1954).—Brenan, F.T.E.A., Chenopodiaceae: 18 (1954).—Tölken in Bothalia **9**: 276 (1967). Type from India.

Main stems prostrate, up to at least 40 cm. long, becoming longer when older and forming loose open mats; lateral branches numerous ascending or erect,

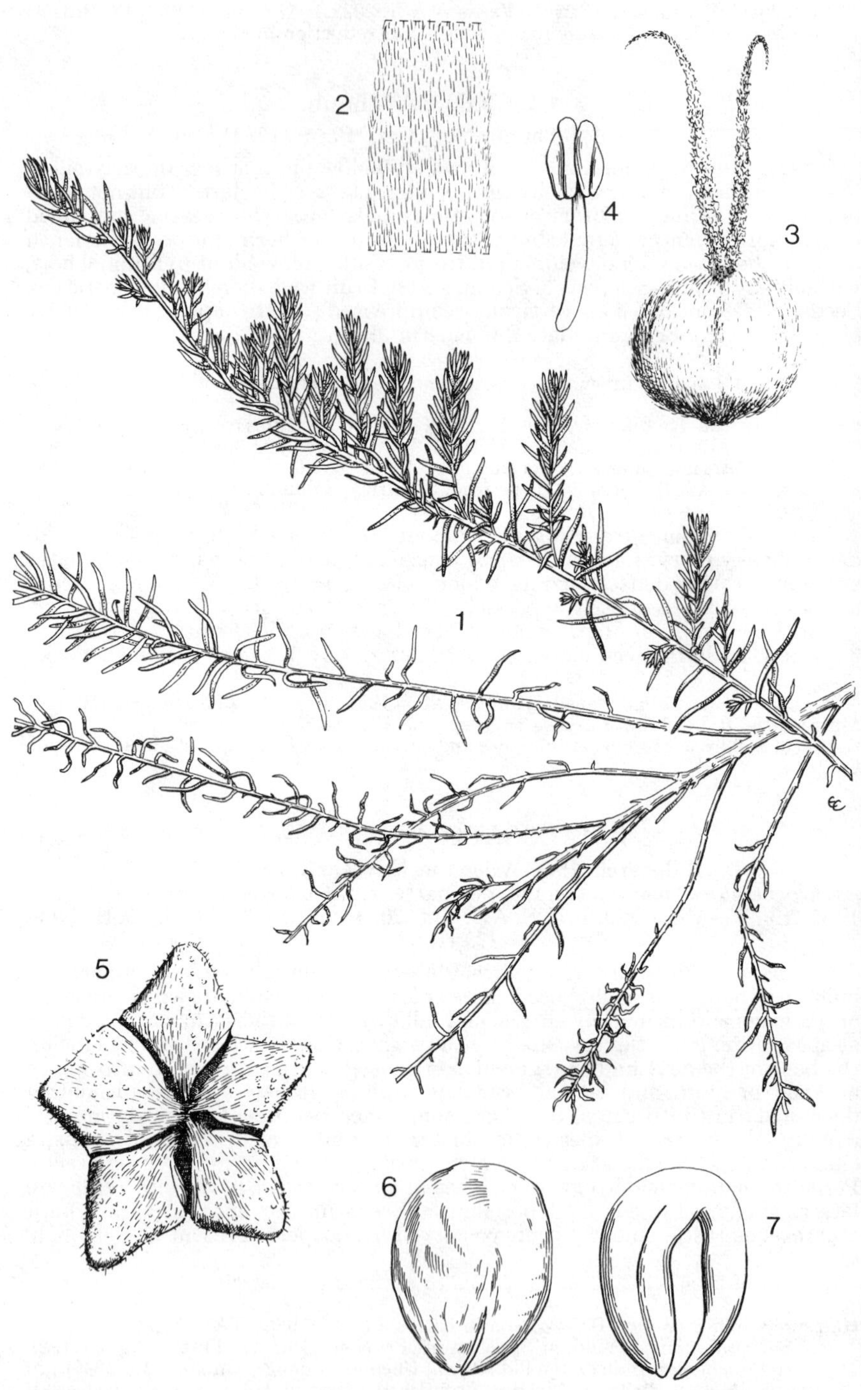

Tab. 36. CHENOLEA DIFFUSA. 1, habit (×⅔); 2, detail of indumentum on leaves (×8); 3, female flower (×20); 1–3 from *Barbosa & Lemos* 8541; 4, stamen (×10) *Dinter* 5997; 5, fruit from above (×10); 6, seed (×10); 7, embryo (×10), 5–7 from *Galpin* 6469.

about 10–30 cm. high, fertile or sterile. Sterile segments about 5–11 mm. long and 3–6 mm. in diam. Flowering spikes about 1–4 cm. long and 4–5 mm. in diam., with the flowers (except for the stigmas) hidden. Flowers deeply embedded in and fused to the segment of the spike above each cluster. Anthers very rarely visible, but sometimes a single often apparently non-functional stamen occurs with the ovary. Fruiting spikes cylindrical or somewhat thicker in middle, built up of numerous closely imbricate, ring-like, thickened and corky fertile segments, their margins 1.5–4 mm. apart, completely hiding the fruits; the segments finally disarticulating and falling away, leaving often a projecting bristle-like axis. Fruiting calyces spongy and thickened, about 3 × 2 mm. Seeds smooth.

Mozambique. N: Goa or St. George's Isl., fr. 31.x.1953, *Gomes e Sousa* 4166 (K). GI: Banamana Salina, 20 km. SW. of Mabate, fl. & fr. x.1973, *Tinley* 2958 (K; LISC). M: Costa do Sol and between Maputo and Matola, fr. 22.ii.1963, *McNae* s.n. (K).

Shores of the Indian Ocean from Somalia southwards to Mozambique and on Madagascar, Sri Lanka, in India from Bombay to Bengal, and on the northern coasts of Australia (P.G. Wilson, op. cit. supra, p. 65). Also in Angola and possibly further north in W. Africa. Salt marshes by the sea and mangrove swamps.

H. indica is readily distinguished from all the other *"glassworts"* in the Flora Zambesiaca area by the combination of perennial habit, thick sterile stem-segments, and thick corky fruiting spikes which ultimately disarticulate into ring-like separate segments or groups of segments, which can float in water.

Although when I wrote up this species for F.T.E.A., Chenopodiaceae, I had not seen male or hermaphrodite flowers (p. 20), I have since seen an occasional small single stamen, which seems doubtfully functional. The regularity of seed-production suggests parthenogenesis.

Macêdo & Macuacua 1132 (K) (GI: 23.viii.1963, Chibuto, Baixo Changana) record that the plant is used as food in times of scarcity. It is interesting to note that *Salicornia* is a palatable and esteemed vegetable in some parts of Europe. P.G. Wilson in Nuytsia 3: 63–69 (1980) distinguished four subspecies of *H. indica*, all except typical subsp. *indica*, confined to Australia and Malaysia. The specific description of *H.indica* above is of subsp. *indica* and does not necessarily apply to the others.

5. SARCOCORNIA A.J. Scott

Sarcocornia A.J. Scott in Bot. Journ. Linn. Soc. 75: 366 (1977).

Perennial herbs or shrubs, glabrous, seemingly leafless, apparently built up of numerous, superposed, more or less tubular segments succulent and usually green or greenish to reddish, ultimately shrivelling and falling away from the stem; each segment at apex forming a little cup, usually with two short teeth or lobes, embracing the base of the next higher segment. Fertile segments aggregated into spikes at ends of stem and lateral branches. Spikes persistent or breaking up irregularly, not regularly disarticulating. Flowers minute 3 (7–12) together in a cluster (cymule), a pair of clusters to each fertile segment, the clusters on opposite sides and flush with the segments. Perianth minutely and irregularly 3-lobed at apex, or more usually on a truncate flattened lateral shield. Stamens usually 1–2 per flower. Seeds with soft membranous testa, minutely hairy or papillose. Embryo folded so that radicle and cotyledonous point downwards. Endosperm absent.

A genus of 15 species, in Europe, Africa, Asia and north and south America; 9 species occur in southern Africa.

The taxonomy within this genus is anything but satisfactory from herbarium specimens alone. Further material, well collected and linked with careful observation in the feild, is most desirable. Tölken (in Bothalia 9: 266–272 (1967)) records a wide range of interspecific hybrids in S. Africa and some are found in the Flora Zambesiaca area. To interpret these with any confidence, field work is essential, with special attention to growth habit ecology, colour of the living plants, and of course the presence of putative parents.

1. Flowers 5–7 at each segment of the flowering spike, each truncate with a lateral shield, through whose centre stamens and stigmas project; barren segments when

fresh glaucous - - - - - - - - - - - - - 4. *decumbens* ★
— Flowers usually 3, rarely 4 or 5, at each segment of the flowering spike, truncate or rounded at apex; barren segments dull to shiny green when fresh 2
2. Perianth dorsoventrally flattened, opening at apex; flowers more or less concealed by segments of spike in fresh condition (not when dry); testa of seeds papillose - - - - - - - - - - - - - - - - - - - 1. *natalensis*
— Perianth with a lateral truncate shield, through whose centre stamens and stigmas project; flowers exp - - - - - - - - - - - - - - - 3
3. Testa of seeds with hairs which are recurved or coiled at apex; flowering spikes tapering - - - - - - - - - - - - - - - - - 2. *perennis*
— Testa of seeds minutely papillose, not hairy, flowering spikes cylindrical - - - - - - - - - - - - - - - 3. *mossambicensis*

★hybrids between *A. decumbens* and *A. natalense* and *A. perennis* will also key out here; see text. p. 153.

1. **Sarcocornia natalensis** (Bunge ex Ungern-Sternb.) A.J. Scott in Bot. Journ. Linn. Soc. **75**: 368 (1977). TAB. **38** fig. B.
Salicornia natalensis Bunge ex Ungern-Sternb., Vers. Syst. Salic.: 62 (1866). Type from S. Africa.
Arthrocnemum natalense (Bunge ex Ungern-Sternb.) Moss in Journ. S. Afr. Bot. **20**: 15 (1954).—Tölken in Bothalia **9**: 277 (1967).

Perennial prostrate or decumbent herb 50—30 cm. high and 20—80 cm. in diam. (fide Tölken, l.c. supra.). Sterile segments 2—6 mm. in diam., shining to dull grey-green or light yellowish-green, fading to brownish-yellow or pinkish-red (Tölken l.c. supra.) Spikes (10) 20—30 (80) mm. long, tapering in flower and fruit. Flowers 3 in a cluster, hidden by the bract below in dried material (Tölken, l.c.) Perianth not truncate laterally, opening at true apex. Seeds brown; testa minutely papillose.

Mozambique. M: Costa do Sol, fl. & fr. 22.ii.1963, *McNae* 11 (K).
Southern Mozambique to S. Africa. Salt-marshes by the sea.

S. natalensis is very distinct from the other species in the Flora Zambesiaca area by the flowers being 3 in each cluster and opening truly apically, without the truncate lateral shield of the other species. The green or greenish colour of dried specimens is noteworthy and there is a marked tendency for the lateral branches to be curved or arcuate and not almost straight. For comments on hybrids see note under *S. decumbens*. All material from the Flora Zambesiaca area appears to the typical *S. natalensis*.

2. **Sarcocornia perennis** (Mill.) A.J. Scott in Bot. Journ. Linn. Soc. **75**: 367 (1977). Type from England.
Salicornia perennis Mill., Gard., ed. 8 (1768).
Arthrocnemum perenne (Mill.) Moss ex Fourcade in Mem. Bot. Surv. S. Afr. **20**: 20 (1941).—Moss in Journ. S. Afr. Bot. **14**: 40 (1948); in Journ. S. Afr. Bot. **20**: 11 (1954).
Arthrocnemum perenne var. *radicans* (Sm.) Moss in Journ. S. Afr. Bot. **14**: 40 (1948); in Journ. S. Afr. Bot. **20**: 11 (1954).—Macnae & Kalk, Nat. Hist. Inhaca I., Moçamb. 13, **119**: 144 (1958).
Salicornia radicans Sm., Engl. Bot. t. 1691 (1807). Type from England.

Prostrate or decumbent herb forming mats 10—25 cm. high. Sterile segments 3—5 mm. in diam., "dull green to shiny green, fading to yellow or reddish-brown" (Tölken, l.c. supra.) Spikes 15—35 (50) mm. long, tapering in flower, cylindrical in fruit. Flowers 3 in a cluster, rarely 4 or 5, not hidden by bract below. Perianth truncate, with a lateral shield in whose centre the perianth opens. Seeds brown; testa clothed with minute hairs recurved or coiled at apex.

Mozambique. Z: mouth of R. Luabo, fl. 16.v.1858, *Kirk* 16 (K).
Also in Europe, NW. Africa and from Mozambique to S. Africa. Muddy salt-marshes by the sea.

Unfortunately the material of *S. perennis* from our area is old, limited and inadequate. In particular it is only in flower and it is thus not possible to verify the important character of hairs on the surface of the testa. *Kirk* 16 is clearly different from *S. decumbens* by the habit and 3-flowered clusters, and from *S. natalensis* in the straight upright lateral branches. More material is much needed. *Kirk* 16 is (apparently) typical *S. perennis*.

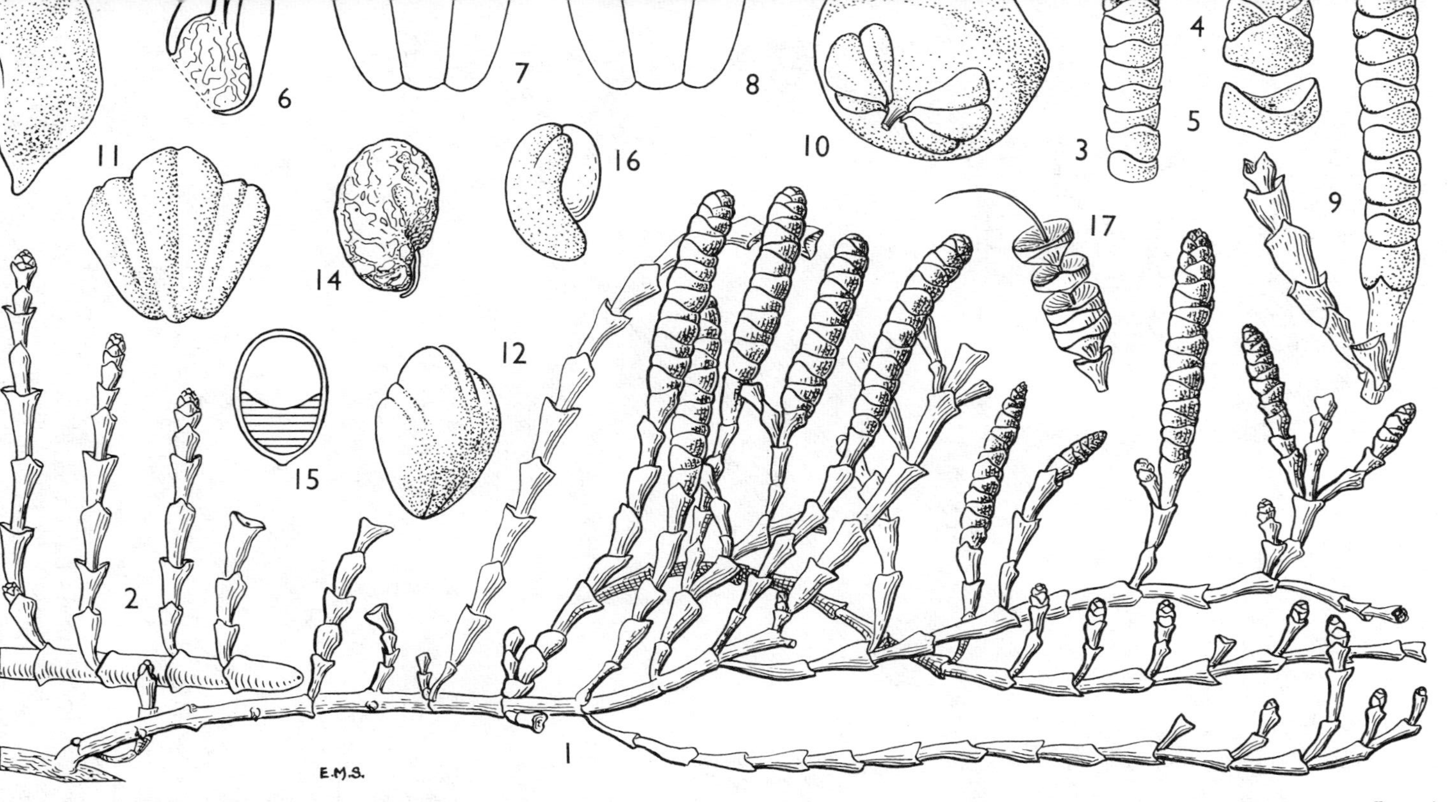

Tab. 37. HALOSARCIA INDICA. 1, part of plant, natural size, *MacNaughtan* 40; 2, sterile lateral shoots, natural size, *Greenway* 4777; 3, young flowering spike (×1½); 4, apex of flowering spike (×3); 5, segment of flowering spike (×3); 6, female flower cut longitudinally (×12), 3–6 from *MacNaughtan* 40; 7,8, groups of 3 female flowers showing different degrees of exertion of stigmas, diagramatic (×12); 9, fruiting spike (×1½); 10, segment of fruiting spike seen obliquely from beneath showing two groups of three fruiting calyces originating in axils of next segment below (×6); 11, group of three fruiting calyces, face view (×12); 12, group of three fruiting calyces, face view (×12); 13, fruiting calyx, lateral view (×12); 14, seed (×24), 9–14 from *Greenway* 4777; 15, seed, diagramatic, transverse section; 16, seed with testa removed, showing embryo to left and endosperm to right (×24), 15–16 from *Greenway* 4777; 17, fruiting spike breaking up (×1½) *Moss* 7072. From F.T.E.A.

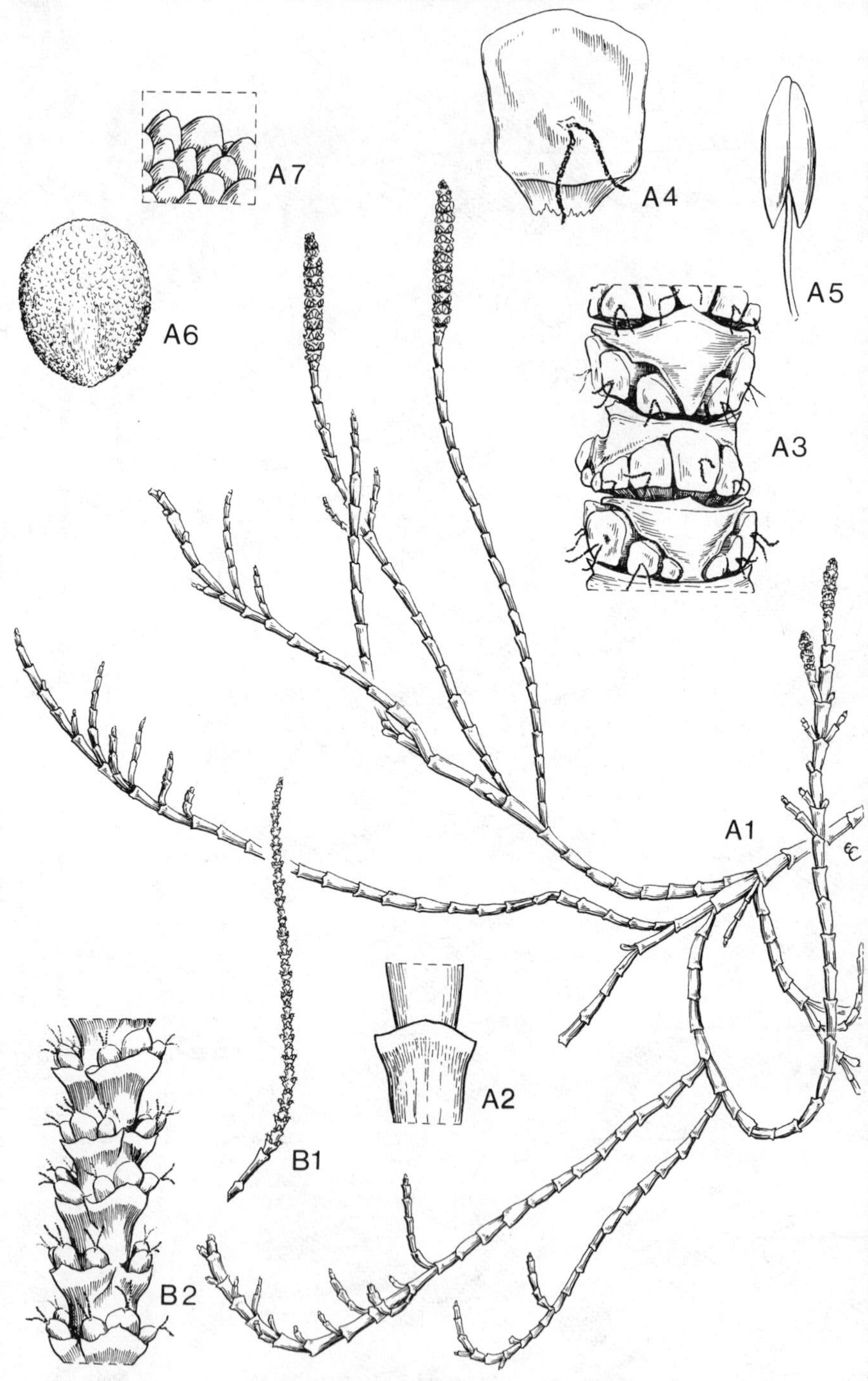

Tab. 38. A.—SARCOCORNIA DECUMBENS. A1, habit (×⅔); A2, sterile stem segments (×8), A1—2 from *McNae* 5; A3, part of inflorescence (×5) *Moss* 27283; A4, flower (×20); A5, anther (×20); A6, seed (×20); A7, detail of surface of testa (×60), A4—7 from *McNae* 5. B.—SARCOCORNIA NATALENSIS. B1, inflorescence (×⅔); B2, part of inflorescence (×5), B1—2 from *McNae* 11.

However, fruiting material may show that it cannot be maintained separate from *S. mossambicensis*. For comments on hybridisation with *S. decumbens*, see note under that species (below).

3. **Sorcocornia mossambicensis** Brenan, sp. nov.★ Type: Mozambique, Inhambane Bay, Mocucuni Isl., fl. & fr. 26.xi.1958, *Mogg* 29306 (LISC, holotype, SRGH).

Prostrate or decumbent herb, with numerous lateral erect or ascending lateral branches 10–40 cm. high. Sterile segments c. 3–4 mm. in diam., green. Spikes 1.3–4.5 cm. long, 2–3.5 mm. in diam., cylindrical. Flowers 3 in a cluster. Perianth truncate with a lateral shield in whose centre the perianth opens. Seeds pale brown; testa minutely papillose.

Mozambique. GI: Rio Pomene Manth, Ponta Barra Falsa, fl. & fr. 21.xi.1958, *Mogg* 28914 (LISC).

The flowers in threes with the perianth of each provided with a lateral truncate shield, distinguish *S. mossambicensis* from all the species in the Flora Zambesiaca area except *S. perennis* from which it differs in having the testa minutely papillose, not hairy, and in having cylindrical not tapering flowering spikes. *S. mossambicensis* is so far restricted to Mozambique (GI). *Mogg* 29305 (K; SRGH), from the same locality as the type, is conspecific. See note also under *S. perennis*.

★ *S. decumbenti* (Tölken) A.J. Scott ut videtur proxima, cymis trifloris differt. Gerba ut videtur perennis, prostrata vel decumbens, e caulibus radicantibus prostratis ramulos numerosos circiter 10–40 cm. longos emittens. Articuli steriles 5–10 mm. longi, 3–4 mm. diametro. Spicae ad apices caulium solitariae sed ob ramulos secundarios saepe profusos numerosae, 1.3–4.5 cm. longae, 2–3.5 mm. diametro, cylindricae. Cymulae triflorae. Perianthi tubus scutulos laterali truncatus. Semina c. 1.25 × 1 mm., testa pallide brunnea, minute papillosa.

A *S. perenni* seminis testa minute papillosa non pilosa, spicis floriferis cylindricis differt.

4. **Sarcocornia decumbens** (Tölken) A.J. Scott in Bot. Journ. Linn. Soc. **75**: 368 (1977). TAB. **38** fig. A.

Arthrocnemum heptiflorum sensu Moss in Journ. S. Afr. Bot. **20**: 18 (1954) quoad specim. afr.-Macnae & Kalk, Nat. Hist. Inhaca I., Moçamb.: 144 (1958).

Arthrocnemum australasicum sensu Moss in Journ. S. Afr. Bot. **20**: 19 (1954) quoad specim. afr. p.p.: vide Tölken, l.c. infra.—Macrae & Kalk, Nat. Hist. Inhaca I., Moçamb.: 144 (1958).

Arthroenemum decumbens Tölken in Bothalia **9**: 293 (1967). Type from S. Africa (Mossel Bay Distr.).

Decumbent, woody herb 10–50 cm. high. Sterile segments 3–6mm. in diam., glaucous fading bluish-red (fide Tölken). Spikes 10–30 (50) mm. long, usually cylindrical, rarely tapering. Flowers (4) 5–7 (9) in a cluster. Perianth truncate, with a lateral shield in whose centre the perianth opens. Seeds pale brown; testa minutely papillose.

Mozambique. M: Inhaca Isl., fl. & fr. 11.vii.1957, *Mogg* 27283 (BM; K; SRGH).

The normally 5–7-flowered clusters (cymules) differentitate *S. decumbens* from other species of the genus in the Flora Zambesiaca area. The specimens in herbaria are generally of dry and dull brownish colour and this is probably correlated with the distinctive glaucous colour of the living plant mentioned by Tölken (l.c., supra).

Tölken (l.c., p. 270) describes hybrids with *S. natalensis* (Mozambique, M: Matola Bridge, fl. 10.vii.1966, *Tölken* 2530, 2532 (K)), differing from *S. decumbens* is either having large tapering spikes or lacking the glaucous colour.

A hybrid with *S. perennis* is also mentioned (p. 272). Mozambique, M: Maputo, fl. & fr. 22.ii.1963, *McNae* 7 (K), "almost indistinguishable" from *S. decumbens* in the herbarium, but when above with green not glaucous segments.

6. SALICORNIA L.

Salicornia L., Sp. Pl. ed. 5: 4 (1754).

Annual herbs, sometimes becoming thinly woody below, glabrous, seemingly leafless, built up of numerous superposed, more or less tubular-segments which are green to reddish and succulent, and ultimately shrivel, each segment at apex forming a little cup, usually with two short teeth, embracing the base of the next higher segment. Fertile segments aggregated into spikes at ends of stem and lateral branches, the latter often very short. Spikes not disarticulating, persistent or breaking up irregularly. Flower minute, hermaphrodite, usually in clusters of three (always in the Flora Zambesiaca area), more or less connate, a pair of clusters to each fertile segment, the clusters on opposite sides and immersed. Perianth minutely 3–4 denticulate, opening in the middle of a truncate flattened lateral shield. Stamens usually 2 per flower. Seeds with thin membranous testa, minutely hairy. Embryo folded so that radicle and cotyledons point downwards. Endosperm absent.

Number of species uncertain, but probably c. 30–40, cosmopolitan, mostly in saline habitats near coasts.

Fruiting spikes cylinrical in outline, markedly thickened, 5–8 mm. in diam., blunt at apex and scarcely tapering, lateral branches suberect - 2. *pachystachya*
Fruiting spikes irregular in outline when dry, less thickened than above, 3–4 mm. in diam., mostly tapering to a narrow apex; lower lateral branches widely spreading and decumbent - - - - - - - - - - - - - 1. *perrieri*

1. **Salicornia perrieri** A. Chev. in. Rév. Bot. Appliq. **2**: 749 (1922).—Macnae & Kalk, Nat. Hist. Inhaca I., Moçamb.: 144 (1958). TAB. **39** fig. B. Type from Madagascar.
Salicornia pachystachya sensu Tölken in Bothalia **9**: 298 (1967) pro parte quoad "second form", non *Salicornia pachystachya* Bunge ex Ungern-Sternb.

Herb 10–25 cm. high; main stem erect with numerous lateral branches, the lower mostly widely spreading and decumbent. Fertile spikes many, 1–5 cm. long, thickening in fruit, but less than in *S. pachystachya*, 3–4 mm. in diam., not cylindrical (at least when dry) and mostly distinctly tapering towards apex.

Mozambique. Z: Kongone, mouth of Zambezi, fl. & fr. 13.xii. 1859, *Kirk* (K). M: Maputo fr. 28.xii.1934, *Wager* (K).
Possibly in Tanzania and Zanzibar; also in Madagascar and Natal. Salt-marshes by the sea.

This is an unsatisfactory taxon, requiring further careful study in the field. I have little doubt that it corresponds with form 2 of *S. pachystachya* as defined by Tölken in Bothalia **9**: 298 (1967) and would have been disposed to follow his view except that the fruiting spikes are (in the herbarium) clearly narrower than in *S. pachystachya*, not regularly cylindrical, and with a strong terndency to taper towards the apex. The record of "*S. herbacea* L." from Mozambique, in P.O.A.C.: 171 (1895) on the authority of Engler, by Baker & Wright in Dyer, Fl. Trop. Afr. **6**, 1: 86 (1909) may be referable here. It almost certainly does not relate to the true *S. herbacea*, a European species.

2. **Salicornia pachystachya** Bunge ex Ungern-Sternb., Vers. Syst. Salicorn. 51 (1866); in Atti Congr. Int. Bot. Firenze: 304, fig. 18 (p. 278) (1876).—Brenan, F.T.E.A., Chenopodiaceae: 21 (1954).—Tölken in Bothalia **9**: 297, fig. 103 (1967) pro parte quoad "first form". TAB. **39** fig. A. Type from Madagascar.

Herb 10–30 cm. high; main stem erect with numerous suberect or ascending lateral branches (forming an angle usually 450 or less with main stem). Fertile spikes many, (0.5) 1–3 (5.5) cm. long, markedly thickening in fruit and 4–8 mm. in diam. Fruiting spikes cylindrical, blunt, rarely somewhat tapering. Perianth exposed in flower and fruit. Seed ellipsoid, c. 1.5–1.7 × 1–1.2 mm.

Mozambique. Z: Quelimane, Micaune Province, fr. xi., *Beirao* 147,076 (K; SRGH). MS?: Save. R., fl. 13.xi.1958, *Mogg* 29200 (K; LISC). GI: Bazaruto Isl., fl. 29.x.1958, *Mogg* 28734 (K; LISC; SRGH). M: Inhaca Isl., fr. 4.iii.1958, *Mogg* 31609 (BM; K).
Coast of E. Africa from S. Kenya to Natal, also in Madagascar. Salt-marshes and mangrove swamps; sea level.

7. SUAEDA Forssk. ex Scop.

Suaeda Forssk. ex Scop., Introd. Hist. Nat.: 33 (1777) excl. verba "Capsula

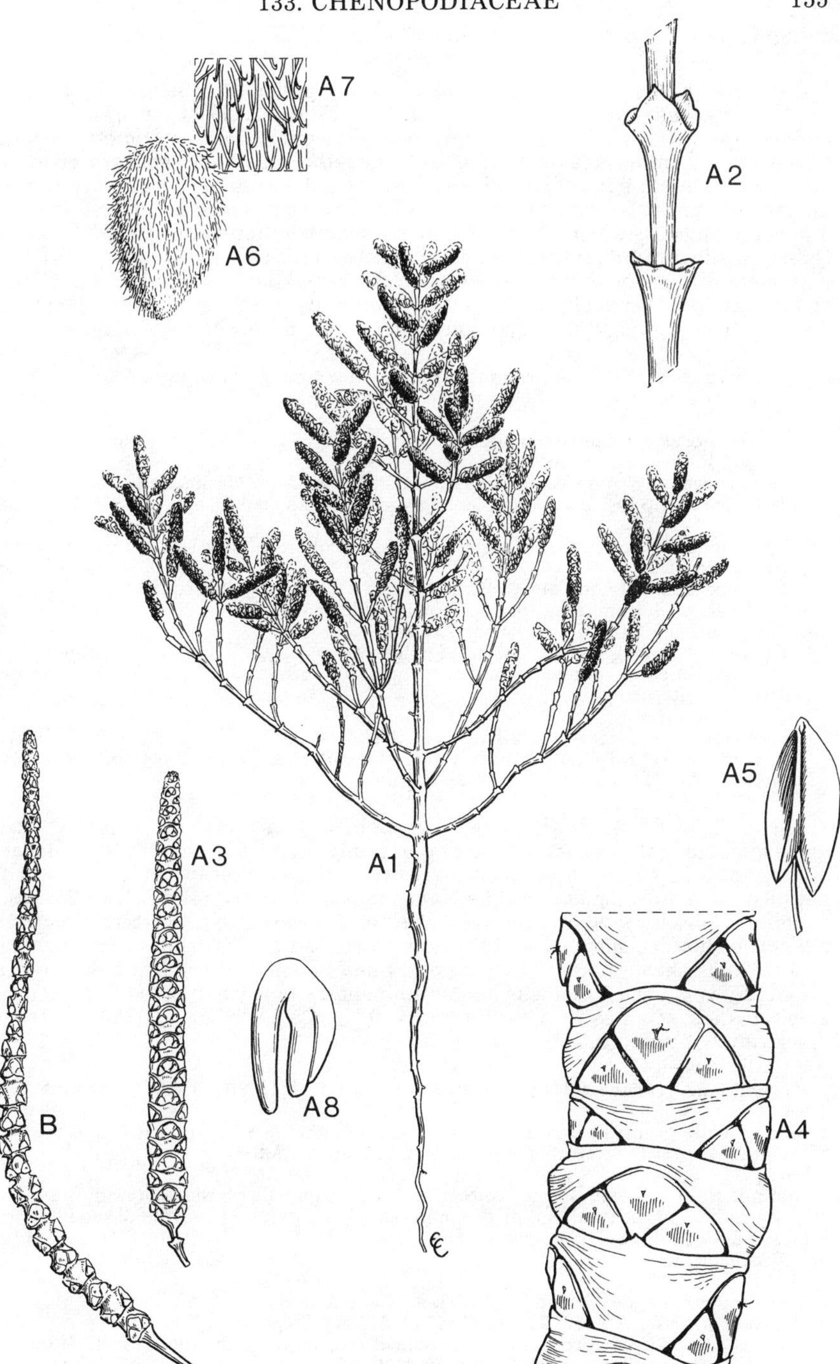

Tab. 39. A.—SALICORNIA PACHYSTACHYA. A1, habit (×$\frac{2}{3}$); A2, sterile segments (×6); A3, inflorescence (×1$\frac{1}{2}$); A4, part of inflorescence (×6), A1—4 from *Mogg* 31609; A5, anther (×40) *McNae* 1; A6, seed (×20); A7, part of surface of testa (×60); A8, embryo (×20), A6—8 from *Mogg* 31609. B.—SALICORNIA PERRIERI. inflorescence (×1$\frac{1}{2}$) from *Tölken* 1096.

quinquevalvis ... polysperma"; vide Brenan, F.T.E.A., Chenopodiaceae: 21 (1954).

Annual or perennial herbs or shrubs, usually glabrous or nearly so. Leaves usually alternate and fleshy (always in the Flora Zambesiaca area), entire, narrow, cylindrical or somewhat flattened. Flowers axillary or sometimes (not in the Flora Zambesiaca area) adnate to the subtending leaf, solitary or more usually clustered, subtended by small scarious bracteoles, hermaphrodite or unisexual. Perianth 5-lobed or-partite, sometimes thickened or inflated in fruit but not winged. Stamens 5, absent or reduced to staminodes in the flowers. Ovary sessile on a broad base, or partly adnate to the calyx. Stigmas 2—5. Fruit with membranous or sometimes spongily thickened indehiscent pericarp. Seeds "horizontal" or "vertical" (i.e. vertically or laterally compressed); testa normally thin, hard and glossy. Embryo spiral. Endosperm absent or scanty.

The delimitation of the species in this genus is far from satisfactory. The material is often inadequate and careful field-work is greatly needed.

1. Seeds laterally compressed ("vertical"); shrub 1.2—6 m. high; leaves mostly 1.3—3.3 (4) cm. long; growing on seacoast - - - - - - - - - - 1. *monoica*
— Seeds vertically compressed ("horizontal"); herbs or low shrubs up to about 60 cm. high; leaves up to 1.2 cm. and plants growing inland (except only for *S. caespitosa*) - 2
2. Stigmas 2, not arising from a disk or cup-like collar, inferior leaves 1—3 cm. long; plants coastal - - - - - - - - - - - - - - - 5. *sp. B*
— Stigmas 3, usually (not always) arising from a disk or cup-like collar; inferior leaves up to about 1.1 cm. long; plants growing inland - - - - - - - - 3
3. Styles arising from a disk or cup-like collar - - - - - - - - - 4
— Styles not arising from a disk or cup-like collar - - - - - - 3. *plumosa*
4. Bracts obtuse or rounded at apex - - - - - - - - - 2. *merxmuelleri*
— Bracts acute at apex - - - - - - - - - - - - 4. *sp. A*

1. **Suaeda monoica** Forssk., ex. J.F. Gmel., Syst. Nat. ed. 13, **2**, 1: 103 (1791).—Baker & C.B. Clarke in F.T.A. **6**, 1: 92 (1909).—Brenan in F.T.E.A., Chenopodiaceae: 23 fig. 6 (1954). Types from Egypt and Arabia.

Shrub 1.2—6 m. high, much branched, glabrous or slightly and inconspicuously pubescent on young parts only. Leaves fleshy, linear to linear-oblong, obtuse to acute at apex, narrowed near the sessile or very shortly petioled base, inferior and middle leaves mostly 1.3—3.3 (4) × 0.15—0.3 cm., superior leaves (bracts) progressively shorter. Flowers green, clustered in upper axils, the clusters aggregated to form interrupted or sometimes dense spikes bracteate as throughout or below only; plants with male and female flowers or only flowers with 5 stamens and a rudimentary ovary expanded at apex into a peltate disk; stigmas 3—4. Seeds "vertical", c. 1.5—1.75 × 1—1.25 mm.; testa black, almost smooth.

Mozambique. N: left bank of R. Ridi, Nangororo, fl. & fr. 30 x.1959, *Gomes e Sousa* 4496 (K).
Also extending northwards along the coast of E. Africa to Egypt, also in Israel and Syria and eastwards to Arabia, Sri Lanka and India (Madras).

Distinct from all other species of *Suaeda* in the Flora Zambesiaca area by its more shrubby habit and the combination of comparatively long (1.3—4 cm.) middle and inferior leaves and 3—4 stigmas. The coastal habitat appears to be peculiar to both these species and *S. caespitosa*.

2. **Suaeda merxmuelleri** Aellen in Mitt. Bot. Staatss. München **4**: 27 (1961); in Merxm., Prodr. Fl. SW. Afr. **32**: 21—22 (1967). TAB. **40**. Type from Namibia.
Suaeda fruticosa sensu Pole-Evans in Bot. Surv. S. Afr. Mem. **21**: 30 (1948).—O.B. Miller in Journ. S. Afr. Bot. **18**: 12 (1952).

Low spreading shrub up to about 0.6 m. high and l m. across. Leaves (middle and inferior) sessile, oblong to elliptic or obovate-oblong, obtuse at apex, 2—4.5 (12) × 1—1.5 mm., pale green or glaucescent, blackish or brownish when dry. Flowers green, solitary or clustered in upper axils, clusters often aggregated to form narrow spiciforn inflorescences; bracts subtending flowers as long as

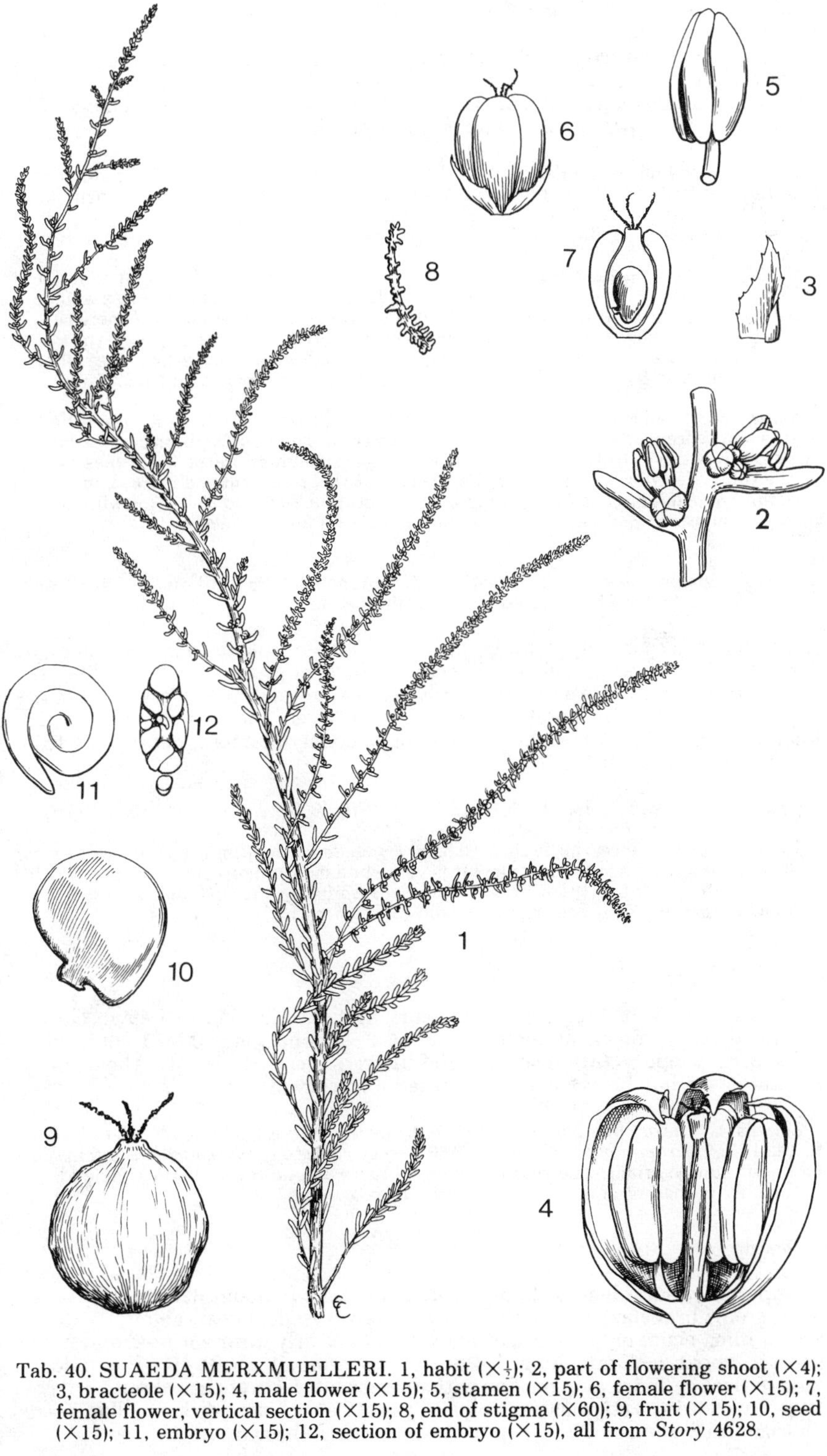

Tab. 40. SUAEDA MERXMUELLERI. 1, habit ($\times\frac{1}{2}$); 2, part of flowering shoot ($\times 4$); 3, bracteole ($\times 15$); 4, male flower ($\times 15$); 5, stamen ($\times 15$); 6, female flower ($\times 15$); 7, female flower, vertical section ($\times 15$); 8, end of stigma ($\times 60$); 9, fruit ($\times 15$); 10, seed ($\times 15$); 11, embryo ($\times 15$); 12, section of embryo ($\times 15$), all from *Story* 4628.

or up to twice as long as flowers (occasionally lowest 3—4 times as long), obtuse. Hermaphrodite flowers about 1.5—2 mm. long; perianth segments about 1.2—1.3 mm. long-cucullate at apex. Stigmas 3, about 0.4—0.5 mm. long, arising from a disk or cup-like enlargement at apex of styles. Seeds, "horizontal" somewhat broader than long, about 1.5 × 1 mm., somewhat beaked on one side; testa blackish-brown, glossy, almost smooth "horizontal".

Botswana. N: Makarikari Pan, fl. 9.ix.1954, *Story* 4628 (K); 2 km. N. of Toromoja school, near Rakops, fl. 21.iv.197l, *Pope* 356 in *Peterhouse* 175 (K; SRGH). SE: Boteti delta area, NE. of Mopipi, fl. 17.iv.1973, *Tyers* in *Peterhouse* 546 (K; SRGH).
Also in Angola, Namibia and S. Africa (Transvaal, Orange Free State, Cape Prov.).

S. merxmuelleri has been from time to time misidentified both in print and in herbaria with *S. fruticosa* (L.) Forssk. ex J.F. Gmel. but differs as follows: the more or less columnar prolongation of the style above the ovary (almost absent in *S. fruticosa*), the more clearly expanded disk or collar at the apex of the style and surrounding the stigmas, the larger perianth segments (1—1.25 mm. long), more cucullate and winged above and more gibbous on back, and also by the longer anthers (1 mm. long or more as against 0.75 mm.).

Smith 2523, from Botswana, N: Mopipi Pan, fl. 20.xi.1978, (K), is at first sight different in appearance from other material of *S. merxmuelleri* in the Flora Zambesiaca area, having larger narrow leaves to 12 × 1—1.5 mm. and longer straighter bracts 4—7 times as long as the subtended flowers. However, I suspect that this represents no more than a state of *S. merxmuelleri* occasioned by an unusual habitat - it was said to be "growing in fast moving water". *Seydel* 844 from Namibia (K!) shows an approach to this.

3. **Suaeda plumosa** Aellen in Mitt. Bot. Staatss. München **4**: 28 (1961); in Merxm., Prodr. Fl. SW. Afr. **32**: 21—22 (1967). Type from Namibia.

Succulent herb about 0.6 m. high. Leaves sessile, oblong to linear-oblong, c. 4—6 × 0.75—1.0 mm., subacute at apex. Flowers greenish-yellow, clustered 1—5 together in middle and upper axils; leaves subtending flowers $1\frac{1}{2}$—2 times as long as flowers. Hermaphrodite flowers not seen. Flowers about 1—2 mm. long. Stigmas 3, 0.5.7 mm. long, not conspicuously plumose, not arising from a basal disk or collar.

Caprivi Strip. Mpilila Isl., fl. & fr. 13.i.1959, *Killick & Leistner* 3378 (K; SRGH).

In general appearance, this is close to *S. merxmuelleri* except for the absence of a clear disk or collar at the base of the styles. It is very similar to and probably conspecific with specimens from Angola (*Ward & Ward* 31, *Barbosa* 9491), From Namibia (*Tölken* 452, *Galpin & Pearson* 7559) and Namaqualand (*Schlechter* 13a).

4. **Suaeda sp. A.**

Herb about 15 cm. high and 20 cm. wide, branched from base. Leaves closely arranged on stem, linear, oblong, about 2—4 mm. long. 0.3—1 mm. wide, subacute at apex. Inflorescence and flowers very similar to those of *S. merxmuelleri* but bracts acutely pointed and curved upwards.

Zambia. B: Liuwe Plain, Paramount Chief's Game Reserve, about 48 km. N. of Kalabo, fl. 14.xi.1959, *Drummond & Cookson* 6466 is very close to *S. merxmuelleri* and may be no more than a variant. The acutely pointed bracts are, however, distinctive, and this is the only gathering of *Suaeda* so far seen from Zambia.

5. **Suaeda sp.B.**

Apparently succulent herb, probably annual, up to about 325 cm. high with erect stems branched especially below. Leaves sessile, linear, about 10—30 × 1—1.5 mm., acute or subacute at apex; bracts clearly differentiated and much shorter, about 2—5 mm. long. Inflorescence terminal, about 10—15 cm. long, more or less cylindrical in outline with profuse lateral branches about 2—3 cm. long. Ovary with 2 stigmas passing into the style without a disk or collar. Seeds "horizontal", about 1.25 × 1 mm.; testa blackish-brown.

Mozambique. M: Campo da Mocidade Portuguesa, fl. & fr. 22.vii.1965, *Caldeira & Marques* 599 (BM; LISC; SRGH).
Salt marshes; sea level.

It is noteworthy that, except for *S. monoica*, this is the only species of *Suaeda* recorded as growing on the coast in the Flora Zambesiaca area. The long, mostly acute leaves, the inflorescence structure and the 2 stigmas are distinctive. It appears close to *S. caespitosa* Wolley-Dod from S. Africa, but that is a perennnial with profuse lateral branches from trailing stems, forming dense mats. The present species may well be undescribed, but better material is needed.

8. SALSOLA L.

Salsola L., Gen. Pl. ed. 5: 104 (1754).

Annual or more often perennial herbs or shrubs, usually pubescent. Leaves usually more or less fleshy, entire, alternate or sometimes opposite, sessile, narrow, often scale-like, sometimes linear-subulate. Flowers axillary, solitary or fascicled, sometimes aggregated in a spiciform way towards ends of branchlets, subtended by two relatively large bracteoles, hermaphrodite. Perianth (4) 5-partite, the lobes nearly always each developing in fruit above the middle a scarious horizontally spreading wing. Stamens 5, rarely fewer. Stigmas 2, very rarely 3. Fruit with membranous or somewhat fleshy pericarp. Seeds normally "horizontal" (i.e. vertically compressed); testa membranous. Embryo spiral. Endosperm absent.

About 150 species, cosmopolitan, especially maritime and in inland steppe regions.

Salsola rabieana C.A. Sm. ex Verdoorn in Bothalia **6**, 1: 218, fig. 3 (219) (1951).—O.B. Miller in Journ. S. Afr. Bot. **18**: 12 (1952).—Aellen in Merxm., Prodr. Fl. SW. Afr. 32: 17, 19 (1967).—Botschantzev in Nov. Syst. Pl. Vasc. **11**: 113, 144 (1974); in Kew Bull. 29, **3**: 600, 607 (1974). TAB. **41**. Type from S. Africa (Orange Free State).

Low spreading shrub about 0.1–2 m. high, with profuse intricate stiff woody branches; stems more or less densely pubescent to puberulous when young. Leaves alternate, scale-like, often closely imbricate, fleshy, very small, 1.25–2 × 1–1.5 mm., with more or less dense short appressed silky pubescence on back, green and thickened in centre, hyaline-margined, obtusely pointed at apex. Flowers yellow-green, solitary in upper axils, about 3 mm. long. Perianth lobes 5, ovate, each with puberulous central thickened green area from which the wings arise in fruit. Fruit (including wings) c. 4–10 mm. in diam., the wings encircling the fruit, usually more or less irregular in shape and size from reniform to obovate or even bifid, to c. 1.5–4 mm. long and c. 2.5–6 mm. wide, glabrous, closely veined and with a sheeny surface.

Botswana. SW: Ghanzi and Kgalagadi Distrs., Nwatle Pan, fl. & fr. ii.1979, *Skarpe* S-318 (K; SRGH). SE: Letlaking Valley, fl. & fr. 16.ii.1960, *Wild* 496 (BM; K).
Also in S. and Namibia. On limestone rocks and slopes in or near pans. Sandy soil in *Acacia* scrub at bottom of valley (*Timberlake* 2037).

Although numerous species are on record for S. Africa and Namibia the most recent workers on the genus, *Aellen* and *Botschantzev*, seem to have drawn their specific limits very finely. There appears to be only two taxa in the Flora Zambesiaca area. The present is best placed under *S. rabieana*, but most closely related to *S. aphylla* L. f., which differs in having broader flattened stigmas (as pointed cut by *Aellen* (l.c. supra)), perianth-segments rounded not acute at apex and by the broader hyaline margins to the leaves. A final taxonomic verdict on this plant must await closer field-studies in S. Africa. The differences from *Salsola* sp. A are given under the latter.

SPECIES INSUFFICIENTLY KNOWN

Salsola sp. A.

Low spreading perennial, with radiating branches; stems more or less densely pubescent when young, later glabrescent. Leaves opposite, scale-like, imbricate on young shoots, small, 2–3 × 2.5–3.5 mm., glabrous or subglabrous, with

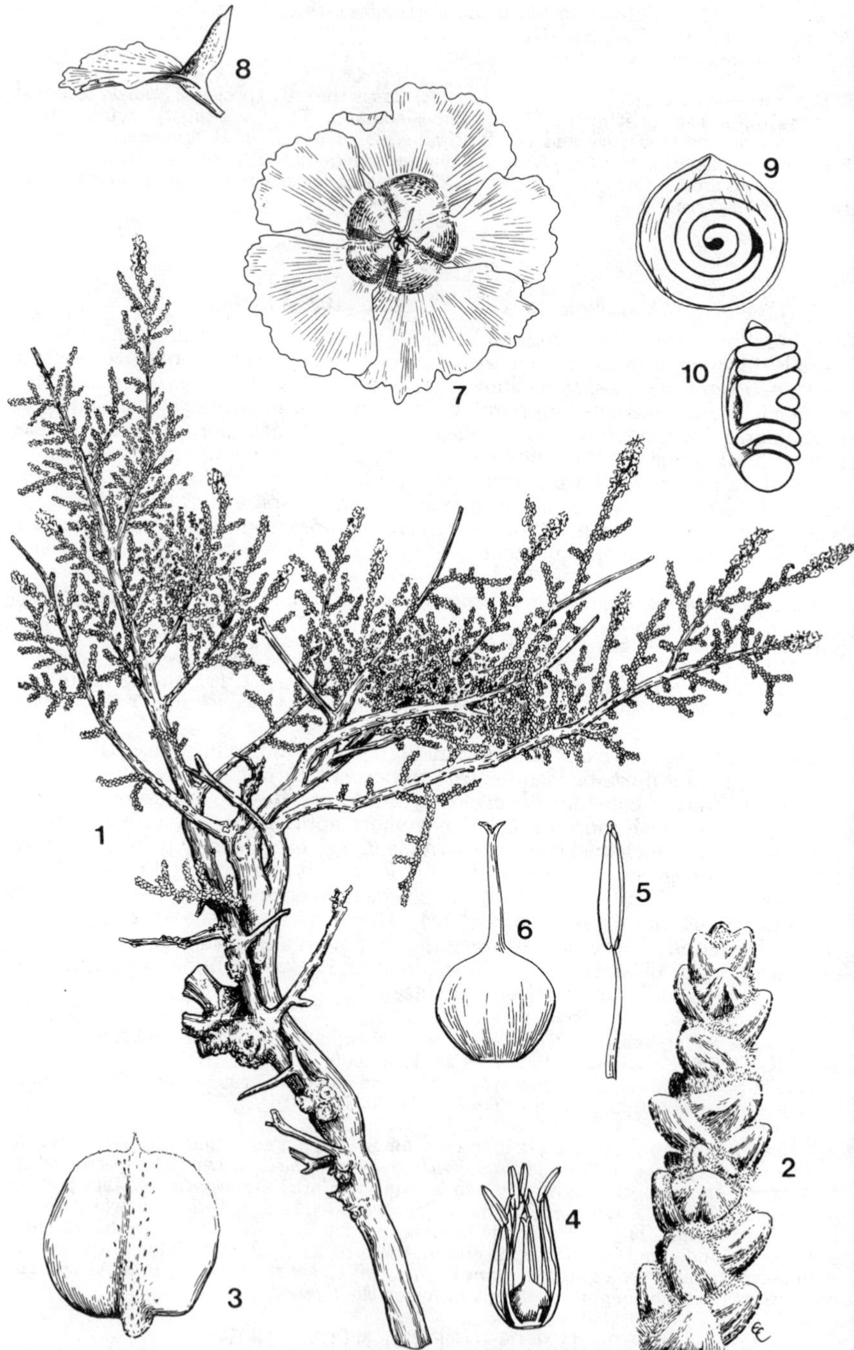

Tab. 41. SALSOLA RABIEANA. 1, habit (×$\frac{1}{3}$); 2, leafy branchlet (×8); 3, bracteole (×15), 1–3 from *Skarpe* S-318; 4, flower with two perianth lobes removed (×6); 5, stamen (×15); 6, ovary (×15), 4–6 from *Wild* 4967; 7, perianth in fruiting stage (×6); 8, single perianth lobe, viewed laterally to show spreading wing (×6); 9, seed (testa transparent) (×15); 10, seed in vertical section (×15), 7–10 from *Skarpe* S-318.

thickened pale midrib and margins, obtusely pointed. Flowers solitary in upper axils, immature only seen. Fruit unknown.

Mozambique. GI: Mocucuni, Inhambane Isl., fl. juv. 20.xi.1958, *Mogg* 29302 (LISC; SRGH).

Clearly different from *S. rabieana* in having opposite glabrous or subglabrous and larger leaves, as well as in its coastal saline habitat. It has not been matched with any described species and further material is needed. *Mogg* 29153 (San Sebastião, 10.xi.1958, LISC; SRGH) is the same.

134. BASELLACEAE

By B.L. Stannard

Subsucculent herbs with slender twining stems, glabrous. Leaves alternate, entire or almost entire, usually petiolate, exstipulate. Inflorescences of spikes, racemes or panicles, axillary or terminal. Flowers small, actinomorphic, hermaphrodite or unisexual; bracts small; bracteoles 2–4, frequently appressed to base of perianth, sometimes winged. Perianth 5-lobed, lobes united at base only or into a 5-lobed tube, imbricate, persistent in fruit. Stamens 5, inserted opposite and at base of perianth lobes; filaments free; anthers dehiscing variously. Ovary superior, unilocular; styles terminal, free or united or 3-fid; stigmas 3; ovule solitary, basal, short stalked. Fruit indehiscent, enveloped by persistent frequently fleshy perianth. Seeds solitary, globose; endosperm present.

A family of five genera from the tropics, occurring mainly in South America but also in Africa and Asia. Only one genus (*Basella*) occurs naturally in the Flora Zambesiaca area. *Anredera cordifolia* (Tenore) Steen. (syns. *Boussingaultia cordifolia* Tenore; *B. baselloides* auct. non H.B.K.; *B. gracilis* Miers incl. f. *pseudobaselloides* Hauman) is found as a cultivated plant. It is similar in habit and appearance to the *Basella* spp. but can be distinguished by its slender, multiflorous, often branched racemes of pedicellate, non-fleshy flowers which open wide at anthesis.

BASELLA L.

Basella L., Sp. Pl. **1**: 272 (1753); in Gen. Pl. ed. **5**: 133 (1754).

Stems long, branching. Leaves petiolate, entire. Inflorescence of axillary spikes or panicles of spikes. Flowers hermaphrodite. Bracteoles 2 connate into cup appressed to perianth. Perianth fleshy. Ovary ovoid, free. Styles 3 or 1 deeply trifid. Stigmas linear. Fruit globose, membranous. Embryo spirally twisted.

A genus of about 5 species, one widespread in Africa and Asia, one endemic to eastern Africa, three endemic in Madagascar.

Leaves up to 15 × 12.5 cm.; petioles up to 9 cm. long. Inflorescences usually simple spikes, long peduncled. Perianth a tube remaining closed at anthesis . . . . 1. *alba*
Leaves up to 4.5 × 3.5 cm.; petioles up to 7 mm long. Inflorescences panicles of spikes. Perianth lobed often almost to base, lobes spreading at anthesis . . 2. *paniculata*

1. **Basella alba** L., Sp. Pl. **1**: 272 (1753).—Moq. in DC., Prodr. **13**, 2: 223 (1849).—Volkens in Engl. & Prantl, Pflanzenfam. 3, **1a**: 126, fig. 73/A-F (1893).—Baker & C. B. Clarke in F.T.A. **6**, 1: 94 (1909).—Peter in Fl. Deutsch Ost.- Afr. 2: 272 (1938).—Hauman in F.C.B. 2: 129 (1951).—F.W.T.A. ed. 1, **1**: 130, fig. 50 (1927) et ed. 2, **1**: 155, fig. 57 (1954).—Verdc. in F.T.E.A., Basellaceae: 1 (1968). TAB. **42** fig. A. Type from specimen cultivated in Europe.

Basella cordifolia Lam., Encycl. Meth. Bot. **1**: 382 (1785).—Moq. in DC., Prodr. **13**, 2: 223 (1849). Type from India.

Basella rubra L., Sp. Pl. **1**: 272 (1753) et ed. 2: 390 (1762).—Hook. f. in Fl. Brit. Ind. 5: 20 (1886).—F.W. Andr., Fl. Pl. Anglo-Egypt. Sudan **1**: 123 fig. 74 (1950).—Cufodontis in Bull. Jard. Bot. Brux. 23. Suppl. 92 (1953). Type from Sri Lanka.

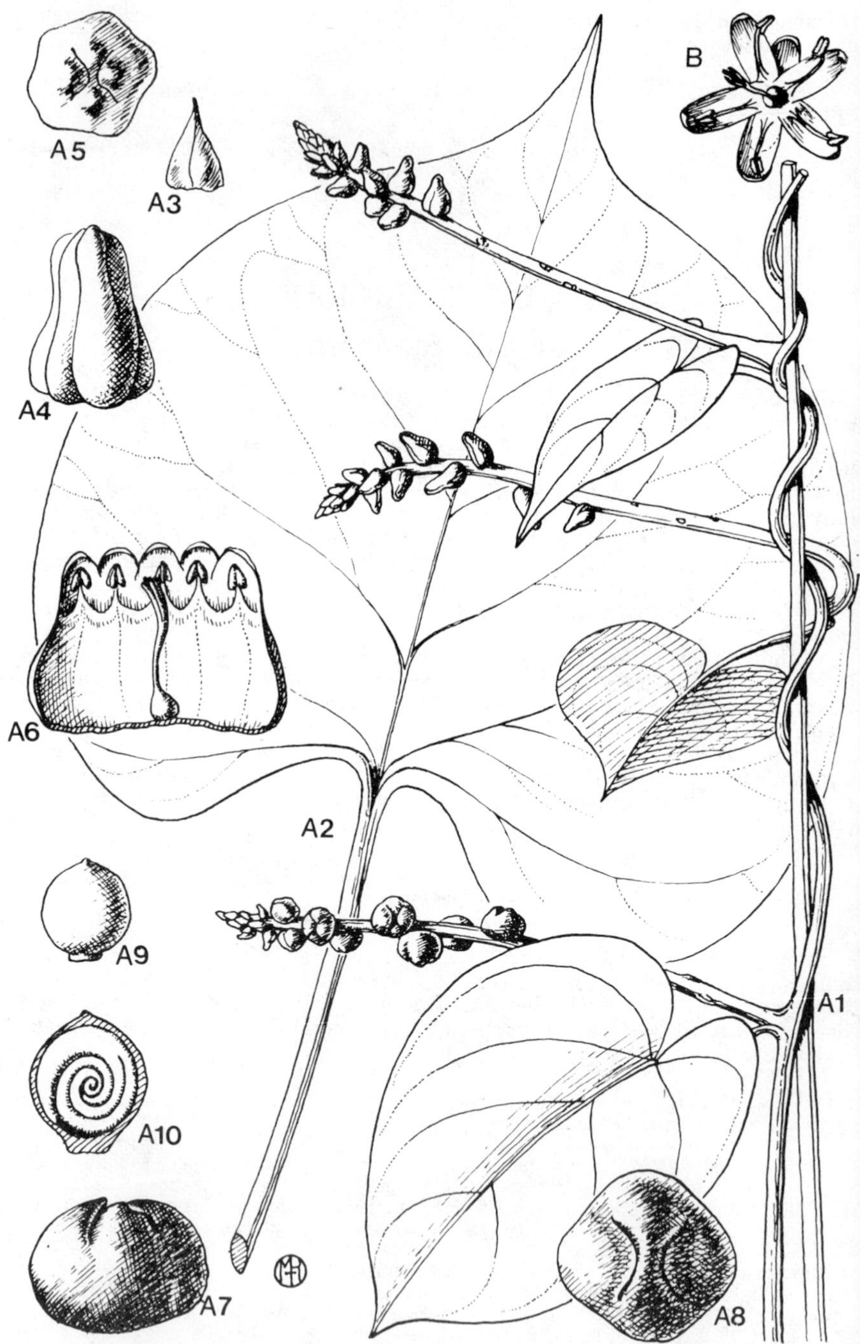

Tab. 42. A.—BASELLA ALBA. A1, habit (×1) *Verdcourt* 1567; A2, leaf (×1) *Fries* 846; A3, bracteole (×4); A4, flower bud, lateral view; A5, flower bud, view from above (×4); A6, flower opened out (×4), A3—6 from *Milne-Redhead & Taylor* 11402; A7, fruit, lateral view (×3); A8, fruit, view from above (×3); A9, seed (×3); 10, vertical section of seed (×4), A7—10 from *Faulkner* 1962. B.—BASELLA PANICULATA. B1, flower (×4) *Milne-Redhead & Taylor* 7244.

Plant 1–8 m. tall; stems much branched, sometimes sparsely leaved, sometimes reddish. Leaves: lamina 2.5–15 × 2.0–12.5 cm., ovate to circular, usually acute or acuminate, sometimes rounded or emarginate at apex, usually cordate at base, entire, lateral nerves 3–4 (5) on either side; petioles 0.5–9 cm. long. Inflorescences axillary, long peduncled, usually unbranched spikes, 1.5–22 (30) cm. long. Flowers 2.5–6 mm. long, white, pink or mauvish. Perianth somewhat fleshy, urceolate, lobes c. one-quarter length of tube, remaining closed. Stamens inserted near apex of tube. Ovary 0.5 mm. tall, ovoid. Style with 3 linear stigmas, 0.5 mm. long. Fruit 4–5 mm. diam., subglobose, black.

Zambia. W: Solwezi Distr., East Lunga Basin, 1200 m., fr. x.1934, *Trapnell* 1652 (K). **Malawi**. N: Chitipa Distr., 4.8 km. N. Mughesse, 1650 m., fl. & fr. 13.ix.1977, *Phillips* 2854 (K; MO). S: Thyolo Mt., 1200 m., fl. 22.ix.1946, *Brass* 17741 (K). **Mozambique**. Z: Posto Agrícola de Mocuba, fr. 21.ii.1943, *Torre* 4803 (LISC).

Probably native to Old World tropics but now pantropical, often cultivated and naturalised. Cultivated as a spinach and for a purple dye strained from the fruits. Some cultivated forms are fleshier and have relatively short petioles. In forest margins and clearings, thickets, often in wet places; 0–2590 m.

2. **Basella paniculata** Volkens in Engl., Bot. Jahrb. **38**: 81 (1905).—Baker & C.B. Clarke in F.T.A. **6**, 1: 94 (1909).—Peter, Fl. Deutsch Ost.-Afr. **2**: 272 (1938).—Verdc. in Kew Bull. **17**: 496 (1964).—Verdc. in F.T.E.A., Basellaceae: 1 (1968). TAB. **42** fig. B. Type from Tanzania.

Perennial climber. Stems branched, often sparsely leaved. Leaves: lamina 2–4.5 × 1.5–3.5 cm., elliptic to circular, acute to shortly acuminate, rounded to cuneate at the base, entire; petioles: 5–7 mm. long. Inflorescences axillary, fleshy panicles of spikes, 1.5–4 cm. long. Flowers 2–3 mm. long, greenish white. Bracteoles 2, c. 2 × 2 mm., united at base, broadly ovate to circular, appressed to perianth. Perianth deeply divided, often almost to base; lobes 1.25–2.5 × 0.75–1.5 mm., ovate to subcircular. Stamens 1.5–2.5 mm. long, inserted level with base of lobes on slightly thickened ring. Ovary ovoid, style with 3 linear stigmas. Fruit globose, 3–5 mm. diam., with longitudinal ridges; receptacles fleshy, enlarged.

Mozambique. GI: Gaza, Guija, old Rd. to Mabalane, 5 km. from Aldera da Barragem, fr. 20.xi.1957, *Barbosa & Lemos* in G.B. 8208 (K; LISC). M: Maputo, Porto Henrique, fl. 23.viii.1944, *Mendonça* 1879 (LISC). M: Maputo, Umbeluzi, near Experimental Station, fr. 28.xi.1944, *Mendonça* 3099 (LISC).

Also in Kenya, Tanzania and South Africa. In dry thickets and forest on sandy or rocky ground; 20–1000 m.

The very different flower structure clearly distinguishes this species from *B. alba*. Further study is required to decide whether they are indeed congeneric or better separated into different genera.

135. PHYTOLACCACEAE

By B.L. Stannard

Trees, shrubs or herbs, sometimes scrambling. Leaves simple, alternate, entire, petiolate to sessile, crystals often visible especially on younger leaves. Stipules absent (in the Flora Zambesiaca area), minute or thorny. Inflorescences terminal or axillary, spicate, racemose or paniculate. Flowers bracteate and bracteolate, hermaphrodite or unisexual (then usually with rudimentary aborted organs), usually actinomorphic. Sepals 4–5, free or connate towards base, imbricate. Petals absent. Stamens (3) 4-many, 1–2-seriate, often inserted on more or less fleshy, annular disk, irregularly arranged or alternate to sepals; filaments slender, free or connate at base; anthers dorsi-or basifixed, dehiscing longitudinally. Ovary usually superior, composed of 1-many free or connate carpels; ovule basal, solitary in each carpel; styles absent or same number as carpels, free or united at base; stigmas linear or capitate. Fruit of 1-many, free or connate carpels, fleshy or dry, sometimes winged; seed subglobose, discoidal or reniform, often compressed; testa membranous or brittle; endosperm present.

A family of about 14 genera and approximately 70 species of tropical and subtropical regions, mainly in the New World.

1. Ovary multicarpellate, carpels free or united - - - - - **1. Phytolacca**
– Ovary unicarpellate - - - - - - - - - - - - - - - 2
2. Ovary with (2) 3–4 stigmas - - - - - - - - - - **4. Lophiocarpus**
– Ovary with 1 stigma - - - - - - - - - - - - - - - 3
3. Flowers more or less zygomorphic; 3 sepals united, 1 free. Fruit dry - - - - - - - - - - - - - - - - - **2. Hilleria**
– Flowers more or less actinomorphic; 4 free sepals. Fruit juicy - - **3. Rivina**

1. PHYTOLACCA L.

Phytolacca L., Sp. Pl. **1**: 441 (1753); in Gen. Pl. ed. **5**: 200 (1754).

Trees, shrubs or herbs, sometimes scrambling. Inflorescences terminal or axillary, racemes, spikes or panicles. Flowers hermaphrodite or unisexual. Sepals free, subequal, sometimes reflexed. Stamens 5-many, functional or rudimentary; filaments sometimes connate at base; anthers dorsifixed. Ovary of 4-many carpels, free or united, functional or rudimentary; styles equal in number to carpels, free or connate at base. Fruit of 4-many, free or connate carpels, fleshy, globose to subglobose. Seed reniform or discoid; testa black.

A genus of about 20 species, mainly in warmer regions, especially America. *P. dioica* L. is found as a cultivated plant in the Flora Zambesiaca area and can be distinguished from the two species below by its tree habit, greater number of stamens ((20) 24–30), broader, glabrous perianth segments and less divergent carpels in the fruit.

Scrambling herb or shrub up to 13 (20) m. Leaf laminas ovate to elliptic. Flowers usually functionally unisexual. Sepals reflexed at anthesis. Upper part of carpels free. Fruit 4–5 (8) lobed, style remains pointing outwards at apex of each carpel, orange to red when ripe - - - - - - - - - - - - 1. *dodecandra*
Bushy herb or subshrub up to 2 m. Leaf laminas narrowly elliptic to elliptic-lanceolate. Flowers hermaphrodite. Sepals spreading at anthesis. Carpels completely fused. Fruit 7–8 (9) lobed, style remains clustered inside depression at apex of fruit, black when ripe - - - - - - - - - - - - - - - - - - 2. *octandra*

1. **Phytolacca dodecandra** L'Herit., Stirp. Nov.: 143, t. 69 (1791).—Heimerl in Engl. & Prantl, Pflanzenfam. 3, **1b**: 11 (1889); in 2 ed., **16 C**: 156, fig. 71 (1934).—Hiern, Cat. Afr. Pl. Welw. **4**: 901 (1900).—Baker & C.H. Wright in F.T.A. **6**, 1: 97 (1909).—H. Walt. in Engl., Pflanzenr. **4**, 83: 42, fig. 15 (1909).—A.W. Hill in Fl. Cap. **5**, 1: 457 (1911).—Engl., Pflanzenw. Afr. **3**, 1: 137, fig. 87 (1915).—F.W.T.A. ed. 1, **1**: 121 (1927); in ed. 2, **1**: 143 (1954).—Peter, Fl. Deutsch Ost.-Afr. **2**: 253 (1938).—Brenan, T.T.C.L.: 449 (1949).—Balle in F.C.B. **2**: 94 (1951).—F. White, F.F.N.R.: 46 (1962).—Nowicke in Ann. Miss. Bot. Gard. **55**: 308 (1968).—Verdc. & Trump, Common Pois. Pl. E. Afr.: 29, fig. 2 (1969).—Polhill in F.T.E.A., Phytolaccaceae: 2, fig. 1 (1971). TAB. **43** fig. B. Type from Ethiopia.

Phytolacca abyssinica Hoffmann in Comm. Götting. **12**: 25, fig. 2 (1796).—Oliver in Trans. Linn. Soc. **29**: 140 (1875).—Engl., Pflanzenw. Ost-Afr. **C**: 175 (1895) nom. illegit. Type as above.

Phytolacca abyssinica var. *apiculata* Engl., Pflanzenw. Ost- Afr. **C**: 175 (1895). Type not indicated.

Phytolacca dodecandra var. *apiculata* (Engl.) Baker & Wright in F.T.A. **6**, 1: 97 (1909).—H. Walt. in Engl., Pflanzenr. **4**, 83: 44 (1909).—Engl., Pflanzenw. Afr. **3**, 1: 137 (1915).—Peter, Fl. Deutsch Ost-Afr. 2: 255 (1938).—Brenan, T.T.C.L. 2: 450 (1949). Type not indicated.

Phytolacca dodecandra var. *brevipedicellata* H. Walt. in Engl., Pflanzenr. **4**, 83: 44 (1909).—Peter, Fl. Deutsch Ost-Afr. 2: 255 (1938).—Brenan, T.T.C.L. 2: 450 (1949). Types from Tanzania and Madagascar.

Herbs or shrubs sometimes scrambling or trailing up to 10–13 m. (20 m.), semi-succulent, sometimes forming dense thickets. Stems usually glabrous, more rarely pubescent. Leaves: lamina 3–14 × 1.5–9.5 cm., ovate to broadly elliptic, base rounded to broadly cuneate, usually oblique, often slightly decurrent, apex

Tab. 43. A.—PHYTOLACCA OCTANDRA. A1, branch with fruits ($\times\frac{2}{3}$) *Noel* 2341; A2, flower (with one sepal and four anthers removed) (×8) *Torre & Correia* 15554; A3, fruit (×5) *Noel* 2341; A4, seed (×8) *Noel* 2341. B.—PHYTOLACCA DODECANDRA. B1, fruit (×5) *Jacobsen* 2241.

acute or rounded with often recurved mucronate point or retuse, glabrous to pubescent; petioles 8–40 mm. long, pubescent to glabrous. Inflorescences terminal and lateral racemes, leaf-opposed, 5–30 cm. long, rhachis pubescent. Flowers dimorphic, usually functionally unisexual, whitish to yellowish green. Pedicels 5–8 (10) mm. long, pubescent. Bracts 1–2.5 mm. long, subulate, pubescent. Bracteoles usually on middle or upper part of pedicels, 0.5–1.5 (3) mm. long, subulate, pubescent. Sepals 2–4 × 1–2 mm., narrowly triangular to oblong ovate, acute, pubescent outside, pubescent to glabrous inside, reflexed at anthesis. Disk 1.5–2 mm. in diam., more or less fleshy. Stamens 10–20 in 1–2 whorls; filaments-filiform, broadening towards base, pubescent to glabrous. Long staminate flowers: usually functionally male but sometimes developing fruit; stamens 3–6.5 mm. long; filaments 3–5 mm. long; anthers 0.5–1 × 0.5–0.75 mm.; carpels 4–5, 1–1.5 × 0.75–1 mm., ovoid, connate at base only; styles 0.5–1.0 mm. long, linear, broadening at base, pubescent to glabrescent; stigmas linear. Short staminate flowers: functionally female; stamens 1.5–3.0 mm. long; filaments 1–2.5 mm. long; anthers 0.5–1.0 × 0.25–0.5 mm., circular to oblong; carpels 4–5, 1.5–2 mm. tall, ovoid, connate at base only, pubescent; styles 0.75–2 mm. long, widening towards base, turned outwards, pubescent to glabrescent; stigmas linear. Fruit usually developed in short-staminate flowers only, 4–5 (8) lobed, 5–10 mm. in diam., ripening orange or red, each carpel 3.5–5 × 2.5–4 mm., fleshy, remains of style pointing outwards at apex of each carpel, pubescent to glabrous. Seeds 2–3 mm. in diam., discoidal, sometimes with ridged edge around circumference; testa shiny black.

Zambia. B: Chikonkwelo Valley, c.1066 m., fl. vi.1952, *Gilges* 114 (K; PRE). N: Mansa Distr., near Samfya Mission, Lake Bangweulu, fl. 28.viii.1952, *White* 3155 (BM; K; PRE). W: Mwinilunga Distr., near River Lwamukunyi, fl. 25.viii.1930, *Milne-Redhead* 968 (K; PRE). C: Mt. Makulu Research St. 13 km. S. of Lusaka, fl. 20.vii.56, *Angus* 1384 (BM; K; PRE). S: Kafue Flats near Mazabuka, fr. 7.x.1930, *Milne-Redhead* 1221 (K; PRE). **Zimbabwe**. N: Lomagundi, Dichwe Farm, Lemon Forest, 2050 m., fl. 5.vii.64, *Jacobsen* 2461 (PRE). W: ? Victoria Falls, fl. 20.iv.1906, *Flanagan* 2998 (PRE). C: Makoni Distr., Rusape, fl. & fr. 11.ix.1952, *Dehn* 96168 (K). E: Mutare Distr., Murahwa's Hill, 1875 m., fl. 31.vii.1963, *Chase* 8046 (LISC). S: Great Zimbabwe, 1097 m., fl. & young fr. 20.vi.1924, *Galpin* 9200 (PRE). **Malawi**. N: Rumphi Distr., Nyika Plateau, Kafwimba rain forest, 1900 m., fl. 15.viii.1975, *Pawek* 10058 (K; PRE). C: Dedza Mt., fl. 1.vii.1980, *Salubeni* 2748 (MO). S: Ntcheu Distr., Chirobwe Mt., fl. 10.ii.1967, *Salubeni* 558 (LISC). **Mozambique**. N: Amaramba, Mandimba, fl. 1.x.1958, *Monteiro* 96 (LISC). Z: Mopeia, fr. 29.vii.1942 Torre 4457 (LISC). T: Angónia, slopes of Mt. Dómuè, 1800 m., fr. 16.x.1943, *Torre* 6046 (LISC). MS: Manica, Maronga, fl. 13.viii.1945, *Simão* 468 (LISC). GI: between Chongoene and Manjacaze, fl. 6.viii.47, *Pedro & Pedrogão* 1660 (PRE).

Widespread in tropical and southern Africa and Madagascar. In broad range of habitats; 600–2800 m.

The reduction of the non-functioning reproductive parts is often not very conspicuous.

2. **Phytolacca octandra** L., Sp. Pl. ed. 2: 631 (1762).—H. Walt. in Engl., Pflanzenr. 4, 83: 58 (1909).—Baker & Wright in F.T.A. 6, 1: 98 (1909).—A.W. Hill in Fl. Cap. 5, 1: 458 (1911).—Heimerl in Engl. & Prantl, Pflanzenfam. ed. 2, **16** C: 158 (1934).—Nowicke, Ann. Miss. Bot. Gard. **55**: 313 (1968).—Polhill in F.T.E.A., Phytolaccaceae: 4 (1971). TAB. **43** fig. A. Type from Mexico.

Bushy herb or subshrub, 1–2 m. tall, semi-succulent. Stems glabrous, sometimes trailing. Leaves: lamina 3.5–13.5 × 1–5 cm., narrowly elliptic to elliptic-lanceolate, base cuneate, decurrent and usually oblique, apex acute-acuminate with often recurved mucronate tip, glabrous; petioles 0.3–2.0 cm. long, glabrous. Inflorescences terminal or lateral, leaf-opposed racemes, 4–14 cm. long, rhachis sparsely and crisply pubescent to glabrescent. Flowers hermaphrodite, white to yellowish green, subsessile or with pedicels up to 1.5 mm. long. Bracts 1.5–2 mm. long, lanceolate, glabrous. Bracteoles c. 1 mm. long, lanceolate to subulate, glabrous. Sepals 5, 2–3 × 1–1.5 mm., ovate-elliptic, apex rounded to acute, spreading at anthesis, glabrous. Disk very thin, almost non-existent. Stamens 8 in one whorl; filaments 1–2 mm. long, filiform, united at base; anthers c. 0.5 x 0.25 mm., oblong elliptic. Ovary of (6) 7–8 (10) united carpels, 1.5–2 mm. in diam., 0.75–1 mm. tall, glabrous; styles c. 0.25–0.5 mm. long, linear. Fruits (6) 7–8 (10) lobed, 5–8 mm. in diam., depressed globose, remains of styles clustered in apical depression, glabrous, black when ripe. Seeds 2–3 mm. in diam., discoidal; testa shiny black.

Zimbabwe. C: Harare Distr., Seki, fl. & fr. 25.ii.1980, *Jardaan* in SRGH 265 452 (K; MO). E: Inyanga Distr., Rhodes Inyanga Orchards, 1980 m., fr. 23.iii.1966, *Simon* 768 (K; LISC). **Mozambique.** MS: Mt. Zuira, Tsetserra, 4 km. from farm, road to Vila Pery, 2100 m., fl. & fr. 2.iv.1966, *Torre & Correia* 15554 (LISC). M: Maputo, fl. & fr. no date given, *Borle* 575 (PRE).

Native to tropical America but now distributed and naturalised locally throughout tropical regions of the world. In forests and forest margins, waste places and as a weed of cultivation; 1550–2100 m.

2. HILLERIA Vellozo

Hilleria Vellozo, Fl. Flum.: 47 (1825); Ic. 1, t. 122 (1835).
Mohlana Mart., Nov. Gen. et Sp. Pl. Bras. **3**: 170 (1832).

Herbs, sometimes woody at base. Leaves ovate or elliptic to broadly lanceolate, acuminate, rounded to cuneate and often unequal at base, petiolate. Inflorescences terminal and axillary racemes. Flowers hermaphrodite, more or less zygomorphic. Bracts single, caudate. Bracteoles in pairs. Sepals 4, one free, 3 united with centre one slightly larger, accrescent at fruiting. Stamens 4–13; filaments filiform; anthers dorsifixed. Ovary subspherical, slightly compressed laterally, unicarpellate; style indistinct or short, stigma capitate. Fruit lenticulate, sometimes slightly asymmetric, pericarp reticulately wrinkled. Seed discoidal; testa black.

A genus of 3 species in South America. One species, *H. latifolia*, common in Africa and Madagascar.

Hilleria latifolia (Lam.) H. Walt. in Engl., Pflanzenr. **4**, 83: 81, fig. 25 (1909).—Engl., Pflanzenw. Afr. **3**, 1: 137, fig. 88 (1915).—F.W.T.A. ed. 1, **1**: 121 (1927) et ed. 2, **1**: 143 (1954).—Heimerl in Engl. & Prantl, Pflanzenfam. ed. 2, **16 C**: 150 (1934).—Brenan, T.T.C.L.: 449 (1949).—Balle in F.C.B. 2: 98, t. 8 (1951).—Cavaco in Fl. Madag., Phytolaccaceae: 4, fig. 1/1–3 (1954).—Nowicke in Ann. Miss. Bot. Gard. **55**: 340 (1968).—Polhill in F.T.E.A., Phytolaccaceae: 6, fig. 2 (1971). TAB. **44**. Type from Mauritius.

Rivina latifolia Lam., Tab. Encycl. Méth. Bot. **1**: 324 (1792). Type from Mauritius.

Hilleria elastica Vellozo, Fl. Flum.: 47 (1825) Ic. 1, t. 122 (1835).—Hiern, Cat. Afr. Pl. Welw. **4**: 900 (1900). Type from Brazil.

Ravina apetala Schumacher & Thonn., Beskr. Guin. Pl.: 84 (1827).—Engl., Pflanzenw. Ost-Afr. C: 174 (1895). Type from Ghana.

Mohlana nemoralis Mart., Nov. Gen. et Sp. Pl. Bras. **3**: 171, t. 290 (1832).—Baker & C.H. Wright in F.T.A. **6**: 95 (1909).—Peter, Fl. Deutsch Ost.-Afr. **2**: 252 (1938). Type from Brazil.

Herbs or sub-shrubs up to c. 1.5 m. tall, stems pubescent on younger parts. Leaves: lamina 4.5–15 × 2.0–6.5 cm., ovate-elliptic to elliptic, pubescent to glabrescent along veins below and sometimes along midrib above; petioles 5–55 mm. long, pubescent on superior surface. Racemes to 15 cm. long in fruit; rhachis pubescent to glabrescent; pedicels c. 2 mm. long in flower, to 4 mm. in fruit, glabrous. Bracts c. 1.5 mm. long, caducous, glabrous to setulose (particularly along margins). Bracteoles minute. Sepals 1.5–2 × 1–1.25 mm. in flower, broadly elliptic to obovate, apex rounded, glabrous, green or white. Stamens 4, approximately same length as sepals. Ovary 1–1.5 mm. tall; style very short; stigma capitate. Fruit 2.0–2.5 mm. in diam., ridged around margin; pericarp very thin, adhering to seed. Seed same shape as fruit; testa hard.

Mozambique. MS: Cheringoma, Inhamitanga, fr., 6.iv.1945, *Simão* 343 (LISC; LMA).

Widespread in tropical Africa, Madagascar and South America. In varied forest habitats; 400–1600 m.

Often considered to have been introduced from South America but its occurrence in apparently little disturbed habitats would indicate that it is possibly native.

Tab. 44. HILLERIA LATIFOLIA. 1, branch with flowers and fruits (×$\frac{2}{3}$) *Garcia* 192; 2, flower (×12); 3, fruit (×8); 4, seed (×8), 2–4 from *Simão* 714.

3. RIVINA L.

Rivina L., Sp. Pl. **1**: 121 (1753) et ed. 2: 177 (1762).
Rivinia L., Gen. Pl., ed. 5: 57 (1754).

Herbs, sometimes subshrubby and woody at base. Stems erect or straggling. Leaves petiolate. Inflorescences racemes, terminal and axillary, suberect, many-flowered. Flowers hermaphrodite, actinomorphic. Sepals 4, more or less free, subequal, oblong-elliptic, rounded or acute at apex, erect or spreading in fruit. Stamens 4; filaments free; anthers dorsifixed, linear. Ovary globose to elliptic, somewhat compressed laterally, unicarpellate; style short, curved; stigma capitate. Fruit globose, orange, red or purple, juicy; pericarp thin, adherent to seed. Seed 1, discoidal; testa crustaceous, black, often pubescent.

Treated by some authors (H. Walt. in Pflanzenr. (1909); Raeder in Ann. Miss. Bot. Gard. **48**: (1961)) as a genus of three species, by others (Heimerl in Pflanzenfam. ed. 2, **16**: (1934); Nowicke in Ann. Miss. Bot. Gard. **55**: (1968)) as one variable species, native to tropical and subtropical America and introduced into the Old World.

Rivina humilis L., Sp. Pl.: 121 (1753).—Heimerl in Engl. & Prantl, Pflanzenfam. 3, **1b**: 8, figs. 2A, B (1889) et ed. 2, 16 C: 147, fig. 67 (1934).—H. Walt. in Engl., Pflanzenr. **4**, 83: 102, fig. 30 (1909).—Cavaco in Fl. Madag., Phytolaccaceae: 2, fig. 1/4, 5 (1954).—Nowicke in Ann. Miss. Bot. Gard. **55**: 332 (1968).—Polhill in F.T.E.A., Phytolaccaceae: 8, fig. 3 (1971). TAB. **45**. Type uncertain.

Herbs or (?)sub-shrubs 0.30—1.2 m. tall. Stems glabrous or slightly pubescent at nodes. Leaves: lamina 2.0—10.0 × 1.0—4.5 cm., elliptic to ovate, cuneate to rounded at base, usually acuminate, glabrescent to pubescent above and below, often particularly along main veins; petioles 8—50 mm. long, thickly pubescent all round (not in the Flora Zambesiaca area) to pubescent along upper side only or glabrescent. Inflorescences to 13 cm. long. Flowers glabrous; pedicels to 5 mm. long. Bract 1, 1—1.2 mm. long, subulate. Bracteoles 2, appressed to sepals, triangular, 0.2—0.25 × 0.2—0.25 mm. Sepals c. 2 × 0.8—1 mm., elliptic, rounded to slightly acute at apex, spreading at anthesis, green, white or pinkish. Stamens 1.5—1.8 mm. long; filaments often persistent in fruiting; anthers 0.4—0.5 mm. long. Ovary 0.5—0.6 mm. in diam. Style 0.4—0.5 mm. long, filiform. Fruit 2.5—3 mm. in diam. Seed c. 2 mm. in diam.

Zimbabwe. N: Mazoe Citrus Estates, fl. 3.xi.1970, *Searle* 70 (PRE; K). E: Mutare 1097 m., fl. & fr. 27.xii.55, *Chase* 5930 (BM). **Mozambique**. M: Maputo, Umbeluzi, edge of R. Umbeluzi, fl. & fr. 2.iii.1964, *Carvalho* 739 (K; LMU; MO).
Introduced from tropical and subtropical America. In forest, river margins, shadier places, grassland; 550—1300 m.

4. LOPHIOCARPUS Turcz.

Lophiocarpus Turcz., Bull. Soc. Nat. Hist. Mosc. **16**: 55 (1843).
Wallinia Moq. in DC., Prodr. **13**, 2: 143 (1849).

Subshrubs or herbs sometimes becoming woody at base, annual and perennial. Leaves alternate, often with fascicles of leaves in axils, filiform or with distinctly broad lamina, sometimes more or less succulent, sessile or petiolate. Inflorescences spikes, terminal and axillary. Flowers hermaphrodite, actinomorphic, usually in clusters of 3. Bracts solitary, sometimes 3-lobed. Bracteoles sometimes present in pairs. Sepals 4—5, more or less equal, membranous. Stamens 4, three alternate with sepals and one opposite; filaments filiform, free. Ovary globose, 1-ovulate; style absent or short; stigmas 4, filiform. Fruit subglobose to globose, ridged, muricate or warty. Seed 1; testa dark brown or black.

A genus of about 3 species confined to southern Africa.

1. Leaves with distinct lamina - - - - - - - - - - - - 1. *latifolius*
— Leaves filiform - - - - - - - - - - - - - - - 2
2. Fruits muricate or warty, not ridged - - - - - - - - 2. *tenuissimus*

Tab. 45. RIVINA HUMILIS. 1, branch with flowers and fruits (×$\frac{2}{3}$); 2, flower (×20), 1–2 from *Balsinhas* 1463; 3, fruit (×8); 4, seed (×8), 3–4 *Balsinhas* 1827.

– Fruits ridged, sometimes muricate or warty between ridges - - - - - - - - - - - - - - - - - - 3. *polystachyus*

1. **Lophiocarpus latifolius** Nowicke in Ann. Miss. Bot. Gard. **56**: 288, fig. 1 (1969) (as *latifolia*). TAB. **46**. Type: Mozambique, Maputo, Matutuine, *Lemos & Balsinhas* 263 (BM, holotype; K, LISC, isotypes).

Subshrubs or herbs with woody base, to 1–1.20 m. tall, glabrous. Leaves: lamina 2–5 × 0.6–2 cm., elliptic to elliptic-lanceolate, cuneate, slightly oblique, decurrent, acute or mucronate, narrow translucent margin (in dried state at least), sessile or with petioles up to 5 mm. long. Inflorescences of longer terminal and shorter, sometimes branching lateral spikes up to 32 cm. long; rhachis glabrous. Flowers: central flower subtended by glabrous, ovate bract c. 1.5–2 × 0.75–0.8 mm. with long central lobe and sometimes 2 slight lateral ones, lateral flowers each subtended by glabrous, ovate bract c. 1–1.25 × 0.6–0.75 mm. and 2 glabrous, elliptic-ovate bracteoles c. 0.75–1 × 0.3–0.5 mm. Sepals 1.25–2 × 1–1.5 mm., oblong to circular, rounded at apex, glabrous. Stamens: filaments 2.5–3.5 mm. long; anthers 0.5–0.75 mm. long, elliptic. Ovary 1–1.5 × c. 1 mm., laterally compressed, glabrous; stigmas 0.5–1.5 mm. long. Fruit c. 1.2–1.5 mm. in diam., obliquely globose, black with raised lighter coloured patterning. Seed shiny black.

Mozambique. GI: c. 12 km. from Chibuto, road to Alto Changane, fl. 12.ii.1959, *Barbosa Lemos* 8389 (K, LISC). M: Maputo, Matutuine between Tinonganine and Floresta de Licuati, fl. 8.xii.1961, *Lemos & Balsinhas* 263 (BM; K; LISC).
Known only from Mozambique, Namibia and S. Africa. On sandy soils.

2. **Lophiocarpus tenuissimus** Hook.f., Ic. Pl. **15**: 50, t. 1463, fig. 10, 11 (1884).—Heimerl in Engl. & Prantl, Pflanzenfam. ed. 2, **16 C**: 163 (1934).—Nowicke in Ann. Miss. Bot. Gard. **55**: 354 (1968).—Neusser & Schreiber in Merxm., Prodr. Fl. SW. Afr. **24**: 1–2 (1969). Type from S. Africa.
Microtea tenuissima (Hook.f.) N.E. Br. in Kew Bull. **1909**: 134 (1909).—Baker & C.H. Wright in F.T.A. **6**, 1: 96 (1909).—Hill in Fl. Cap. **5**, 1: 455 (1911).

Herbs, slightly woody towards base, 10–40 cm. tall, branching, stems glabrous or pubescent towards base. Leaves sessile, single or more usually in fascicles, 8–40 × 0.5–0.7 mm., subsucculent filiform, glabrous (rarely papillate). Inflorescence usually terminal, up to c. 20 cm. long; rhachis glabrous. Flowers yellowish-green to white, clustered in threes, glabrous, central flower subtended by a glabrous, ovate to cuspidate bract c. 0.8–1 × 0.5–0.6 mm., sometimes 3-lobed with larger triangular central lobe; lateral flowers each subtended by a glabrous, ovate, faintly lobed bract 0.7–0.8 × 0.3–0.4 mm. and usually 2 glabrous, ovate bracteoles 0.5–0.6 × 0.25–0.3 mm. Sepals 1–1.2 × 0.3–0.7 mm., lanceolate to ovate, rounded or acute at apex, one sepal sometimes larger than others. Stamens: filaments c. 0.7–1.5 mm. long; anthers 0.3–0.4 mm. long, elliptic. Ovary 0.4–0.6 × 0.2–0.5 mm., glabrous; stigmas 0.15–0.5 mm. long. Fruit 1–1.2 mm. in diam., muricate or warty.

Botswana. N: Ngamiland, nr. Toteng, Lake Ngami, 930 m., fl. & fr. 25.iii.1961, *Richards* 14847 (K). SW: 27 km. N. of Kang, fl. 18.ii.1960, *Wild* 5035 (K, MO) SE: Ilatamabele-Mosu area nr. Soa Pan, fl. & fr. 7.1.1974, *Ngoni* 260 (K). **Zambia**. S: Katombora, 914 m., fl. & fr. 3.iv.56, *Robinson* 1400 (K). **Zimbabwe**. W: Bulawayo, fl. & fr. xii.1897, *Rand* 75 (BM). S: Beitbridge, Chiturupazi, fl. 22.ii.1961, *Wild* 5329 (COI; K; MO).
From Southern Africa. On sandy soil.

Four specimens examined from southern Botswana (*Ngoni* 260, *Rand* 75, *Rogers* 6553, *Wild* 5053) and retained here under *L. tenuissimus* have distinctly papillose leaves. Further study is required to establish whether or not these represent a new taxon.

3. **Lophiocarpus polystachyus** Turcz. in Bull. Soc. Imp. Nat. Mosc. **16**: 56 (1843).—Benth. & Hook., Gen. Pl. **3**, 1: 50 (1880).—Engl. Pflanzenw. Afr. **3**, 1: 139 (1915).—Heimerl in Engl. & Prantl. Pflanzenfam. ed. 2, **16 C**: 163 (1934).—Nowicke in Ann. Miss. Bot. Gard. **55**: 353 (1969).—Schreiber in Merxm., Prodr. Fl. SW. Afr., Phytolaccaceae: 24: 1 (1969). Type from S. Africa.
Wallinia polystachya (Turcz.) Moq. in DC., Prodr. **13**, 2: 143 (1849). Type as above.
Lophiocarpus burchellii Hook.f. in Benth. & Hook., Gen. Pl. 3, 1: 50 (1880).—Hook.,

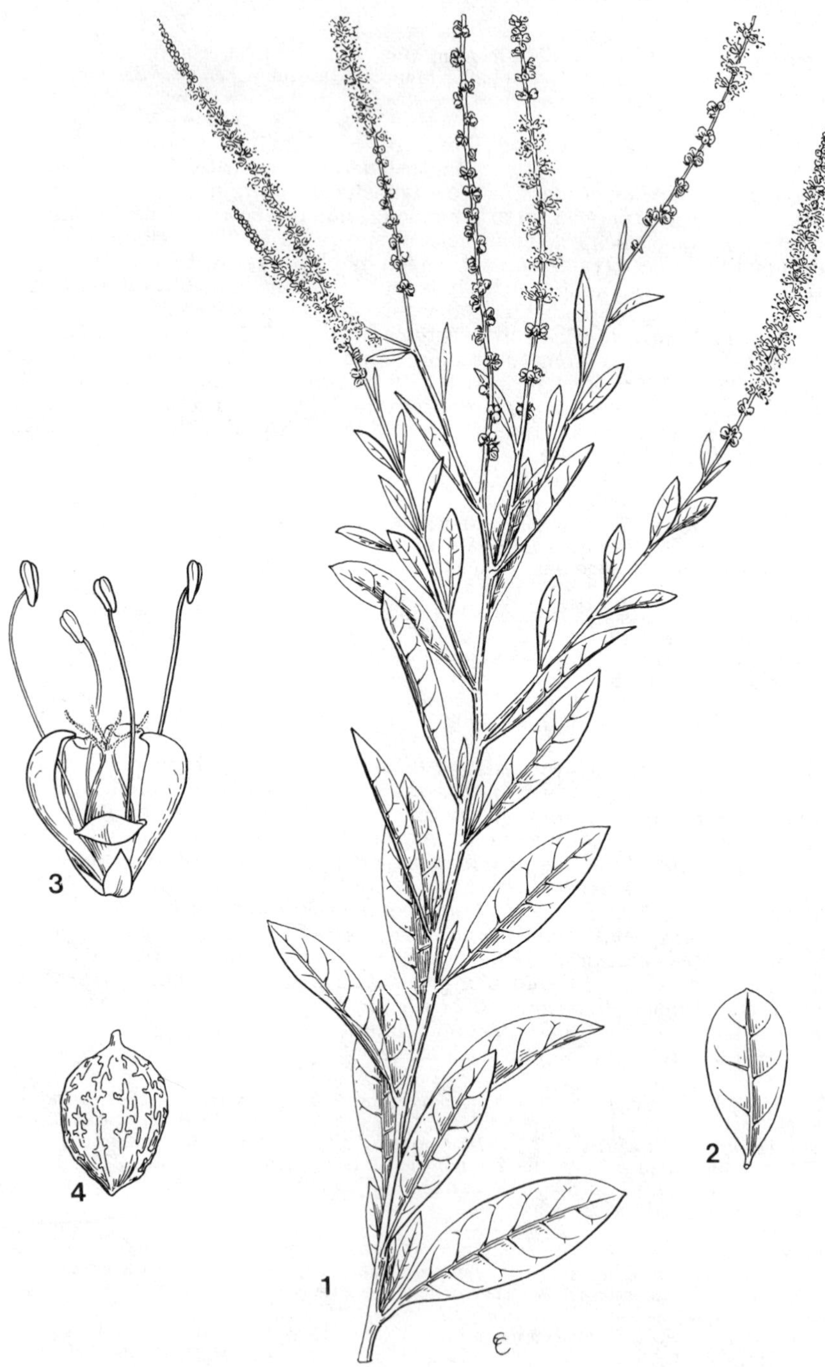

Tab. 46. LOPHIOCARPUS LATIFOLIUS. 1, branch with flowers (×⅔) *Lemos & Balsinhas* 263; 2, leaf to show variation (×⅔) *Torre* 3945; 3, flower (×10) *Lemos & Balsinhas* 263; 4, fruit (×10) *Torre* 3945.

Ic. Pl. **15**: 49, t. 1463 (1884).—Engl., Pflanzenw. Afr. **3**, 1: 139 (1915).—Heimerl in Engl. & Prantl, Pflanzenfam. ed. 2, **16 C**: 163 (1934). Type from S. Africa.

Microtea burchellii (Hook.f.) N.E. Br. in Kew Bull. **1909**: 135 (1909).—Hill in Fl. Cap. **5**, 1: 456 (1912).—Burtt Davy, F.P.F.T. **1**: 173 (1926). Type as above.

Microtea polystachya (Turcz.) N.E. Br. in Kew. Bull. **1909**: 135 (1909).—Hill in Fl. Cap. **5**, 1: 455 (1912). Type from S. Africa.

Microtea gracilis A.W. Hill in Kew Bull. **1910**: 56 (1910).— Hill in Fl. Cap. **5**, 1: 456 (1912).—Burtt Davy, F.P.F.T. **1**: 173 (1926). Type from S. Africa.

Lophiocarpus gracilis (A.W. Hill) Engl. in Pflanzenw. Afr. **3**, 1: 138 (1915).—Heimerl in Engl. & Prantl, Pflanzenfam. ed. 2, **16 C**: 163 (1934). Type as above.

Subshrubs or herbs with woody base to 0.5 m. tall, usually much branched, stems ascending, glabrous. Leaves fasciculate, (sometimes some leaves single and alternate), 5—30 × 0.4—2 mm., filiform, mucronate, subsucculent, glabrous, sub-sessile or with petiole to 2 mm. long. Inflorescence terminal, to 25(30) cm. long; rhachis glabrous. Flowers greenish-yellow, clustered in threes, glabrous; central flower subtended by a bract 1—1.5 × 0.8—1 mm., ovate, acuminate, sometimes faintly 3-lobed, membranous, glabrous; lateral flowers each subtended by a bract 0.8—1 × 0.6—0.8 mm., ovate, membranous, glabrous and usually 2 bracteoles 0.5— 0.75 × 0.3—0.5 mm., ovate, membranous, glabrous. Sepals 1—1.5 × 0.5—1 mm., oblong to broadly elliptic, apex rounded. Stamens: filaments 1—1.3 mm. long; anthers c. 0.3—0.5 mm. long, oblong to broadly elliptic. Ovary 0.5—1.3 mm. in diam., glabrous; stigmas 0.8—1 mm. long. Fruit 1—1.5 mm. in diam., ridged, sometimes muricate or warty between ridges.

Botswana. SW: Tshabong, hill above Government Camp, 1000 m., fl & fr. 2.iii.1977, *Mott* 1100 (K).

From southern Africa. In savanna, on dunes, along dry river beds, mainly on sandy soils.

INDEX TO BOTANICAL NAMES